Random Signal Processing

Random Signal Processing

Shaila Dinkar Apte

CRC Press
Taylor & Francis Group
Boca Raton London New York

CRC Press is an imprint of the
Taylor & Francis Group, an **informa** business

CRC Press
Taylor & Francis Group
6000 Broken Sound Parkway NW, Suite 300
Boca Raton, FL 33487-2742

© 2018 by Taylor & Francis Group, LLC
CRC Press is an imprint of Taylor & Francis Group, an Informa business

International Standard Book Number-13: 978-1-4987-8199-2 (Hardback)
978-1-138-74627-5 (Paperback)

Library of Congress Cataloging-in-Publication Data

Names: Apte, Shaila Dinkar, author.
Title: Random signal processing / Shaila Dinkar Apte.
Description: Boca Raton : CRC Press, 2018. | Includes bibliographical references and index.
Identifiers: LCCN 2017013790| ISBN 9781498781992 (hardback : acid-free paper) | ISBN 9781315155357 (ebook)
Subjects: LCSH: Signal processing. | Random variables.
Classification: LCC TK5102.9 .A678 2018 | DDC 621.382/2--dc23
LC record available at https://lccn.loc.gov/2017013790

Visit the Taylor & Francis Web site at
http://www.taylorandfrancis.com

and the CRC Press Web site at
http://www.crcpress.com

My beloved husband late Shri Dinkar

My grand children

*Aarohi, Shriya and Shreyas
and Shruti*

My students from

*W.C.E., Sangli.
R.S.C.O.E., Pune*

Contents

Preface

It gives me great satisfaction in handing over the book *Random Signal Processing* to all my beloved UG, PG, and Ph.D. students, who are eagerly waiting for it to be published. The difficulties encountered by my students while carrying out their research work inspired me to write a student-friendly book that will be rich in technical content. Random signal processing is essential for all UG, PG, and Ph.D. students of electronics engineering, electrical engineering, computer engineering, and instrumentation engineering disciplines. The subject has a large number of diverse applications. A thorough knowledge of signals and systems and digital signal processing is essential for better understanding of the subject.

Random signal processing involves a number of complex algorithms that require in-depth domain knowledge. If the concept is explained with concrete examples and programs, the reader will take interest in the field. The reader will experience a great joy when he/she actually experiences the outcome of the experiment. Speech signals are naturally occurring and hence, random in nature. Research in this field is rather difficult. Image processing is also a random signal and requires good practical knowledge for implementation of the algorithms. Despite significant progress in this area, there still remain many things that are not well understood. There is a vast scope in advanced techniques used for signal processing especially for speech and image processing. There are many MATLAB® programs for applying the different techniques for speech and image processing. I have personally observed while working with many of my students that any beginner working in this field requires gathering a lot of material to become competent. Finally, when we reach a level of expertise, it is what others have already achieved. It is felt that if some concrete preliminary guidance is available in this field, a person may go very far. This is my motivation for writing this book. The idea is that a newcomer in this field of research must know from where to start, how to start, and how to proceed using this book.

There was a constant demand from my students and from well wishers to write a book on random signal processing that is self-explanatory with lots of practical examples for speech and image processing. This book has concrete examples to illustrate the concepts. For better visualization, we have included the output of many MATLAB programs in the text. The interpretation of the results is also included, which will definitely help a reader to better understand the subject. This will enable instructors to incorporate the exploration of examples in their classroom teaching. It will also help the reader to try out different experiments on their own for better understanding.

The book is intended to provide a rigorous treatment of random signal processing at the PG and Ph.D. level. It will provide a solid foundation for specialized courses in signal processing at the UG and PG level. It will also serve as a basic practical guide for Ph.D. students and provide research directions for further work. A large number of MATLAB programs will provide a start for researchers. The results of a number of experiments conducted in our lab are included for motivating and guiding the students.

I have enjoyed whatever little work I have done in signal processing research. Although I had undertaken this project for newcomers, it was my inner wish to go through the subject thoroughly for my own benefit. I wish the same for the users of this book. I feel that they will enjoy working with advanced techniques in signal processing and obtain great satisfaction when they put in extra effort from their heart.

Organization of the Book

The book contains 10 chapters organized as follows. Chapter 1 covers the basic mathematical background about random signals and processes essential for understanding further chapters. Chapter 2 covers the statistical parameter measurements of random variables, density, and distribution functions, and joint functions for more than one random variable. Chapter 3 deals with multiple random variables and introduces random processes. Chapter 4 introduces detection and estimation theory used in communication. Chapter 5 is devoted to speech processing and applications. Chapter 6 deals with spectra estimation including higher order spectra. Chapter 7 covers statistical speech processing. Chapter 8 is on applications of DCT and WT for speech and image processing. It also discusses the spectral parameters of speech. Chapter 9 deals with image processing techniques. Chapter 10 is devoted to case studies related to handwritten character recognition and writer verification systems. The chapter details are as follows.

Chapter 1 discusses set theory. The concept of probability, conditional probability, and Bayes theorem will be discussed. We will introduce the random variable. The probability distribution function and density functions will be discussed for discrete as well as continuous random variables.

Chapter 2 deals with the properties of random variables. Statistical properties of random variables like estimation of mean, variance, standard deviation, higher order moments like skew, kurtosis, etc. will be explained with numerical examples. There are some functions called moment generating functions and characteristic functions used for finding higher order moments of a random variable. These functions are explained and the use of these functions for finding the moments is illustrated with numerical problems. Energy and power signals are defined. Estimation of energy spectra density (ESD) and power spectral density (PSD) is described. Use of correlograms and autocorrelation is illustrated with examples. Some numerical examples are solved for illustrating the concepts.

Chapter 3 discusses the random variable as it is a useful concept for evaluating the parameters for the variable such as mean and standard deviation. In some situations, it is necessary to extend the concept to more than one random variable. In the engineering field, many interesting applications such as autocorrelation, cross-correlation, and covariance can be handled by the theory of two random variables. The focus is on two random variables.

Chapter 4 describes the main components of the system, namely transmitter, channel, and receiver. It starts with the basics of communication theory and deals with signal transmission through the channel, distortionless transmission, bandwidth of the signal, and relation between bandwidth and rise time, Paley–Wiener theorem, ideal LPF, HPF, and BPF. The next section deals with optimum detection methods. These methods include weighted probability test, maximum likelihood criterion, Bayes criteria, minimax criteria, Neyman–Pearson criteria, and receiver operating characteristics. Estimation theory is described in a further section. We will study three types of estimators, namely Bayesian estimator, maximum likelihood estimator (MLE), and minimum mean square estimation (MMSE). Cramer–Rao inequality will be explained.

Chapter 5 introduces basic processing concepts and algorithms used for speech signals. Speech signals are generated by nature. They are naturally occurring and hence modeled as random signals. There are several models put forth by researchers based on their perception of the speech signal. Here, the speech signal is processed using statistical parameters. Even though speech has a random component, if we consider a short segment of speech, say about 10 ms of speech, we can assume that over the small segment, the speech parameters will remain constant. Hence, parametric models are also used. Different parameters extracted are pitch frequency, formants, cepstral coefficients, Mel frequency cepstral coefficients, linear prediction coefficients, etc.

Chapter 6 introduces the classification of signals into two classes, namely energy signals and power signals. The exact calculation of the spectrum is not possible. Hence, the need for estimation of the spectrum for random signals like speech will be emphasized. Different methods for spectrum estimation such as the periodogram method, the Welch method and the Blackman–Tukey method will be discussed. The signal models AR, MA, and ARMA are introduced. We will explain power spectrum estimation using these parametric methods such as using MA model parameters, AR model parameters, and ARMA model parameters. We will also deal with other spectrum evaluation methods like minimum variance method and eigenvalue algorithm method. The cepstrum domain is discussed and its use for pitch period measurement of speech signal is explained. Finally, the higher order spectrum estimation called cumulant spectra is described. The evaluation of cumulant spectra and higher order spectrum estimation is discussed in detail with illustrations.

Chapter 7 deals with statistical speech processing methods. It deals with statistical parameters such as mean, standard deviation, skew, and kurtosis. The dynamic time warping required due to natural changes in time duration when the same speech is uttered a number of times is discussed in the next section. The speech recognition methods use statistical modeling of speech such as the hidden Markov modeling (HMM) and Gaussian mixture models. These are described in detail in Section 7.3. We then proceed to discuss statistical sequence recognition and pattern recognition. Speech recognition uses vector quantization and HMM models. These are explained in short. HMM models require acoustic probability estimation. We then proceed to parametric speech processing. The pitch synchronous analysis is dealt with in detail.

Chapter 8 deals with small segment analysis of speech and pitch synchronous analysis of speech. It discusses transform domain speech processing, namely processing using short time Fourier transform, wavelet transform, etc. The statistical analysis of speech has limitations due to the length of the speech segment available. Practically, we have only limited duration of speech available for analysis. The statistical estimate of parameters approaches the actual value only when the length of the speech segment is very large. Again, the probability density function is not known. We have to be satisfied with only the estimate of statistical parameters. We may extract speech parameters such as pitch frequency, formants, LPC, MFCC, etc. and use these parameters for speech recognition, speaker recognition, etc. This is termed as the parametric approach. This parametric approach is not directly valid, as speech varies with time. Speech parameters also vary with time. The only possibility of using the parametric approach is for short segment analysis of speech.

Chapter 9 describes image formation, image storage, and image representation. The filtering techniques used for images, namely low pass filters and high pass filters; edge enhancement filter, high frequency emphasis filter, and contrast enhancement filters in the spatial domain are discussed in detail. We will describe different spatial masks for image smoothing, edge detectors such as Sobel and Prewitt. Edge detection using derivative of gradients, that is, Laplacian mask, will be explained. Laplacian of Gaussian (LOG) is discussed. We then go through different image transformations such as logarithmic and piecewise linear transformations for image enhancements. We describe different statistical parameters of image. The different processing techniques will be explained like image compression, noise cancellation, and image resizing using transform domain techniques like DCT and WT.

Chapter 10 discusses two different case studies. One related to Marathi handwritten character recognition and the other related to writer verification using a handwritten document.

In organizing the material as described here, we have tried to explain the concepts via concrete examples and a lot of MATLAB programs. This has been done to provide a source of motivation to the reader.

MATLAB® is a registered trademark of The MathWorks, Inc. For product information, please contact:

The MathWorks, Inc.
3 Apple Hill Drive
Natick, MA 01760-2098 USA
Tel: 508-647-7000
Fax: 508-647-7001
E-mail: info@mathworks.com
Web: www.mathworks.com

Acknowledgments

This book is the outcome of great inspiration by my students from Walchand College of Engineering, Sangli and from Rajarshi Shahu College of Engineering, Pune. I sincerely thank all the students who worked under my guidance for their UG, PG, or Ph.D. project. Their painstaking efforts and sincerity encouraged me to face further challenges. We have carried out a lot of experiments in speech and image processing that have made my understanding perfect and enriched the knowledge of my students. The experimental work and the encouraging results further prompted me to undertake this project. The constant demand by my students for the book has accelerated the work.

I had great help and inspiration from my family members, especially my husband late Mr. Dinkar, my son Mr. Anand, daughter-in-law Dr. Aditi, my daughter Mrs. Amita, and son-in-law Mr. Prasad.

The idea of publishing the book came to reality when I received an invitation to write a book from CRC Press/Taylor & Francis. Lastly, I wish to thank all who helped me directly and indirectly for their support during the long process of bringing out this book.

Shaila D. Apte

About the Author

Dr. Shaila Dinkar Apte is currently working as a professor at Rajarshi Shahu College of Engineering, Pune, teaching postgraduate students and as reviewer for the *International Journal of Speech Technology*, Springer, *International Journal of Digital Signal Processing*, Elsevier. She is currently guiding five Ph.D. candidates. Eight candidates have completed their Ph.D. and about seventy candidates have completed their M.E. dissertations under her guidance. Almost all dissertations were in the area of signal processing. She has a vast teaching experience of 35 years in electronics engineering, and enjoys great popularity among students. She has been teaching digital signal processing and advanced digital signal processing for the last 22 years. She had also been an assistant professor at Walchand College of Engineering, Sangli for 27 years; a member of the board of studies for Shivaji University and a principal investigator for a research project sponsored by ARDE, New Delhi.

Dr. Apte completed her M.Sc. (electronics) from Mumbai University in 1976 and acquired first rank. She then received her M.E. (electronics) in 1991 from Walchand College of Engineering, Sangli and Ph.D. in electronics engineering in 2001 from Walchand College of Engineering, under Shivaji University, Kolhapur. Her Ph.D. thesis involved work on speaker modeling using optimal mother wavelets.

She has published 40 papers in reputed international journals, more than 40 papers in international conferences, and about 15 papers in national conferences. She has a patent granted to her for generation of a mother wavelet from speech signals and has applied for three Indian patents related to emotional speech synthesis, emotion detection from speech, and context-dependent human-like emotional speech synthesis with prosody. She has published many books: *Digital Signal Processing* (second reprint of second edition), *Advanced Digital Signal Processing*, and *Speech and Audio Processing* published by Wiley India; *Signals and Systems: Principles and Applications* published by Cambridge University Press. Her areas of interest include emotional speech synthesis, emotion detection, and context-based personalized speech synthesis. E-mail: sdapte@rediffmail.com.

1

Introduction to Random Signals

LEARNING OBJECTIVES

- Introduction to set theory
- Probability
- Conditional probability
- Bayes' theorem
- Random variables
- Probability distribution function
- Standard probability density functions

The theory of signals and systems is applied for processing of signals like speech, radio detection and ranging (RADAR), sound navigation and ranging (SONAR), earthquake, electrocardiogram (ECG), and electroencephalogram (EEG). All these are naturally occurring signals and hence have some random components. The primary goal of this chapter is to introduce the principles of random signals for an in-depth understanding of the processing methods. We will start with an introduction to set theory. The concept of probability, conditional probability, and Bayes' theorem will be discussed. We will introduce the random variable. The probability distribution function and density functions for discrete and continuous random variables will be discussed.

1.1 Introduction to Set Theory

We will first introduce the concept of set theory. Let us define a set. A set is defined as a collection of objects. The objects will then be called as the elements of the set. The examples of the sets can include students in a class, students scoring more than 60% marks, and all capital letters. Let C be the set of capital letters. If "A" belongs to set C but "a" does not belong to set C, then this can be given as $A \in C$, but $a \notin C$.

Let us now understand how to specify a set. A set C of all capital letters will be specified by a tabular method or by a rule-based method.

$$\text{Tabular method: } C = \{A, B, C, \ldots, Z\} \tag{1.1}$$

Rule-based method: C = {capital letters}   (1.2)

A set is said to be countable if it has countable number of elements in it corresponding to integers 1, 2, 3, etc., and it is said to be uncountable if it has elements that are not countable. If a set contains no element, it is said to be an empty set or a null set and is denoted by the symbol ϕ. Let us consider examples of sets of each type.

$A = \{1, 3, 5, 7\}$ this is an example of countable set

$B = \{1, 3, 5, 7, \ldots\}$ this is an example of countable set and is infinite

$C = \{2, 4, 6, 8, \ldots\}$ this is an example of countable set and is infinite   (1.3)

$D = \{0.1 < c < 2.1\}$ is an example of rule-specified set, uncountable and infinite

A is subset of B, $A \subset B$

Sets B and C are disjoint or mutually exclusive sets as they have no common elements.

We define the universal set for every random problem. Consider the problem of rolling a die. Here, there are six possibilities for the event. All these six outcomes form a universal set. $S = \{1, 2, 3, 4, 5, 6\}$. Let $A = \{1, 3, 5\}$. This is a subset of S. For any universal set with N number of elements in it, there are 2^N possible subsets.

Example 1

Find the possible subsets for a universal set with six elements in it.

Solution: There will be 2^N possible subsets. Here, N is equal to 6, so possible subsets will be 64. Let us list out these possible subsets.

$S_1 = \{1\}, S_2 = \{2\}, S_3 = \{3\}, S_4 = \{4\}, S_5 = \{5\}, S_6 = \{6\}$

$S_7 = \{1,2\}, S_8 = \{2,3\}, S_9 = \{3,4\}, S_{10} = \{4,5\}, S_{11} = \{5,6\}, S_{12} = \{6,1\},$
$S_{13} = \{1,3\}, S_{14} = \{1,4\}, S_{15} = \{1,5\}, S_{16} = \{2,4\}, S_{17} = \{2,5\}, S_{18} = \{2,6\},$
$S_{19} = \{3,5\}, S_{20} = \{3,6\}, S_{21} = \{4,6\}$

$S_{22} = \{1,2,3\}, S_{23} = \{2,3,4\}, S_{24} = \{3,4,5\}, S_{25} = \{4,5,6\}, S_{26} = \{5,6,1\},$
$S_{27} = \{6,1,2\}, S_{28} = \{1,2,4\}, S_{29} = \{1,2,5\}, S_{30} = \{1,2,6\}, S_{31} = \{1,3,4\},$
$S_{32} = \{1,3,5\}, S_{33} = \{1,3,6\}, S_{34} = \{1,4,5\}, S_{35} = \{1,4,6\}, S_{36} = \{1,5,6\},$
$S_{37} = \{2,3,5\}, S_{38} = \{2,3,6\}, S_{39} = \{3,4,6\}.$ (1.4)

$S_{40} = \{1,2,3,4\}, S_{41} = \{2,3,4,5\}, S_{42} = \{3,4,5,6\}, S_{43} = \{4,5,6,1\}, S_{44} = \{5,6,1,2\},$
$S_{45} = \{6,1,2,3\}, S_{46} = \{1,2,4,5\}, S_{47} = \{1,2,4,6\}, S_{48} = \{1,2,5,6\}, S_{49} = \{1,3,4,5\},$
$S_{50} = \{1,3,4,6\}, S_{51} = \{1,3,5,6\}, S_{52} = \{1,4,5,6\}, S_{53} = \{1,2,3,5\}, S_{54} = \{2,3,4,6\},$
$S_{55} = \{2,4,5,6\},$

$S_{56} = \{1,2,3,5,6\}, S_{57} = \{1,2,3,4,5\}, S_{58} = \{2,3,4,5,6\}, S_{59} = \{3,4,5,6,1\},$
$S_{60} = \{4,5,6,1,2\}, S_{61} = \{5,6,1,2,3\}, S_{62} = \{6,1,2,3,4\}, S_{63} = \{1,5,6\}, S_{64} = \{1,2,3,4,5,6\}$

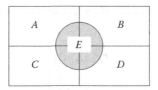

FIGURE 1.1
Venn diagram for the universal set and five different subsets in it.

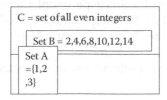

FIGURE 1.2
Venn diagram for Example 3.

Venn diagram To work with sets, we use geometrical or diagrammatic representation called the Venn diagram. Venn diagram uses an enclosed area to represent the elements of the set. The universal set is depicted as a rectangle (shown in Figure 1.1). The figure shows five sets, namely, A, B, C, D, and E. The intersection of the two sets A and E is represented as the common area for the two sets. Set E has common elements with all other sets.

Example 2

Specify the following sets with the rule-based method.

$$A = \{1, 2, 3\}, B = \{2, 4, 6, 8, 10, 12, 14\}, C = \{2, 4, 6, 8, ...\}$$

Solution: Set A has only three elements 1, 2, and 3. We can represent this set using the rule-based method. Similarly, sets B and C can be represented using the rule-based method.

$$A = \{1 < x < 3\}, \text{ where } x \text{ is an integer}$$

$$B = \{2x \text{ for } 4 < x < 7\}, \text{ where } x \text{ is an integer}$$

$$B \text{ is a set of even integers between 2 and 14}$$

$$C = \{\text{set of even integers}\}$$

(1.5)

Example 3

Draw the Venn diagram for the sets A, B, and C given in Example 2.

Solution: The universal set for this example is the set of all integers. The Venn diagram for three sets with a universal set is shown in Figure 1.2. Set A has a common element 2 with sets C and B. Set B is a subset of set C.

1.1.1 Union and Intersection

The union of the two sets is defined as the set of elements of two sets taken together. It consists of A or B or A and B. The union is represented by

$$C = A \cup B$$

(1.6)

The intersection of the two sets is defined as the set of elements common for both the sets. The intersection is denoted as

$$D = A \cap B \tag{1.7}$$

The complement of any set A is the set of all elements of the universal set, which are not included in set A. Let us first define the universal set. The universal set is defined as the set that consists of all elements for a particular situation or experiment. Let us consider the example of a rolling die. Here, for this experiment, there are only six possible elements or outcomes, namely, number 1 on the face, number 2 on the face, number 3 on the face, number 4 on the face, number 5 on the face, and number 6 on the face. We write this as the universal set $S = \{1, 2, 3, 4, 5, 6\}$. Let us define set A as $A = \{1, 2, 3\}$.

Now, we know that $A \subset S$; the complement of A is denoted by

$$\bar{A} = \{4, 5, 6\} = S - A \tag{1.8}$$

1.1.2 Algebra of Sets

Let us study the algebraic rules for the sets.

The three laws governing the union and intersection operations are as follows.

1. The Commutative law: It states that

$$A \cap B = B \cap A$$
$$A \cup B = B \cup A \tag{1.9}$$

2. The Distributive law: It states that

$$A \cap (B \cup C) = (A \cap B) \cup (A \cap C)$$
$$A \cup (B \cap C) = (A \cup B) \cap (A \cup C) \tag{1.10}$$

3. The Associative law: It states that

$$(A \cup B) \cup C = A \cup (B \cup C) = A \cup B \cup C$$
$$(A \cap B) \cap C = A \cap (B \cap C) = A \cap B \cap C \tag{1.11}$$

1.1.3 De Morgan's Laws

The law states that the complement of a union (intersection) of two sets, say A and B, is equal to the intersection (union) of the complements of sets A and B. We can write it as

$$\overline{(A \cup B)} = \bar{A} \cap \bar{B}$$
$$\overline{(A \cap B)} = \bar{A} \cup \bar{B} \tag{1.12}$$

These identities can be easily proved using the Venn diagrams.

1.1.4 Duality Principle

The duality principle states that if we replace the operator of union with the intersection and operator of intersection by the union, the identity is preserved. This is evident from De Morgan's laws stated earlier.

Example 4

Two sets are given by $A = \{-6, -4, -0.5, 0, 1.6, 8\}$ and $B = \{-0.5, 0, 1, 2, 3\}$. Find $A - B, B - A, A \cup B, A \cap B$.

Solution: Note that $A - B$ consists of all elements of A, which are not present in B. $B - A$ consists of all elements of B, which are not present in A.

$$A - B = \{-6, -4, 1.6, 8\}, B - A = \{1, 2, 3\}, \text{Note: } A - B \neq B - A$$

$$A \cup B = \{-6, -4, -0.5, 0, 1, 1.6, 2, 3, 8\}, A \cap B = \{-0.5, 0\} \tag{1.13}$$

Concept Check

- Define a set.
- When will you call the set as countable or uncountable?
- Define the universal set.
- Define $A - B$. Will $A - B = B - A$?
- Find the possible number of subsets if the universal set contains only two elements.
- What is a Venn diagram?
- Define the union and intersection of the two sets.
- Define De Morgan's laws.
- Define a duality principle.

1.2 Probability

Consider the experiment of throwing a die. The result of the throw is random in nature, in the sense that we do not know the result or the outcome of the experiment until the outcome is actually available. The set of all possible outcomes is called as a sample space and is a universal set for the given experiment. The randomness in the outcome is lost once the outcome is evident. The result of the experiment is a 1 dot, 2 dots, or 3 dots, and so on. The outcome for this experiment is a discrete random variable. For the experiment of throwing a die, the universal set can be written as $S = \{1, 2, 3, 4, 5, 6\}$.

We will define the term event. When any experiment is performed to generate some output, it is called an event. The output obtained is called as the outcome of the experiment. The possible set of outcomes of the experiment is called as a sample space or an event space. Many times, there is uncertainty or randomness involved in the outcome of the experiment. The term probability is closely related to uncertainty. In the case of the experiment of throwing of a die, the uncertainty is involved in the outcome. When a die is thrown, we do not know what will be the outcome. Using the notion, we may define a relative frequency of occurrence of some event. Consider the example of a fair coin. Let it be tossed n times. Let n_H times head appear. The relative frequency of occurrence of head R_H is defined as

$$R_H = \frac{\text{Number of times head occurs}}{\text{Total number of times the coin is tossed}} = \frac{n_H}{n} \tag{1.14}$$

The relative frequency approaches a fixed value called probability when n tends to infinity. We define probability as

$$\text{Probability} = \lim_{n \to \infty} \frac{n_H}{n} \tag{1.15}$$

Using the definition of probability as a frequency of occurrence, we can easily conclude that the probability P is always less than 1. When $P = 1$, the event becomes a certain event.

Let us consider the example of an event. When a die is thrown, say number "1" comes up. This is an event. This experiment spans a discrete random space as the output of the experiment is discrete, and the total number of outcomes is finite and equal to 6. Discrete random spaces are easier to deal with as the events are countable. Continuous sample spaces are generated when the outcomes of the experiment are real numbers. Examples include acquisition of speech samples and measurement of temperature.

To understand the concept of relative frequency, we will consider one example.

Example 5

Let there be 100 resistors in a box, all are of the same size and shape. There are 20 resistors of 10 Ω, 30 resistors of 22 Ω, 30 resistors of 33 Ω, and 20 resistors of 47 Ω. Let us consider the event of drawing one resistor from the box. Find the probability of drawing 10, 22, 33, and 47 Ω resistors.

Solution: The concept of relative frequency suggests that

$$p(10\,\Omega) = \frac{20}{100},\ p(22\,\Omega) = \frac{30}{100},$$
$$p(33\,\Omega) = \frac{30}{100},\ \text{and}\ p(47\,\Omega) = \frac{20}{100} \tag{1.16}$$

Total probability sums to 1.

Example 6

If a single (fair) die is rolled, determine the probability of each of the following event.

1. Obtaining the number 3
2. Obtaining a number >4
3. Obtaining a number <3 and obtaining a number ≥2

Solution:

1. Obtaining the number 3
   Because the die we are using is fair, each number has an equal probability of occurring; therefore,

$$P(3) = \frac{1}{6}$$

2. Obtaining a number >4
   In this case, the set of possible outcomes is given by (5, 6). Note that these are all mutually exclusive events. Therefore,

$$P(5 \cup 6) = P(5) + P(6) = \frac{1}{6} + \frac{1}{6} = \frac{2}{6} = \frac{1}{3} \qquad (1.17)$$

3. Obtaining a number <3

$$A = \{\text{a number} < 3\} = \{1, \ 2\} \ P(A) = \frac{1}{3} \qquad (1.18)$$

$$B = \{\text{a number} \geq 2\} = \{2, \ 3, \ 4, \ 5, \ 6\}$$

$$P(B) = \frac{5}{6} \qquad (1.19)$$

1.2.1 Conditional Probability

In the case of random experiments, sometimes the probability of occurrence of one event may depend on the other event. The probability of occurrence of such an event is termed as the conditional probability. Consider one example. Let a card be drawn from a pack of cards. Let the card drawn be the heart card. Let us now define another event of drawing the second card if the first card drawn is heart card. The probability of occurrence of second event depends on the first event, i.e., if the first card drawn is heart card, the probability of second event is 12/51. This is because when the first card drawn is the heart card, the number of heart cards remaining in the pack of cards is 12, and the total number of cards remaining in a pack of cards is 51 as the card drawn in the first draw is not replaced. If the first card drawn is not a heart card, then the probability of drawing the second card as the heart card is 13/51.

We define the conditional probability as follows. Probability of occurrence of event A given that event B has occurred is called conditional probability and is given by

$$P(A/B) = \frac{P(A \cap B)}{P(B)} \qquad (1.20)$$

(*Note:* Conditional probability is a defined quantity and cannot be proven.)

Example 7

Let the event A be drawing a heart card and event B be drawing a heart card in the second draw. Find $P(A \cap B)$.

Solution: In the example of a pack of cards, $(A \cap B)$ represents the intersection of the sample space for the two events, namely, the heart card is drawn in the first draw and the heart card is drawn in the second draw. $P(A)$ is the probability of occurrence of event A, namely, drawing the heart card in the first draw. $P(A) = \frac{1}{4}$ and $P(B/A) = \frac{12}{51}$, as explained earlier.

$$\text{So, } P(A \cap B) = P(B/A) \times P(A) = \frac{12}{51} \times \frac{1}{4} = \frac{12}{204} \tag{1.21}$$

Example 8

Let there be balls in a box. There are balls of two colors, namely, red and blue. There are three red balls and two blue balls. Find the probability of picking up a red ball first and a blue ball on the second pick.

Solution: Let us define the event A as picking up a red ball on the first pick. Event B is picking up a blue ball. We will find the probabilities.

$$P(A) = \frac{3}{5}, P(B) = \frac{2}{5}, P(B/A) = \frac{2}{4}, \text{as after the first draw, the number of balls remaining is 4.}$$

$$P(A \cap B) = P(B/A) \times P(A) = \frac{2}{4} \times \frac{3}{5} = \frac{3}{10} \tag{1.22}$$

If the two events A and B are mutually exclusive, $P(A \cap B) = 0$ and $P(B/A) = 0$.

Example 9

Let there be three events with probabilities given by $P(A) = 44/100$, $P(B) = 62/100$, and $P(C) = 32/100$. Let the joint probabilities be given by $P(A \cap B) = 28/100, P(A \cap C) = 0$, and $P(B \cap C) = 24/100$. Find the conditional probabilities $P(A/C)$, $P(A/B)$, and $P(B/C)$.

Solution:

$$P(A/C) = \frac{P(A \cap C)}{P(C)} = \frac{0}{\frac{32}{100}} = 0 \tag{1.23}$$

$$P(B/C) = \frac{P(B \cap C)}{P(C)} = \frac{\frac{24}{100}}{\frac{32}{100}} = \frac{3}{4} \tag{1.24}$$

$$P(A/B) = \frac{P(A \cap B)}{P(B)} = \frac{\frac{28}{100}}{\frac{62}{100}} = \frac{14}{31} \tag{1.25}$$

Example 10

Consider a box with five 100 Ω resistors and two 1000 Ω resistors. We remove two resistors in succession. What is the probability that the first resistor is 100 Ω and the second is 1000 Ω?

Solution:
Let A = {draw a 100 Ω resistor}
Let B = {draw a 1000 Ω resistor in the second draw}

$P(A) = 5/7$, we assume that the first resistor drawn is not replaced.
$P(B/A) = 2/6 = 1/3$. As there are only four resistors of 100 Ω and two of 1000 Ω

$$P(A \cap B) = P(B / A)P(A)$$

$$= \frac{1}{3} \times \frac{5}{7} = \frac{5}{21} \tag{1.26}$$

Example 11

A dodecahedron is a solid object with 12 equal faces. It is frequently used as a calendar paperweight with a month name placed on each face. If such a calendar is randomly placed on a desk, the outcome is taken to be the month of the upper face.

1. What is the probability of the outcome being February?
2. What is the probability of the outcome being January or April or August?
3. What is the probability that the outcome will be a month with 31 days?
4. What is the probability that the outcome will be a month in the third quarter of the year?

Solution:

1. What is the probability of the outcome being February?
   Let A be the event "outcome being February"

$$P(A) = \frac{1}{12} \tag{1.27}$$

2. What is the probability of the outcome being January or April or August?
   Let B be the event "outcome being January or April or August"

$$P(B) = \frac{1}{12} + \frac{1}{12} + \frac{1}{12} = \frac{1}{4} \tag{1.28}$$

3. What is the probability that the outcome will be a month with 31 days?
   We will find months with 31 days. They are listed in set C.
   C = {January, March, May, July, August, October, December}

$$P(C) = \frac{7}{12} \tag{1.29}$$

4. What is the probability that the outcome will be in the third quarter of the year?
   Let D be the event "outcome will be a month in the third quarter of the year"
   B = {July, August, September}

$$P(D) = \frac{3}{12} = \frac{1}{4} \tag{1.30}$$

Example 12

Two fair, six-sided dice are thrown. Find the probability of

1. Throwing a sum of 11
2. Throwing two 7s
3. Throwing a pair

Solution:

1. Throwing a sum of 11
   Let A be the event of "sum of 11."

$$A = \{(5,\ 6), (6,5)\}$$

$$P(A) = \frac{2}{36} = \frac{1}{18} \tag{1.31}$$

2. Throwing two 7s
   Let B be the event of "two 7s."

$$B = \{\phi\}$$
$$P(B) = 0 \tag{1.32}$$

3. Throwing a pair
   Let C be the event of "throwing a pair." There will be six pairs out of 36 possible outcomes.

$$P(C) = \frac{1}{6} \tag{1.33}$$

Example 13

In the network of switches shown in Figure 1.3, each switch operates independently of all others and each switch has a probability of 0.5 of being closed. What is the probability of a complete path through the network?

Solution:
Let A be the event of a complete path through the network.
   The path in the network will be completed by the following switch combination.

$$\{(S1, S2, S3),\ (S1, S4),\ (S1, S2, S4),\ (S1, S3, S4),\ (S1, S2, S3, S4)\}$$

Therefore, the probability of a complete path through the network is

$$P(A) = \frac{1}{2} \times \frac{1}{2} \times \frac{1}{2} + \frac{1}{2} \times \frac{1}{2} + \frac{1}{2} \times \frac{1}{2} \times \frac{1}{2} + \frac{1}{2} \times \frac{1}{2} \times \frac{1}{2} + \frac{1}{2} \times \frac{1}{2} \times \frac{1}{2} \times \frac{1}{2} = \frac{11}{16} \tag{1.34}$$

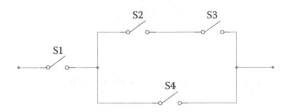

FIGURE 1.3
Combination of connecion of four switches.

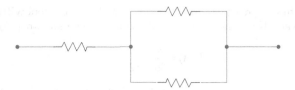

FIGURE 1.4
Configuration for connecting three resistors in random manner.

Example 14

A technician has three resistors, with values 100, 300, and 900 Ω. They are connected randomly in the configuration, as shown in Figure 1.4; now, answer the following questions.

1. What is the probability that the equivalent resistance of this network will be 390 Ω?
2. What is the probability that the equivalent resistance will be >390 Ω?
3. What is the probability that the equivalent resistance will be <1000 Ω?

Solution:
All possible resistor configurations will form a set.

$$S = \{(100 + 300||900), (100 + 900||300), (300 + 100||900), (300 + 900||100),$$

$$(900 + 100||300), (900 + 300||100)\}$$

1. Probability *P(A)* of getting equivalent resistance of 390 Ω can be found by tracking the combinations that give 390 Ω. They are 300 + 100||900 and 300 + 900||100.

$$P(A) = \frac{2}{6} = \frac{1}{3}. \tag{1.35}$$

2. Probability *P(B)* of getting equivalent resistance >390 Ω can be found by tracking combinations that give equivalent resistance of >390 Ω. They are 900 + 100||300, 900 + 300||100.

$$P(B) = \frac{1}{3} \tag{1.36}$$

3. What is the probability *P(C)* that the equivalent resistance will be <1000 Ω? This will happen for all resistor combinations.

$$P(C) = 1 \tag{1.37}$$

Example 15

A fair, six-sided die is thrown. If the outcome is odd, the experiment is terminated. If the outcome is even, the die is tossed a second time. Find the probability of

1. Throwing a sum of 7
2. Throwing a sum of 2

Solution:
Let us find the sample space for the experiment.

$$S = \{(2,1),(2,2),(2,3),(2,4),(2,5),(2,6),(4,1),(4,2),(4,3),(4,4),(4,5),(4,6),$$

$$(6,1),(6,2),(6,3),(6,4),(6,5),(6,6)\}$$

Note that if the first die has an outcome of 1, 3, or 5, the experiment will be terminated. Let *A* be the event "throwing a sum of 7." Possible combinations are (4,3), (2,5), and (6,1).

$$P(A) = \frac{3}{36} = \frac{1}{12} \tag{1.38}$$

Let *B* be the event "throwing a sum of 2" here, but in this case, the only possibility is (1,1); however, this is to be discarded, as when the first die gives an odd output, the experiment is terminated.

$$P(B) = \frac{0}{36} = 0 \tag{1.39}$$

1.2.2 Bayes' Theorem

Bayes' theorem is also called as the theorem of inverse probability. Consider an example. Let us start with the definition of conditional probability. Let there be one event *A* with sample space *S*. Let there be *n* number of events B_n, which are mutually exclusive and which have the total sample space of *S*. The definition of conditional probability says that

$$P(A/B_n) = \frac{P(A \cap B_n)}{P(B_n)} \quad \text{if } P(B_n) \neq 0$$

$$\text{or } P(B_n/A) = \frac{P(A \cap B_n)}{P(A)} \quad \text{if } P(A) \neq 0 \tag{1.40}$$

Let us equate the two equations stated in Equation 1.26. We get the statement of Bayes' theorem.

$$P(B_n/A) = \frac{P(A/B_n)P(B_n)}{P(A)} \tag{1.41}$$

We know that event *A* spans the sample space *S*, which is spanned collectively by all *n* events, namely, B_n.

$$A \cap S = S = A \cap \left(\sum_{n=1}^{N} B_n \right) = \sum_{n=1}^{N} (A \cap B_n);$$

$$P(A) = \sum_{n=1}^{N} P(A \cap B_n) = \sum_{k=1}^{n} P(A/B_n)P(B_n) \tag{1.42}$$

$$= P(A/B_1)P(B_1) + P(A/B_2)P(B_2) + \cdots P(A/B_N)P(B_N)$$

We can write Equation 1.27 as

$$P(B_n/A) = \frac{P(A/B_n)P(B_n)}{P(A)} = \frac{P(A/B_n)P(B_n)}{\sum_{n=1}^{N} P(A/B_n)P(B_n)} \tag{1.43}$$

Example 16

Let us define the events as follows.

> A—Picking up a diamond card
> B—Picking up a spade card
> C—Picking up a club card
> D—Picking up a heart card
> E—Picking up a king

Events A, B, C, and D are mutually exclusive. That is, their sample space is exclusively different and there is no overlapping. Consider the events A and E. They are not mutually exclusive. There is some overlapping space between the two events. The event spaces are illustrated in Figure 1.5.

Find the inverse probability, namely, if the king is drawn, what is the probability that it is the diamond card? We are now interested in $P(A/E)$.

Solution:
$P(A/E)$ is the probability of picking a diamond card given that it is a king.

Let us find $P(A/E)$.

$$P(A/E) = \frac{P(A \cap E)}{P(E)}$$
(1.44)

$$P(A \cap E) = P(A/E) \times P(E) = \frac{1}{4} \times \frac{4}{52} = \frac{1}{52}$$

Now,

$$P\left(\frac{A}{E}\right) = \frac{P(A \cap E)}{P(E)}$$
(1.45)

And,

$$P(E) = P(E \cap A) + P(E \cap B) + P(E \cap C) + P(E \cap D)$$
(1.46)

So,

$$P(A/E) = P(E/A) \times \frac{P(A)}{P(E \cap A) + P(E \cap B) + P(E \cap C) + P(E \cap D)}$$
(1.47)

Note that we have used Equation 1.45. This is known as Bayes' theorem.

Let us calculate $P(A/E)$.

$$P(A/E) = \frac{\dfrac{1}{13} \times \dfrac{1}{4}}{\dfrac{1}{52} + \dfrac{1}{52} + \dfrac{1}{52} + \dfrac{1}{52}} = \frac{\dfrac{1}{52}}{\dfrac{4}{52}} = \frac{1}{4}$$
(1.48)

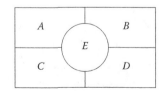

FIGURE 1.5
Sample spaces for the events.

1.2.2.1 *Alternative Statement of Bayes' Theorem*

Bayes' theorem has application in finding the class membership value based on the value of selected feature. Let us consider a problem of class selection based on a single feature, say "x." Let $P(x)$ denote the probability distribution of x in the entire population. Let $P(C)$ denote the probability that the sample belongs to class C. $P(x/C)$ is a conditional probability for feature value x given that a sample comes from class C. The goal of the experiment is to find the inverse probability, namely, probability that a sample belongs to class C given that the feature value is x, i.e., $P(C/x)$.

We know that

$$P(x/C) = \frac{P(x \cap C)}{P(C)} \text{ and } P(x/C) \times P(C) = P(x \cap C) = P(C \cap X) \qquad (1.49)$$

Hence,

$$P(x/C) \times P(C) = P(C \cap X) = P(C/x) \times P(x) \qquad (1.50)$$

We can write Bayes' theorem as

$$P(x/C) \times P(C) = P(C/x) \times P(x)$$

$$P(C/x) = [P(x/C) \times P(C)]/P(x) \qquad (1.51)$$

Example 17

Let A and B be the nonmutually exclusive events. Draw the Venn diagram for the two events and verify the relation: $P(A \cap B) = P(A) + P(B) - P(A \cup B)$.

Solution: Let us draw the Venn diagram to show nonmutually exclusive events. Figure 1.6 shows a Venn diagram for such events.

The sample space spanned by events A and B, namely, $P(A)$ and $P(B)$, is shown by two circles in the diagram. The symbol $P(A \cap B)$ represents the area common to two events. The symbol $P(A \cup B)$ is a space spanned by two events taken together. It can be easily verified from the Venn diagram that $P(A \cap B) = P(A) + P(B) - P(A \cup B)$. Obviously, in the case of two mutually exclusive events, $P(A \cap B) = P(A) + P(B)$, as the common area or the intersection is zero.

(*Note*: $P(A \cup B)$ is called the joint probability for the two events A and B.)

Example 18

Consider a binary symmetric channel with *a priori* probabilities $P(B_1) = 0.6$, $P(B_2) = 0.4$. *A priori* probabilities indicate the probability of transmission of symbols 0 and 1 before

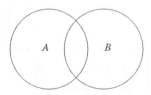

FIGURE 1.6
Venn diagram for two mutually nonexclusive events A and B.

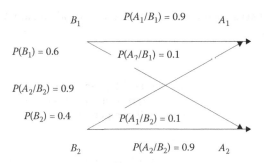

FIGURE 1.7
Binary symmetric channel and probabilities for Example 18.

the experiment is performed, i.e., before transmission takes place. The conditional probabilities are given by $P(A_1/B_1) = 0.9$; $P(A_2/B_1) = 0.1$ and $P(A_2/B_2) = 0.1$; $P(A_2/B_2) = 0.9$. Find received symbol probabilities. Find the transmission probabilities for correct transmission and transmission with error.

(*Note*: Conditional probabilities are the same for symmetrical links. Hence, they are termed symmetric channels. Let A_1 and A_2 indicate received symbols and B_1 and B_2 transmitted symbols, as shown in Figure 1.7.)

Solution: The received symbol probabilities can be calculated as

$$P(A_1) = P(A_1 \, / \, B_1) \times P(B_1) + P(A_1 \, / \, B_2) \times P(B_2)$$
$$= 0.9 \times 0.6 + 0.1 \times 0.4 = 0.58 \tag{1.52}$$

$$P(A_2) = P(A_2 \, / \, B_1) \times P(B_1) + P(A_2 \, / \, B_2) \times P(B_2)$$
$$= 0.1 \times 0.6 + 0.9 \times 0.4 = 0.42 \tag{1.53}$$

The probability of reception at both the received symbols is contributed from the symbols transmitted from both the inputs as indicated in Figure 1.7.

The probabilities for correct symbol transmission are given by

$$P(B_1/A_1) = \frac{P(A_1/B_1) \times P(B_1)}{P(A_1)}$$
$$= 0.9 \times 0.6/0.58 \approx 0.931 \tag{1.54}$$

$$P(B_1/A_2) = \frac{P(A_2/B_2) \times P(B_2)}{P(A_2)}$$
$$= 0.9 \times 0.4/0.42 \approx 0.857 \tag{1.55}$$

Now, let us calculate probabilities of error.

$$P(B_1/A_2) = \frac{P(A_2/B_1) \times P(B_1)}{P(A_2)}$$
$$= 0.1 \times 0.6/0.42 \approx 0.143 \tag{1.56}$$

$$P(B_2/A_1) = \frac{P(A_1/B_2) \times P(B_2)}{P(A_1)}$$
$$= 0.1 \times 0.4/0.58 \approx 0.069 \tag{1.57}$$

The application of Bayes' theorem is mostly in the medical domain. We will go through a couple of examples.

Example 19

What is the probability that a person has a cold given that he/she has a fever? The probability of a person having cold $P(C) = 0.01$, probability of fever in a given population is $P(f) = 0.02$, and probability of fever given that a person has a cold $P(f/C) = 0.4$.

Solution:

$$P(C/f) = \frac{P(f/C) \times P(C)}{P(f)}$$

$$= 0.4 \times 0.01 / 0.02 = 0.2 \qquad (1.58)$$

Example 20

Consider detecting HIV virus using ELISA test. The probabilities are specified as follows. Probability of having HIV is $P(H) = 0.15$. Probability of not having HIV is $P(\bar{H}) = 0.85$. Probability for testing positive given that person has HIV is $P(\text{Pos}/H) = 0.95$ and probability for testing positive given that person is not having HIV is $P(\text{Pos}/\bar{H}) = 0.0.02$. Find the probability that a person has HIV given that the test is positive.

Solution:

$$P(H/\text{Pos}) = \frac{P(\text{Pos}/H) \times P(H)}{P(H) \times P(\text{Pos}/H) + P(\bar{H}) \times P(\text{Pos}/\bar{H})}$$

$$= \frac{0.15 \times 0.95}{0.15 \times 0.95 + 0.85 \times 0.02} = 0.893 \qquad (1.59)$$

Note that we have used Equation 1.45.

1.2.3 Mutually Exclusive and Independent Events

Let us first define independent events. The two events are said to be statistically independent if the probability of occurrence of one event is not affected by the occurrence of the other event. This can be mathematically written as

$$P(A/B) = P(A) \text{ or } P(B/A) = P(B) \qquad (1.60)$$

We know that

$$P(A/B) = \frac{P(A \cap B)}{P(B)}$$

$$P(A)P(B) = P(A \cap B) \qquad (1.61)$$

The probability of joint occurrence of the events is equal to the product of their probabilities if the events are statistically independent. Condition stated in Equation 1.60 is sufficient for deciding the independence of the two events. Condition stated in Equation 1.60 can be used as a test to decide the independence of the two events. If the two events are statistically independent, the problems involving the probabilities get greatly simplified.

We will now define mutually exclusive events. The two events are said to be mutually exclusive if their joint probability is zero. The Venn diagram will show that the intersection is zero. This can be mathematically stated as

$$P(A \cap B) = 0 \qquad (1.62)$$

We can conclude that if the two events are statistically independent, they cannot be mutually exclusive.

Let us consider an example to illustrate the concepts.

Example 21

Let event A be the drawing of a heart card from a pack of cards. Event B is drawing a king and event C is drawing a jack card. Find which two events are statistically independent and which two events are mutually exclusive.

Solution: Let us find the probabilities of the events and probabilities of the joint events.

$$P(A) = \frac{13}{52}, \, P(B) = \frac{4}{52}, \, P(C) = \frac{4}{52} \qquad (1.63)$$

$$P(A \cap B) = \frac{1}{52}, \, P(B \cap C) = 0, \, P(A \cap C) = \frac{1}{52}$$

Events B and C are mutually exclusive.

$$P(A)P(B) = \frac{1}{52} = P(A \cap B), \, P(B)P(C) = \frac{1}{169} \neq P(B \cap C),$$

$$P(A)P(C) = \frac{1}{52} = P(A \cap C) \qquad (1.64)$$

We can conclude that events A and B are not statistically independent. Events A and C are statistically independent.

Concept Check

- Define the term event.
- Define probability in terms of relative frequency of occurrence.
- What is conditional probability?
- Define a sample space for a random experiment.
- State the relation between conditional probability and the intersection of the sample spaces of the two events.
- When will you classify the events as mutually exclusive?
- State Bayes' theorem in two different ways.
- What will a Venn diagram for mutually exclusive events look like?
- When will you call a channel as symmetrical?

1.3 Random Variable

We have considered the examples of random experiments like tossing of a coin and throwing a die. The outcome of these experiments was random in the sense that the outcome cannot be predicted. The sample space spanned by the outcomes is also known. These are called as discrete random outcomes, with sample space consisting of only finite values. A random variable can be defined as the characterization of the outcome of the random experiment, which will associate a numerical value with each outcome. Consider the example of throwing a die. Let us denote the random variable associated with the outcome of the experiment as "X". The range of "X" is from 1 to 6. Every time a die is thrown, a value corresponding to the number on the face is associated with the outcome. If number "1" appears, "X" = 1. If the random variable takes on only discrete values, the random variable is called a discrete random variable. In contrast, if the random variable takes on real values, it is called continuous random variable. Here, the range of random variable is infinite.

1.3.1 Cumulative Distribution Function (CDF)

Let us now define the concept of distribution function. CDF is defined as the probability that the random variable "X" takes a value less than or equal to some allowed value "x." The value is allowed if it belongs to a sample space. CDF is defined by the following equation.

$$F_X(x) = P(X \leq x) \tag{1.65}$$

Example 22

Consider a fair die. Plot a CDF versus "x".

Solution: We consider a fair die. This indicates that the probability of getting any value on the face is equally likely. There are six possible outcomes. Hence, the probability of getting each face is 1/6, as the total probability is 1. We will now find CDF for each value of "x."

$$
\begin{aligned}
F_X(1) &= P(X \leq 1) = \frac{1}{6} \\
F_X(2) &= P(X \leq 2) = P(1) + P(2) = \frac{2}{6} \\
F_X(3) &= P(X \leq 3) = P(1) + P(2) + P(3) = \frac{3}{6} \\
F_X(4) &= P(X \leq 4) = P(1) + P(2) + P(3) + P(4) = \frac{4}{6} \\
F_X(5) &= P(X \leq 5) = P(1) + P(2) + P(3) + P(4) + P(5) = \frac{5}{6} \\
F_X(6) &= F(X \leq 6) = P(1) + P(2) + P(3) + P(4) + P(6) = 1
\end{aligned}
\tag{1.66}
$$

A plot of CDF is shown in Figure 1.8.

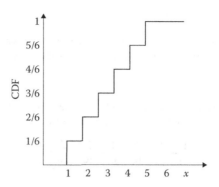

FIGURE 1.8

Plot of CDF for a toss of a single die.

The CDF $F_X(x)$ obeys the following properties.

1. $0 \le F_X(x) \le 1$
2. $\lim_{x \to \infty} F_X(x) = 1$
3. $\lim_{x \to -\infty} F_X(x) = 0$
4. $F_X(x)$ is a nondecreasing function of x
5. $P(X = x) = F_X(x) - \lim_{\varepsilon \to 0} F_X(x - \varepsilon)$
6. $F_X(x) = P(X \le x) = 1 - P(X > x)$

The earlier properties can be easily verified by the reader. The first property holds good because $F_X(x)$ is a probability. Second and third properties also follow from the concept of probability. For a certain event, i.e., the face value of a die if ≤6 is 1 and probability for no face value is 0, the fourth property is true because the probability is a positive value. Fifth property says that for a discrete random variable, the probability of getting a specific value is finite. We will see that for a continuous random variable, the probability of getting a specific value is 0. The sixth property indicates that the events $X \le x$ and $X > x$ are mutually exclusive.

Example 23

Consider a pair of dice. Find the probability of getting the sum of the faces as <6. Also find the probability for getting the sum of faces equal to 10.

Solution: We understand that the probability of getting any particular combination of faces is $1/6 \times 1/6 = 1/36$. The possible outcomes can be listed as shown in Table 1.1.

Total outcomes are 36. To find the probability of getting the sum of faces as <6, we have to count such combinations for which the sum of faces is ≤6. It can be verified that such a probability is 15/36.

To find the probability for getting the sum of faces equal to 10, let us list such combinations (5,5), (6,4), and (4,6). The probability is 3/36.

TABLE 1.1

Possible Outcomes for the Throw of a Pair of Dice

(1,1)	(2,1)	(3,1)	(4,1)	(5,1)	(6,1)
(1,2)	(2,2)	(3,2)	(4,2)	(5,2)	(6,2)
(1,3)	(2,3)	(3,3)	(4,3)	(5,3)	(6,3)
(1,4)	(2,4)	(3,4)	(4,4)	(5,4)	(6,4)
(1,5)	(2,5)	(3,5)	(4,5)	(5,5)	(6,5)
(1,6)	(2,6)	(3,6)	(4,6)	(5,6)	(6,6)

Example 24

Consider two unfair dice. The probability of getting a face value of 2 is 2/7 and the probability of getting a face value 3 is 2/7. Find the probability of getting a sum of 6 for two faces.

Solution:
To get a sum of 6 for two faces, we have the following combinations:
 (1,5), (5,1), (2,4), (4,2), and (3,3). Let us find the probability.

$$\frac{1}{49}+\frac{1}{49}+\frac{2}{49}+\frac{1}{49}+\frac{2}{49}=\frac{7}{49}=\frac{1}{7} \tag{1.67}$$

Example 25

A random variable has a distribution function given by

$$F_X(x)=0 \quad -\infty < x \le -10$$

$$=\frac{1}{6} \quad -10 \le x \le -5$$

$$=\frac{x}{15}+\frac{1}{2} \quad -5 < x < 5$$

$$=\frac{5}{6} \quad 5 \le x < 10$$

$$=1 \quad 10 \le x < \infty$$

Draw the CDF. Find $P(X \le 4)$ and $P(-5 \le X \le 4)$.

Solution: We will first plot the CDF. Figure 1.9 shows a plot of CDF.
 Let us find

$$P(X \le 4)=\frac{4}{15}+\frac{1}{2}=\frac{23}{30}=F_X(4) \tag{1.68}$$

$$P(-5 \le X \le 4)=F_X(4)-F_X(-5)=\frac{23}{30}-\frac{1}{6}=\frac{18}{30}=\frac{3}{5} \tag{1.69}$$

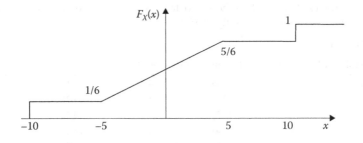

FIGURE 1.9
Plot of CDF for Example 25.

1.3.2 Probability Density Function (pdf)

pdf is defined as the derivative of the CDF for a continuous random variable.

$$f_X(x) = \frac{d}{dx}(F_X(x)) \tag{1.70}$$

In the case of a discrete random variable, pdf is defined in terms of the values of the probabilities at the discrete sample values. It can be stated as

$$
\begin{aligned}
f_X(x) &= P(X = x_i) \quad \text{if } x = x_i \;\; i = 1, 2, \ldots, n \\
&= 0 \qquad\qquad\quad \text{if } x \neq x_i
\end{aligned}
\tag{1.71}
$$

Example 26

Consider the experiment of throwing a die. There are six possible outcomes that are equally probable. Plot $f_X(x)$.

Solution: As all events are equally probable, the probability of each event is 1/6. Let us plot $f_X(x)$. Figure 1.10 shows a plot of pdf.

Example 27

Plot pdf for the CDF specified for Example 12.

Solution: The plot of pdf is a plot of the derivative of the graph shown in Figure 1.9. Figure 1.11 shows a plot of pdf.

The nature of pdf drawn can be explained as follows. Referring to Figure 1.4, we see that the value of the derivative at $x = -10$ is 1/6. The slope of the CDF graph is 0 for values

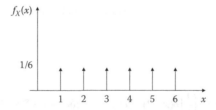

FIGURE 1.10
Plot of pdf.

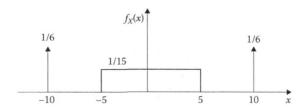

FIGURE 1.11
Plot of pdf for CDF specified in Example 27.

of x from -10 to -5 and for values of x from 5 to 10. For values of x between -5 and 5, the slope of CDF graph is

$$\frac{\frac{5}{6}-\frac{1}{6}}{10}=\frac{4}{6\times10}=\frac{1}{15} \tag{1.72}$$

(Refer to Figure 1.9 and Figure 1.11.)
   We can list the properties of pdf as follows.

1. $f_X(x)\geq0$

2. $\displaystyle\int_a^b f_X(x)dx = P(a\leq X \leq b)$

3. $\displaystyle\int_{-\infty}^{\infty} f_X(x)dx = 1$

The pdf being a probability must be a nonnegative quantity. If we integrate the pdf between some limits, it gives the probability for the variable to be between the limits specified. The integration of pdf over the entire range is nothing but the probability of a certain event. Hence, it is equal to 1.

Example 28

The pdf for different x values is given as follows: $x = 1$, pdf $= 0.2$; $x = 2$, pdf $= 0.1$; $x = 3$, pdf $= 0.3$; $x = 4$, pdf $= 0.3$; $x = 5$, pdf $= 0.1$. Draw the pdf and its corresponding CDF.

Solution: Let us first draw the pdf as shown in Figure 1.12.
   To draw the CDF, we have to integrate the pdf. CDF is shown in Figure 1.13.
   Note that we have to add the probability values at each step. At 1, there is a step of 0.2. Then it remains constant until 2. At 2, it has a step of 0.1 and it reaches 0.3. It remains constant until 3. At 3, it has a step of 0.3 and it reaches 0.6. It remains constant until 4. At 4, it has a step of 0.3 and it reaches 0.1. It remains constant until 5. At 5, it has a step of 0.1 and it reaches 1.0. The total value of probability must reach 1.

Example 29

The pdf for a random variable is given by

$$f_X(x)=\left\{\frac{1}{6}[\delta(x-2)+2\delta(x-3)+2\delta(x-4)+\delta(x-5)]\right\}$$

Draw the pdf and its corresponding CDF.

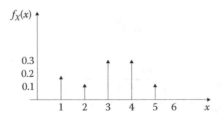

FIGURE 1.12
Plot of pdf for Example 28.

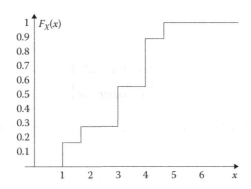

FIGURE 1.13
Plot of CDF for Example 28.

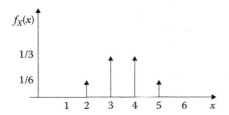

FIGURE 1.14
Plot of pdf for Example 29.

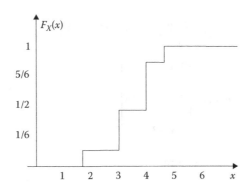

FIGURE 1.15
Plot of CDF for Example 29.

Let us first draw the pdf as shown in Figure 1.14.

To draw the CDF, we have to integrate the pdf. CDF is shown in Figure 1.15.

Note that we have to add the probability values at each step. At 2, there is a step of 1/6. Then it remains constant until 3. At 3, it has a step of 1/3 and it reaches 1/2. It remains constant until 3. At 4, it has a step of 1/3 and it reaches 5/6. It remains constant until 5. At 5, it has a step of 1/6 and it reaches 1. The total value of probability must reach 1.

Example 30

Find the density function such that,

$$f_X(x) = \begin{cases} cx & 0 < x < 3 \\ 0 & \text{otherwise} \end{cases}$$

Compute the probability $P(1 < x < 2)$. Find the distribution function $F_X(x)$.

Solution:

1. The pdf is specified. To find c, we will equate the total probability to 1.

$$\int_0^3 f_X(x)dx = 1$$

$$\int_0^3 cx\,dx = \frac{cx^2}{2}\Big|_0^3 = \frac{9c}{2} = 1 \Rightarrow c = \frac{2}{9} \tag{1.73}$$

$$f_X(x) = \begin{cases} \dfrac{2}{9}x & 0 < x < 3 \\ 0 & \text{otherwise} \end{cases}$$

The plot of pdf is shown in Figure 1.16.

2. Probability that $P(1 < x < 2)$

$$P(1 < x < 2) = \int_1^2 \frac{2}{9}x\,dx = \frac{2}{9} \times \frac{x^2}{2}\Big|_1^2 = \frac{2}{9}\left(2 - \frac{1}{2}\right) = \frac{1}{3} \tag{1.74}$$

3. Find the distribution function

$$F_X(x) = \int_0^x \frac{2}{9}x\,dx = \frac{2}{9} \times \frac{x^2}{2}\Big|_0^x = \frac{2}{9} \times \frac{x^2}{2} = \frac{x^2}{9}$$

$$F_X(x) = \begin{cases} 0 & x < 0 \\ \dfrac{x^2}{9} & 0 \le x \le 3 \\ 1 & x > 3 \end{cases} \tag{1.75}$$

Plot of CDF is shown in Figure 1.17.

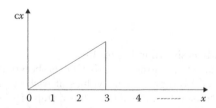

FIGURE 1.16
Plot of pdf for Example 30.

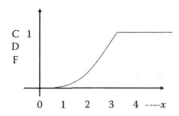

FIGURE 1.17
Plot of CDF for Example 30.

Example 31

Consider the function

$$f_X(x) = \begin{cases} \dfrac{1}{2}e^{-x} + \dfrac{1}{2}\delta(x-3) & \text{for } x \geq 0 \\ 0 & \text{for } x < 0 \end{cases}$$

1. Verify that the function represents a density function.
2. Calculate $P(X = 1)$, $P(X = 3)$, and $P(X \geq 1)$.

Solution:
To verify $f_X(x)$ represents a valid density function, we have to verify that

$$\int_0^\infty \left(\frac{1}{2}e^{-x} + \frac{1}{2}\delta(x-3) \right) dx = \frac{1}{2}(-e^{-x})\downarrow_0^\infty + \frac{1}{2}\delta(x-3)$$

put $x = 3$ for second term   (1.76)

$$= \frac{1}{2}[1+1] = 1$$

$$P(x = 1) = f_X(1) = \frac{e^{-1}}{2} = 0.1839$$

$$P(x = 3) = f_X(3) = \frac{e^{-3}}{2} + \frac{1}{2} = 0.5248$$

(1.77)

$$P(X \geq 1) = \int_1^\infty \left[\frac{1}{2}e^{-x} + \frac{1}{2}\delta(x-3) \right] dx$$

$$= -\frac{1}{2}e^{-x}\downarrow_1^\infty + \frac{1}{2} = 0.1839 + 0.5 = 0.6839$$

Concept Check

- Define a random variable.
- How will you define a discrete random variable?
- How will you define a continuous random variable?

- Define CDF.
- State the properties of CDF.
- Define pdf. How is it related to CDF?
- How will you find pdf if CDF is specified?
- List the properties of pdf.

1.4 Standard Distribution Functions

There are standard distribution functions defined for continuous random variables, namely, uniform probability distribution, Gaussian distribution, Rayleigh distribution, and exponential distribution. Discrete random variables use Poisson distribution and binomial distribution. We will first describe standard distribution functions for continuous random variables.

1.4.1 Probability Distribution Functions for Continuous Variables

1. *Uniform probability distribution*: Uniform pdf over a range of values between a and b is defined as

$$f_x(x) = K \quad \text{if } a \leq x \leq b$$
$$= 0$$

(1.78)

Let x be a uniformly distributed random variable between a and b. Let us plot the pdf. Figure 1.18 shows a pdf.

Let us find the value of K.

$$\int_a^b f_X(x)\,dx = \int_a^b K\,dx = 1$$

$$[Kx]_a^b = 1$$

(1.79)

$$K = \frac{1}{b-a}$$

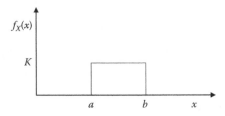

FIGURE 1.18
Plot of pdf for a uniform distribution.

2. *Gaussian random variable:* The pdf associated with Gaussian distribution function is given by

$$f_X(x) = \frac{1}{\sigma\sqrt{2\pi}} \exp\left[\frac{-(x-\mu)^2}{2\sigma^2}\right] \quad -\infty \leq x \leq \infty$$

(1.80)

$$\sigma > 0 \text{ and } -\infty < \mu < \infty$$

The Gaussian pdf is completely specified by σ and μ.

It is found that many physical situations follow the Gaussian distribution function. As the mean value increases, the entire curve for Gaussian distribution function shifts toward right. If the variance increases, the curve becomes wider. To find the probability that the random variable lies between $-\infty$ and some fixed value, we have to integrate the square function that is computationally involved task. To help the user in this regard, the tables are available to read the integral value. If the mean and variance change, the integration values in the table will change. The problem can be simplified if only a single table is prepared for the mean value of 0 and variance of 1. The random variable can be normalized to get a mean value of 0 and variance of 1. Let us illustrate the procedure for normalization.

The function $(x-\mu)/\sigma$ can be replaced by another variable y as $y = (x-\mu)/\sigma$. After replacing the function by a normalized random variable, we obtain the CDF for a normalized random variable y as

$$F_Y(y) = \int_{-\infty}^{y} \frac{1}{\sqrt{2\pi}} \exp\left[\frac{-(y)^2}{2}\right] dy \quad -\infty \leq x \leq \infty$$

(1.81)

Now, the table can be listed for the normalized random variable to find the integral value. We can then find the value for the actual variable x by putting the obtained value for y. The probability distribution function listed in Equation 1.47 is called the normal distribution function.

A MATLAB® program to plot Gaussian pdf is as follows.

```
clear all;
sigma=1;
mean=0;
x=(-10:0.1:10);
a=(x-mean).*(x-mean);
y=(1/(sigma*sqrt(2*pi)))*exp(-(((x-mean).*(x-mean))/(2*sigma^2)));
plot(x,y);xlabel('x');ylabel('pdf');title('Gaussian pdf');
```

A Gaussian pdf is plotted in Figure 1.19. If the variance sigma increases, the width of the curve increases as shown in Figure 1.20. If the width of the curve is large, it is said to have broad shoulders as shown in Figure 1.19. If the variance is small, the distribution is said to be peaked distribution as shown in Figure 1.21. If the mean is positive and it increases, the curve shifts toward right. If the mean is negative and it increases, the curve shifts toward left as shown in Figure 1.21.

Example 32

Find the probability of the event $(X \leq 5.5)$ for a Gaussian variable having a mean value of 3 and variance of 2.

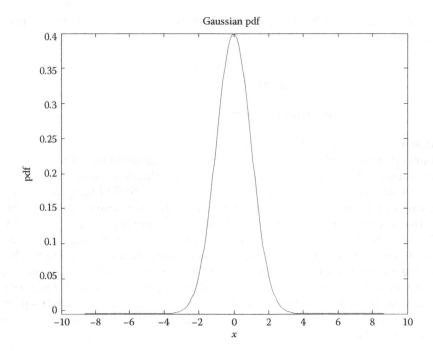

FIGURE 1.19
Plot of Gaussian pdf for value of sigma = 1 and mean = 0.

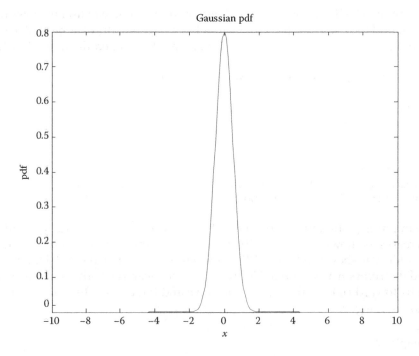

FIGURE 1.20
Plot of Gaussian pdf for value of sigma = 0.5 and mean = 0.

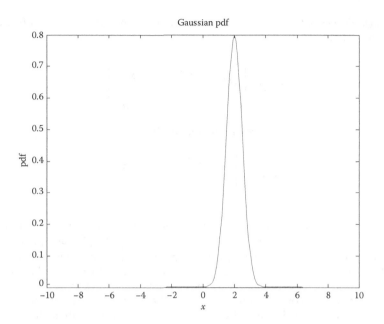

FIGURE 1.21
Plot of Gaussian pdf for value of sigma = 0.5 and mean = 2.

Solution:
We use the equation $y = (x - \mu)/\sigma$ and find the value of y.

$$y = \frac{(5.5 - 3)}{2} = \frac{2.5}{2} = 1.25$$

$$F_Y(y) = P(y \leq 1.25)$$

(1.82)

We will now use a table for a normalized distribution to find the value of CDF. Referring to the normal distribution function table in Appendix A, we can read $F(1.25) = P(Y \leq 1.25) = 0.8944$.

If we want to find the probability of the event $(X \geq 5.5)$, i.e., $(1 - P(Y \leq\leq 5.5))$, then referring to the table, we find the probability of the event as $(1 - 0.8944) = 0.1056$.

Example 33

Find the probability of the event $(X \leq 1)$ for a Gaussian variable having a mean value of 0 and variance of 1.

Solution:
We use the equation $y = (x - \mu)/\sigma$ and find the value of y. The variable is already normalized.

$$y = \frac{(1 - 0)}{1} = 1$$,

$$F_Y(y) = P(y \leq 1)$$

(1.83)

We will now use a table for a normalized distribution to find the value of CDF. Referring to the normal distribution function table in Appendix A, we can read $F_Y(1) = P(Y \leq 1) = 0.8413$.

1. *Rayleigh density function:* Rayleigh density function and distribution function can be stated as

$$f_X(x) = \begin{cases} \dfrac{2}{b^2}(x-a)e^{-(x-a)^2/2b^2} & x \geq a \\ 0 & x < a \end{cases}$$ (1.84)

$-\infty < a < \infty$ and $b > 0$

$$F_X(x) = \begin{cases} 1 - e^{-(x-a)^2/2b^2} & x \geq a \\ 0 & x < a \end{cases}$$ (1.85)

The Rayleigh occurs in several physical situations. Consider the example of a rifle aimed at a target. Let the distance of the target be r given by $r = \sqrt{x^2 + y^2}$. Let x and y represent the x and y coordinates of the distance. If x and y are statistically independent random variables that are Gaussian distributed with 0 mean and unit variance, then the pdf of r is said to be Rayleigh distributed. A second example of the Rayleigh distribution arises in the case of random complex numbers with real and imaginary components denoted by x and y, respectively. If the real and imaginary components are statistically independent Gaussian random variables with identical mean, then the absolute value of the complex number is Rayleigh distributed. Rayleigh distribution is not symmetrical. It is said to be skewed on one side. We will measure the skew for this variable in Section 2.1.

A MATLAB program to plot a Rayleigh density function for a particular value of b is given here. Raylpdf defines a Rayleigh pdf in MATLAB. Y = raylpdf(X, B) returns the rayleigh pdf with parameter B = 0.5 for range of values of X.

```
clear all;
x=[0:0.01:5];
b=0.5;
p=raylpdf(x,b);
plot(x,p); title('Rayleigh density function pdf'); xlabel('value of
x');ylabel('pdf');
```

Figure 1.22 shows a plot of Rayleigh density function for $b = 0.5$ and Figure 1.23 shows a plot of Rayleigh CDF for $b = 0.5$. A MATLAB program to plot CDF is as follows.

```
clear all;
x=[0:0.01:5];
b=0.5;
p=raylcdf(x,b);
plot(x,p); title('Rayleigh CDF'); xlabel('value of x');
ylabel('CDF');
```

1. *Exponential distribution function:* Exponential density function and distribution function can be stated as

$$f_X(x) = \begin{cases} \dfrac{1}{b}e^{-(x-a)/b} & x \geq a \\ 0 & x < a \end{cases}$$ (1.86)

$-\infty < a < \infty$ and $b > 0$

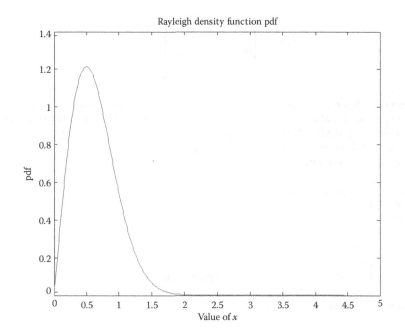

FIGURE 1.22
Plot of Rayleigh density function for $b = 0.5$.

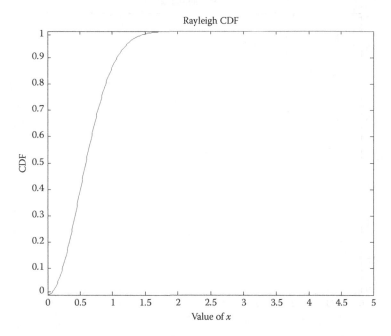

FIGURE 1.23
Plot of Rayleigh distribution function for $b = 0.5$.

$$F_X(x) = \begin{cases} 1-e^{-(x-a)/b} & x \geq a \\ 0 & x < a \end{cases}$$

$$-\infty < a < \infty \text{ and } b > 0$$

(1.87)

It is used to describe the size of a raindrop when large measurements in term conditions are done. It is also used to describe the fluctuations in the strength of the RADAR return signal. A MATLAB program to plot exponential pdf is shown. Figure 1.24 shows an exponential pdf graph. Similarly, we can plot exponential CDF, as shown here, and the plot of CDF is shown in Figure 1.25.

```
% To plot exponential pdf
clear all;
d=0.25;
c=0;
x=(0:0.1:10);
y=(1/d)*exp(-(x-c)/d);
plot(x,y);xlabel('x'); ylabel('pdf');title('exponential pdf');

% plot of exponential CDF
clear all;
d=0.25;
c=0;
x=(0:0.1:10);
y=1-exp(-(x-c)/d);
plot(x,y);xlabel('x'); ylabel('CDF'); title('exponential CDF');
```

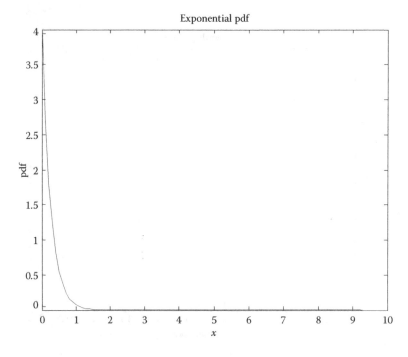

FIGURE 1.24
Plot of exponential pdf.

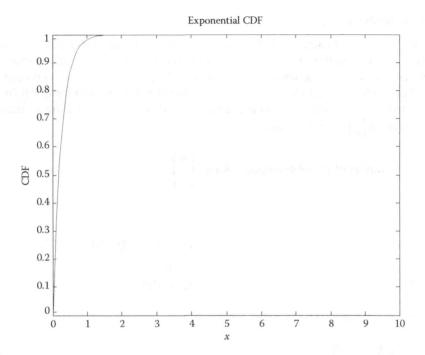

FIGURE 1.25
Plot of exponential CDF.

1.4.2 Probability Distribution Functions for Discrete Variables

We will discuss the standard distribution functions for discrete random variables. Before going through the details of distribution function, we will introduce some terminology like permutation and Bernoulli's trials.

1.4.2.1 Permutations

To understand this term, we will consider one simple example of a pack of cards. Let us say we have four successive draws. After every draw, the card drawn is not replaced. In this case, the sample space reduces after each draw. It becomes 51, 50, and 49 after first, second, and third draws, respectively. Here, the sequence in which the cards are drawn is important. Consider that there are n elements in the pack and there are n possible outcomes in the first draw, then $(n-1)$ possible outcomes for second draw, $(n-2)$ outcomes in third draw, etc. If r elements are drawn, the number of possible sequences or outcomes is $n(n-1)(n-2)\cdots(n-r)$.

We can write the possible outcomes for r draws as

$$\text{Number of possible outcomes} = P_r^n$$

$$= n(n-i)(n-2)\ldots(n-r+1) \tag{1.88}$$

$$\text{Permutations} = \frac{n!}{(n-r)!}$$

This number is called the number of permutations of r elements taken from n elements when the order of sequence of draw is important.

1.4.2.2 Combinations

Let us now define combinations. When the order of sequence of draw is not important, we get the possible number of combinations. In this case, the number is less than the permutation because different ordering of same elements is counted as one combination. The number of permutations reduced is equal to the number of permutations of r things taken from r elements, i.e., $r!$. Hence, the total number of combinations of r things taken from n things denoted as $\binom{n}{r}$ can be written as

$$\text{Number of possible combinations} = \binom{n}{r}$$

$$= \frac{P_r^n}{P_r^r} \qquad (1.89)$$

$$= n(n-i)(n-2)\ldots(n-r+1) / r!$$

$$= \frac{n!}{(n-r)!\,r!}$$

This number is also called the binomial coefficient.

1.4.2.3 Bernoulli's Trials

Consider an experiment where there are only two possible outcomes for every trial. For example, in the experiment of tossing a coin, there are two possible outcomes, namely, getting "Head" or getting "Tail." If we denote the probability of getting one outcome as P, then the probability of getting the other outcome is exactly $1 - P$. If the experiment is repeated a number of times, it is called Bernoulli's trials. When the experiment is repeated N times, we ask a question, what is the probability of getting "Head" exactly k times? Let us calculate this probability. We will first find the binomial coefficient indicating the number of times (k) we get the first outcome out of N number of trials. It is exactly the number of possible sequences that result in the same output. It is given by $\binom{N}{k} = \frac{N!}{(N-k)!\,k!}$. Let us now calculate the probability of getting the first event exactly k times out of N trials. It can be written as

$$P(\text{first oucome } k \text{ times out of } N) = \binom{N}{k} P^k (1-P)^{N-k}$$

$$= \frac{N!}{(N-k)!\,k!} P^k (1-P)^{N-k} \qquad (1.90)$$

Discrete random variables use either binomial distribution or Poisson distribution.

1.4.2.4 Binomial Distribution

Let us first discuss binomial density function. It is given by

$$f_X(x) = \sum_{k=0}^{N} \frac{N!}{(N-k)!\,k!} P^k (1-P)^{N-k} \delta(x-k) \qquad (1.91)$$

The term $\begin{pmatrix} N \\ k \end{pmatrix} = \dfrac{N!}{(N-k)!\,k!}$ is called the binomial coefficient. This density function is used in the case of Bernoulli's trials. When the density function is integrated, we get binomial distribution function given by

$$F_X(x) = \sum_{k=0}^{N} \frac{N!}{(N-k)!\,k!} P^k (1-P)^{N-k} u(x-k) \qquad (1.92)$$

Let us plot these functions for $P = 0.2$ and $N = 6$. A MATLAB program to plot binomial pdf and CDF is given. Figure 1.26 shows a plot of binomial pdf and Figure 1.27 shows a plot of binomial CDF.

```
clear all;
x=0:20;
y=binopdf(x,20,0.25);
stem(x,y);title('Binomial pdf function for N=20 and P=0.25');xlabel('x');
ylabel('pdf value');
figure;
clear all;
x=0:20;
y=binocdf(x,20,0.25);
stem(x,y);title('Binomial CDF function for N=20 and P=0.25');xlabel('x');
ylabel('CDF value');
```

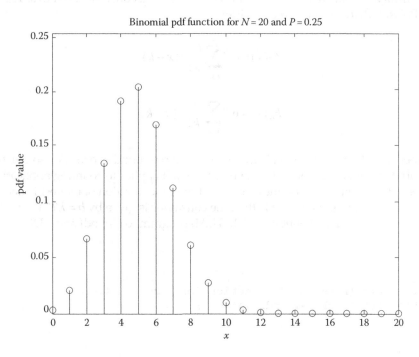

FIGURE 1.26
Plot of binomial pdf for $N = 20$ and $P = 0.25$.

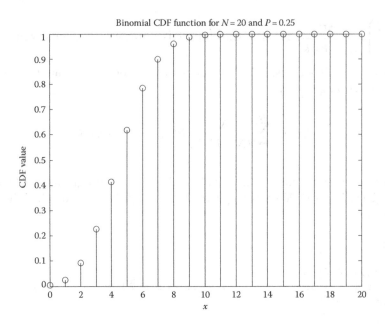

FIGURE 1.27
Plot of binomial CDF for $N = 20$ and $P = 0.25$.

1.4.2.5 Poisson Distribution

Poisson distribution can be considered as a special case of binomial distribution when N tends to infinity, P tends to zero such that NP product is a constant equal to b. The pdf and CDF for Poisson distribution can be written as

$$f_X(x) = e^{-b} \sum_{k=0}^{\infty} \frac{b^k}{k!} \delta(x-k) \qquad (1.93)$$

$$F_X(x) = e^{-b} \sum_{k=0}^{\infty} \frac{b^k}{k!} u(x-k) \qquad (1.94)$$

The number of customers arriving at any service station is found to obey Poisson distribution. This distribution finds application in queuing theory. It applies to counting applications such as number of telephone calls in some duration. If the time duration of interest is denoted as T and if the events occur at the rate of λ, then the constant b is given by $b = \lambda T$. We will plot the pdf and CDF for Poisson distribution. A MATLAB program to plot pdf and CDF is as follows.

```
clear all;
x=0:20;
y=poisspdf(x,0.7);
stem(x,y);title('Poisson pdf function for N=20 and lamda
=0.7');xlabel('x');ylabel('pdf value');
clear all;
x=0:20;
y=poisscdf(x,0.7);
stem(x,y);title('Poisson CDF function for N=20 and lamda
=0.7');xlabel('x');ylabel('CDF value');
```

Figure 1.28 shows a plot of Poisson pdf and Figure 1.29 shows a plot of Poisson CDF.

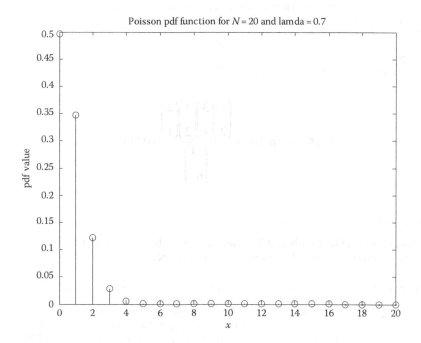

FIGURE 1.28
Plot of Poisson pdf for $N = 20$ and $\lambda = 0.7$.

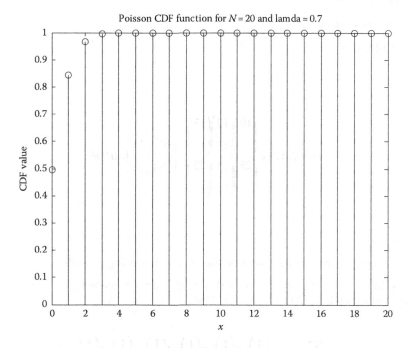

FIGURE 1.29
Plot of Poisson CDF for $N = 20$ and $\lambda = 0.8$.

Example 34

A box contains five red, three green, four blue, and two white balls. What is the probability of selecting a sample size of six balls containing two red, one green, two blue, and one white ball?

Solution:
In this case, the probability is given by

$$P(2R,1G,2B,1W) = \frac{\binom{5}{2}\binom{3}{1}\binom{4}{2}\binom{2}{1}}{\binom{14}{6}} = 0.1198 \tag{1.95}$$

Example 35

A box contains 10 black balls and 15 white balls. One ball at a time is drawn at random; its color is noted and the ball is then replaced in the box for the next draw.

1. Find the probability that the first white ball is drawn in the third draw.

Solution: The events are independent because the ball is replaced in the box, and thus the sample space does not change. Let B denote drawing a black ball and W denote drawing a white ball. The total number of balls in the sample space is 25. We will find the probability that the first white ball is drawn in the third draw. Hence, we have

1st draw $\rightarrow B$

2nd draw $\rightarrow B$

3rd draw $\rightarrow W$

Thus,

$$P(B,B,W) = \frac{\binom{10}{1}\binom{10}{1}\binom{15}{1}}{\binom{25}{1}\binom{25}{1}\binom{25}{1}} = \left(\frac{10}{25}\right)^2\left(\frac{15}{25}\right) = 0.096 \tag{1.96}$$

Example 36

Find the expected value of the points on the top face in tossing a fair die.

Solution: In tossing a fair die, each face shows up with a probability 1/6. Let X be the points showing on the top face of the die. Then,

$$E(X) = \sum_i x_i P_i = 1\left(\frac{1}{6}\right)+2\left(\frac{1}{6}\right)+3\left(\frac{1}{6}\right)+4\left(\frac{1}{6}\right)+5\left(\frac{1}{6}\right)+6\left(\frac{1}{6}\right) = 3.5 \tag{1.97}$$

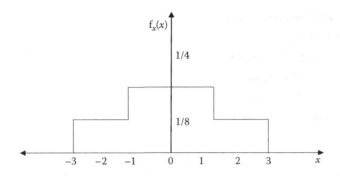

FIGURE 1.30
pdf for a variable.

Example 37

Consider the random variable X with the density function shown in Figure 1.30. Find $E[X]$ and variance of the random variable?

Solution:
The expected value of X is given as

$$E[X] = \int_{-3}^{-1} x\frac{1}{8}dx + \int_{-1}^{1} x\frac{1}{4}dx + \int_{1}^{3} x\frac{1}{8}dx = 0 \tag{1.98}$$

The mean square value $E[X^2]$ is given as

$$E[X^2] = 2\left[\int_{0}^{1} x^2\left(\frac{1}{4}\right)dx + \int_{1}^{3} x^2\left(\frac{1}{8}\right)dx\right] = \frac{7}{3} = 2.3333 = \sigma^2 \tag{1.99}$$

Because the mean is 0, the mean square value is equal to the variance $\sigma_x^2 = 7/3 = 2.3333$.

Concept Check

- Define a uniform pdf.
- Define a Gaussian pdf.
- Why is the variable used in Gaussian distribution normalized?
- How will you find probability of the event $(X \le x)$ for a Gaussian variable having a mean value of m and variance of v?
- What is the effect of change of mean and variance on the nature of the distribution plot in the case of Gaussian random variable?
- Define Rayleigh distribution.
- Give one example for Rayleigh distribution.
- Define exponential distribution function.

- Name the pdfs for a discrete random variable.
- Define permutations.
- Define the term combinations.
- What are Bernoulli's trials?
- Define a binomial distribution.
- Define Poisson distribution.

1.5 Central Limit Theorem, Chi-Square Test, and Kolmogorov–Smirnov Test

We will try to understand the procedure for the computer generation of a random variable. To understand this, we need to know the rules for the sum of random variables. If $Y = \frac{1}{\sqrt{M}}[X_1 + X_2 + \cdots X_M]$ denotes the sum of M random variables, then the density function for Y is given by

$$f_Y(y) = f_{X_1}(x_1) \times f_{X_2}(x_2) \times \cdots f_{X_M}(x_M) \tag{1.100}$$

The density function of Y is the $M-1$-fold convolution of M density function. The distribution function of Y can be found by integrating the density function.

1.5.1 Central Limit Theorem

The theorem states that the probability distribution function of a sum of a large number of random variables approaches a Gaussian distribution. The theorem does not guarantee that the density function will always be Gaussian. Under certain conditions, the density function is also Gaussian [1]. Let the mean values of M random variables be defined as $\mu_1, \mu_2, \ldots \mu_M$. Then the mean of sum of random variables denoted as Y can be written as

$$\mu_Y = \frac{1}{\sqrt{M}}[\mu_1 + \mu_2 + \cdots + \mu_M] \tag{1.101}$$

Consider the random variables as statistically independent. The variance of the sum variable can be written as

$$\sigma_Y^2 = \frac{1}{M}[\sigma_1^2 + \sigma_2^2 + \cdots + \sigma_M^2] \tag{1.102}$$

The theorem is found to hold good for special cases of dependent variables as well.

1.5.2 Computer Generation of a Gaussian Distributed Random Variable

Many times there is a need to simulate a system for estimating the performance before its actual implementation. The simulation of a system usually requires simulation of a random variable having a Gaussian distribution. A function like rand is available, for example, with MATLAB, for the generation of random variable with uniform distribution.

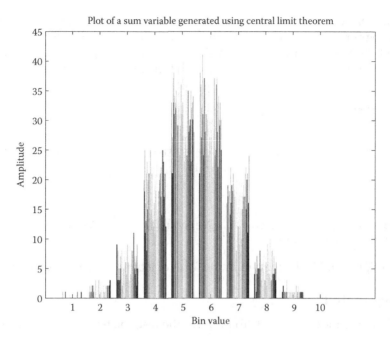

FIGURE 1.31

Histogram plot of the Gaussian-generated variable.

Let us take some 10 uniformly distributed random variables. We will use the addition of these variables and plot its density function. It can be shown to be a Gaussian density function to a good approximation using central limit theorem. Let us write a MATLAB program to generate a Gaussian variable as the sum of 12 uniformly distributed random variables. Figure 1.31 shows the histogram plot of the generated variable. We can easily verify that the envelope of the histogram approaches Gaussian pdf.

```
clear all;
x1=rand(100);
x2=rand(100);
x3=rand(100);
x4=rand(100);
x5=rand(100);
x6=rand(100);
x7=rand(100);
x8=rand(100);
x9=rand(100);
x10=rand(100);
y=(1/sqrt(10))*(x1+x2+x3+x4+x5+x6+x7+x8+x9+x10);
z=hist(y,10);
bar(z);title('plot of a sum variable generated using Central limit
theorem');xlabel('bin value');ylabel('amplitude');
```

Let us try to understand the meaning of histogram. We have used 10 random variables with uniform distribution from 1 to 100. These are generated using a rand function. The generated values of random variable are divided into, for example, 10 bins: one from 1 to 10, the other from 11 to 20, and so on. Now, the number of random variables lying in the range 1–10 is counted, 11–20 is counted, and so on. The histogram plots this count on the

y axis and the bin value on the x axis. If we increase the number of random variables to 1000, the histogram plot will further approach the Gaussian distribution.

Let us plot the histogram for a speech signal. Consider a speech segment of size 4012 samples. Let us write a MATLAB program to plot the histogram of a signal.

```
clear all;
fp=fopen('watermark.wav');
fseek(fp,224000,-1);
a=fread(fp,2048);
subplot(2,1,1);plot(a);title('plot of voiced part of a signal');
xlabel('sample no.');ylabel('amplitude');
b=hist(a,100);
figure;
bar(b);
title('plot of a histogram for a speech segment');xlabel('bin
value');ylabel('amplitude');
```

The plot of a signal and its histogram in the form of a bar graph are shown in Figure 1.32.

1.5.3 Chi-Square Test

It is also called goodness-of-fit test. To perform the test, the test statistic is calculated. The test statistic is defined as

$$\chi^2 = \sum_{i=1}^{N} \frac{(g_i - f_i)^2}{f_i} \tag{1.103}$$

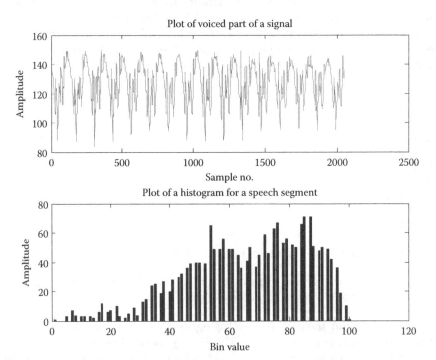

FIGURE 1.32
Plot of a signal and its histogram in the form of a bar graph.

When a histogram is drawn, the data values are divided into bins. The frequency of occurrence of the data in the bin interval is computed and plotted against the bin. To find the test statistic, we find the squared distance between the two frequency values, namely, observed frequency and frequency as per the required distribution. We will illustrate the procedure with a simple example of a uniform distribution.

Example 38

The experiment is conducted with a fair die. The die is tossed, for example, 120 times. The following data are obtained as shown in the table. Find the test statistic to check if it obeys a uniform distribution.

Face No.	1	2	3	4	5	6
Observed frequency	18	20	22	19	21	20

Solution: For uniform distribution, the face values are equi-probable. The expected frequency for each face is 20. Let us find the test statistic.

$$\chi^2 = \sum_{i=1}^{N} \frac{(g_i - f_i)^2}{f_i}$$

$$= \frac{(18-20)^2}{20} + \frac{(20-20)^2}{20} + \frac{(22-20)^2}{20} + \frac{(19-20)^2}{20} + \frac{(21-20)^2}{20} + \frac{(20-20)^2}{20} \tag{1.104}$$

$$= 0.5$$

This value of test statistic is less than the 5% level equal to $120 \times 0.05 = 6$. Hence, the distribution is uniform.

1.5.4 Kolmogorov–Smirnov Test

This test is used when the set of observations is from a continuous distribution. It detects the deviation form a specified distribution. Given a set of observations, we determine the following statistic.

$$D = \max |F(x) - G(x)| \tag{1.105}$$

In the equation, $G(x)$ is a sample distribution and $F(x)$ is a standard distribution against which the sample distribution is compared. If the value of D is less than the critical value specified in Kolmogorov–Smirnov (K–S) table, then the hypothesis that an unknown distribution is a known standard distribution is accepted.

Usually the signal of unknown distribution is to be analyzed. When the pdf of the random signal is not known, the mean and variance, for example, cannot be calculated as they require the mathematical expression of pdf. To solve this problem, the unknown signal is used to find the histogram and to compare with the standard known density functions discussed earlier. The extent to which the match occurs between unknown density/distribution and known density/distribution is evaluated as discussed, using chi-square test or K–S test. If the hypothesis for the known distribution can be accepted, then the mathematical expressions for the known density/distribution are used for computation of signal statistical properties.

Let us find the distance between speech signal histogram and the Gaussian distribution obtained using 12 uniformly distributed random variables. A MATLAB program is as follows.

```
clear all;
fp=fopen('watermark.wav');
fseek(fp,224000,-1);
a=fread(fp,2048);
subplot(2,1,1);plot(a);title('plot of voiced part of a signal');
xlabel('sample no.');ylabel('amplitude');
b=hist(a,100);
subplot(2,1,2);
bar(b);
title('plot of a histogram for a speech segment');xlabel('bin
value');ylabel('amplitude');
x1=rand(100);
x2=rand(100);
x3=rand(100);
x4=rand(100);
x5=rand(100);
x6=rand(100);
x7=rand(100);
x8=rand(100);
x9=rand(100);
x10=rand(100);
x11=rand(100);
x12=rand(100);
y=(1/sqrt(12))*(x1+x2+x3+x4+x5+x6+x7+x8+x9+x10+x11+x12);
z=hist(y,100);
figure;
bar(z);title('plot of a sum variable generated using Central limit
theorem');xlabel('bin value');ylabel('amplitude');
sum=0.0;
for i=1:100,
d(i)=abs(b(i)-z(i));
sum=sum+d(i);
end
disp(sum);
```

The distance obtained is 2004. Hence, the distribution for speech segment does not match that of Gaussian distribution.

Concept Check

- State central limit theorem.
- Define a procedure for computer generation of Gaussian random variable.

- Explain the procedure to find the histogram for a random variable.
- What is a chi-square test?
- Define the K–S test.
- How will you select a standard pdf for a given distribution?

Summary

We started with the introduction to probability theory. In this chapter, we have described different standard distribution functions and their use for signal analysis. The highlights for the chapter are as follows.

1. The term event is defined. When the experiment is performed to generate some output, it is called as an event. The output of the experiment is called as the outcome of the experiment. $\text{Probability} = \lim_{n \to \infty} \dfrac{n_H}{n}$ is defined in terms of relative frequency. The conditional probability is given by $P(A / B) = \dfrac{P(A \cap B)}{P(B)}$. Bayes' theorem is introduced, which is the theorem of inverse probability.

2. A random variable is defined. A random variable can be defined as the characterization of the outcome of the random experiment, which will associate a numerical value with each outcome. CDF is defined as the probability that the random variable "X" takes a value less than or equal to some allowed value "x". The pdf is defined as the derivative of the CDF. The properties of CDF and pdf are listed.

3. The standard distribution functions are described for continuous and discrete random variables. Uniform, Gaussian, Rayleigh, and exponential distributions are discussed for continuous random variables. Permutation and combination is defined. Bernoulli's trials are described. The binomial distribution and Poisson distribution are discussed for discrete random variables.

4. Central limit theorem is defined. The theorem states that the probability distribution function of a sum of a large number of random variables approaches a Gaussian distribution. Use of histogram to find the density function of the unknown signal is explained. The chi-square test is also known as goodness-of-fit test. To perform the test, the test statistic is calculated. The K–S test is used when the set of observations is from a continuous distribution. It detects the deviation from a specified distribution. When the unknown signal is to be analyzed, it is used to find the histogram and is compared with the standard known density functions discussed. The extent to which the match occurs between unknown density/distribution and known density/distribution is evaluated using chi-square test or K–S test. If the hypothesis for the known distribution can be accepted, then the mathematical expressions for the known density/distribution are used for the computation of signal statistical properties.

Key Terms

Probability
Conditional probability
Statistically independent variables
Mutually exclusive events
Events
Ensemble
Outcome of the experiment
Random variable
Continuous random variable
Discrete random variable
Uniform random variable
Gaussian random variable
Rayleigh distribution
Exponential distribution
Combinations
Permutations
Binomial distribution
Bernoulli's trials
Poisson distribution
pdf
CDF

Multiple-Choice Questions

1. The conditional probability is defined as

   a. $P\left(\dfrac{A}{B}\right) = \dfrac{P(A \cap B)}{P(B)}$

   b. $P\left(\dfrac{B}{A}\right) = \dfrac{P(A \cup B)}{P(B)}$

   c. $P\left(\dfrac{A}{B}\right) = \dfrac{P(A \cap B)}{P(A)}$

   d. $P\left(\dfrac{B}{A}\right) = \dfrac{P(A \cap B)}{P(B)}$

2. Bayes' theorem states that

   a. $P\left(\dfrac{C}{x}\right) = \dfrac{P\left(\dfrac{x}{C}\right) \times P(x)}{P(C)}$

b. $P\left(\dfrac{C}{x}\right) = \dfrac{P(x \cup C) \times P(C)}{P(x)}$

c. $P\left(\dfrac{C}{x}\right) = \dfrac{P(x \cap C) \times P(C)}{P(x)}$

d. $P\left(\dfrac{C}{x}\right) = \dfrac{P\left(\dfrac{x}{C}\right) \times P(C)}{P(x)}$

3. The CDF is defined as
   a. $F_X(x) = P(x \leq X)$
   b. $F_X(x) = P(X \leq x)$
   c. $F_X(x) = P(X \leq x)$
   d. $F_X(x) = P(X \geq x)$

4. Poisson distribution is a standard distribution for
   a. A deterministic random variable
   b. A random variable
   c. A continuous random variable
   d. A discrete random variable.

5. K–S test detects the deviation of unknown distribution
   a. From a standard distribution
   b. From a uniform distribution
   c. From other distribution
   d. From other unknown distribution

6. If the mean of the Gaussian distribution is positive and it increases,
   a. The curve shifts toward right
   b. The curve shifts toward left
   c. The curve shrinks
   d. The curve does not shift

7. The central limit theorem states that
   a. Probability distribution function of a sum of a large number of random variables approaches a Gaussian distribution
   b. Probability distribution function of a sum of a large number of random variables approaches a Rayleigh distribution
   c. Probability distribution function of a sum of a large number of random variables approaches uniform distribution
   d. Probability distribution function of product of a large number of random variables approaches a Gaussian distribution

8. Histogram is a plot of
   a. Frequency of occurrence versus the probability of a random variable
   b. Frequency of occurrence versus the random variable
   c. Probability versus the pdf of a random variable
   d. Probability versus the CDF of a random variable

Review Questions

1. Define the term probability. Define conditional probability.
2. State Bayes' theorem in two different ways. What will a Venn diagram for mutually exclusive events look like?
3. Define a random variable. Define a CDF and pdf for a random variable. How will you obtain pdf if CDF is specified?
4. How will you draw the histogram for the samples of unknown random variable?
5. How will you fit a standard distribution to unknown random variable? Explain chi-square test and K–S test.
6. Describe the characteristics of Gaussian random variable. What is the need to normalize the Gaussian variable? Describe a method to obtain the probability value from a table for normal distribution.
7. Define pdf and CDF for a Rayleigh distribution and exponential distribution.
8. Define the terms permutation and combination. What are Bernoulli's trials?
9. Define the characteristics of Poisson random variable.
10. Define the characteristics of a binomial distribution.
11. Define the standard distributions used for discrete random variables.

Problems

1. Let there be 80 balls in a box, all of same size and shape. There are 20 balls of red color, 30 balls of blue color, and 30 balls of green color. Let us consider the event of drawing one ball from the box. Find the probability of drawing a red ball, a blue ball, and a green ball.
2. Let the event A be drawing a red ball in problem 1 and event B be drawing a blue card in the second draw without replacing the first ball drawn. Find $P(A \cap B)$.
3. Consider a binary symmetric channel with *a priori* probabilities $P(B_1) = 0.8$, $P(B_2) = 0.2$. *A priori* probabilities indicate the probability of transmission

of symbols 0 and 1 before the experiment is performed, i.e., before transmission takes place. The conditional probabilities are given by $P(A_1/B_1) = 0.8$, $P(A_2/B_1) = 0.2$, and $P(A_2/B_2) = 0.2$, $P(A_2/B_2) = 0.8$. Find received symbol probabilities. Find the transmission probabilities for correct transmission and transmission with error.

4. Probability of having HIV is $P(H) = 0.2$. Probability of not having HIV is $P(\bar{H}) = 0.8$, probability for testing positive given that person has HIV is $P(\text{Pos}/H) = 0.95$ and probability for testing positive given that person is not having HIV is $P(\text{Pos}/\bar{H}) = 0.02$ Find the probability that a person has HIV given that the test is positive.

5. Consider a pair of dice. Find the probability of getting the sum of the dots on the two faces as <5. Also find the probability for getting the sum of faces equal to 1.

6. A random variable has a distribution function given by

$$F_X(x) = 0 \qquad -\infty < x \leq -8$$

$$= \frac{1}{6} \qquad -8 \leq x \leq -5$$

$$= \frac{x}{15} + \frac{x}{2} \qquad -5 < x < 5$$

$$= \frac{5}{6} \qquad 5 \leq x < 8$$

$$= 1 \qquad 8 \leq x < \infty$$

Draw the CDF. Find $P(X \leq 4)$ and $P(-5 \leq X \leq 5)$.

Plot the pdf for a CDF specified in problem 5.

7. Find the probability of the event $(X \leq 4)$ for a Gaussian variable having a mean value of 2 and variance of 1.

8. Find the total number of combinations of five things taken from 10 things.

9. Find the total number of permutations of five things taken from 10 things.

10. Write a MATLAB program to generate a Gaussian random variable. Use a rand command to generate 12 random variables with uniform distribution (use central limit theorem).

11. The experiment is conducted with a fair die. The die is tossed 240 times. The following data are obtained, shown in the table below. Find the test statistic to check if it obeys a uniform distribution.

Face No.	1	2	3	4	5	6
Observed frequency	36	40	44	38	42	40

Answers

Multiple-Choice Questions

1. (a)
2. (d)
3. (b)
4. (d)
5. (a)
6. (a)
7. (a)
8. (b)

Problems

1. $P(\text{red ball}) = 20 / 80 = 25\%$, $P(\text{blue ball}) = 30 / 80 = 37.5\%$

    $P(\text{green ball}) = 30/80 = 37.5\%$

2. $P(A) = 2/8$ and $P(B / A) = \dfrac{30}{79}$ So,

    $P(A \cap B) = \dfrac{60}{632}$

3. Received symbol probabilities

    $P(A_1) = 0.68, P(A_2) = 0.32$

    The probabilities for correct symbol transmission

    $P(B_1/A_1) \approx 0.94, P(B_2/A_2) \approx 0.5$

    Probabilities of error

    $P(B_1/A_2) \approx 0.5$

    $P(B_2/A_1) \approx 0.06$

4. $P(H/Pos) = 0.9223$

5. $10/36$ and $5/36$

6. $P(X \leq 4) = 23/30$, $P(-5 \leq X \leq 4) = 18/30$

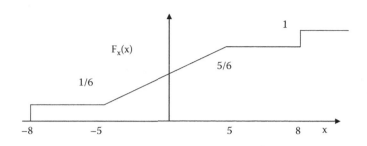

7. $y = 2, P(y \leq 2) = 0.5793$

8. Number of possible combinations: 252

9. Number of permutations: 30,240

10. Refer to program in text.

11. Value of test statistic is 1, which is less than the 5% level equal to $240 \times 0.05 = 1.2$. Hence, the distribution is uniform.

2

Properties of Random Variables

LEARNING OBJECTIVES

In this chapter, we will learn about the following:

- Properties of random variables
- Statistical properties of random variables
- Moments, higher order moments
- Moment generating functions
- Characteristic functions
- Transformation of a random variable
- Estimation of energy spectral density (ESD) and power spectral density (PSD)
- Correlogram
- Autocorrelation
- Cross-correlation

In this chapter, we will discuss the properties of random variables. Statistical properties of random variables like estimation of mean, variance, standard deviation, and higher order moments including skew and kurtosis will be explained with numerical examples. Moment generating functions and characteristic functions used for finding higher order moments of a random variable will also be explained and the use of these functions for finding moments will be illustrated with numerical problems. Energy and power signals will be defined. The chapter also includes a description of how energy spectral density (ESD) and power spectral density (PSD) are estimated. Use of correlogram and autocorrelation is illustrated with examples. Solved numerical examples are included to illustrate the concepts explained in the chapter.

2.1 Statistical Properties of Random Variables

We come across many natural signals in our day-to-day life like speech, electro cardio gram and electro ensephalo gram signals. The naturally occurring signals have some random component. These signals are random signals. The random signal can be defined as the signal for which the next samples can not be predicted using the previous samples by writing down some mathematical equation. For such signals, the probability density function can not be exactly defined. It can only be estimated or approximated using standard density functions. The hurdles in the estimation of properties of these signals are as follows. The duration of the signals may range from minus infinity to plus infinity. Signals can be interfaced to a computer for processing via an analog to digital converter (ADC). However, the memory capacity of the computer is finite or limited. Therefore, only signals of finite duration can be used for processing; in some cases, only finite data values are available. Based on these finite samples, the statistical properties of the signals are predicted or estimated. If we were to calculate the actual properties, strictly speaking, we would have to use infinite number of samples. The actual data does exist over the infinite interval. However, we cannot process infinite samples due to the limitations of the processing capacity of computers. We can still process, as we do now, finite samples and try to predict or estimate the statistical properties from finite samples. Let us define some important statistical properties of a random variable—mean, variance, higher order moments, and central moment are some important statistical properties—and discuss the methods for predicting these properties. We will start with the mean or the expected value. The random variable can be discrete or continuous. A discrete random variable takes on the discrete values such as 1, 2, 3 etc. The continuous random variable will take all continuous values such as 0.9234, 1.012 etc.

1. *Mean or expected value*: The mean or expected value of a continuous random variable is defined as

$$\mu = E[X] = \int_{-\infty}^{\infty} x f_X(x) \, dx, \tag{2.1}$$

where $f_X(x)$ is the probability density function (pdf) of the random variable. In case of a discrete random variable, mean is defined as

$$\mu = E[X] = \sum_i x_i P(x_i), \tag{2.2}$$

where $P(x_i)$ refers to the probability of occurrence of value x_i. The mean obtained using this equation may not belong to a sample space. There are two problems that have to be faced when we are trying to find the value of mean for a continuous random variable. First, the signal is not available for infinite duration, and second, the pdf for the random variable is not known. We fit the pdf for an unknown signal and try to find the mean using the mathematical expression for that standard pdf in the equation for mean. Therefore, we cannot say that the mean is calculated; rather, the mean is estimated by approximating or fitting a standard distribution to a random variable. It is evaluated over the finite interval over which the signal is actually available. We try to get closer to the actual mean value by processing only finite number of samples. This is why the mean is also termed as the expected value. For the discrete random variable, the probability of occurrence of each value of the variable may not be known.

To solve the problem of unknown pdf, a number of standard distribution functions have been defined along with the mathematical expression for the pdfs—uniform distribution, Gaussian distribution, Rician distribution, Rayleigh distribution, exponential distribution, and so forth. To fit the standard distribution to any unknown signal, the following steps will be taken.

1. Plot the histogram of the signal. For this, divide the input signal range into a number of bins and find the number of signal values occurring in the bin range (frequency of occurrence) for every bin. Plotting the frequency of occurrence for a bin versus the bin value gives us the histogram.

2. Use the K–L test or chi-square test discussed in Section 1.5 to fit a distribution for the unknown continuous/discrete random signal respectively.

3. Now use the equations of the matching pdf to find the mean and other statistical parameters.

2. *Moments*: The nth moment, denoted as m_n, of a continuous random variable, is defined as

$$m_n = E[X^n] = \int_{-\infty}^{\infty} x^n f_X(x)\,\mathrm{d}x. \tag{2.3}$$

The mean is the first-order moment. This can be seen by comparing Equations 2.1 and 2.3. When $n = 1$ in Equation 2.3, we obtain Equation 2.1, which is the definition of mean. When $n = 2$, we obtain the mean square value of the variable given by

$$m_2 = E[X^2] = \int_{-\infty}^{\infty} x^2 f_X(x)\,\mathrm{d}x. \tag{2.4}$$

3. *Central moment*: The nth order central moment is defined as

$$m_n^C = E[(X-\mu)^n] = \int_{-\infty}^{\infty} (x-\mu)^n f_X(x)\,\mathrm{d}x. \tag{2.5}$$

Variance: The second-order central moment is called the *variance* of the random variable X and is given by

$$m_2^C = \text{Variance}\,(\sigma^2)$$

$$= E[(X-\mu)^2]$$

$$= \int_{-\infty}^{\infty} (x-\mu)^2 f_X(x)\,\mathrm{d}x. \tag{2.6}$$

$$E[X-\mu]^2 = E[X^2 - 2\mu X + \mu^2]$$

$$= E[X^2] - 2\mu \times E[X] + \mu^2$$

$$= E[X^2] - 2\mu^2 + \mu^2$$

$$= E[X^2] - \mu^2.$$

Hence, $\sigma^2 = E[X^2] - \mu^2$.

$$\text{as } = E[X] = \mu \text{ and } E[\mu^2] = \mu^2. \tag{2.7}$$

Standard Deviation: The symbol σ represents the standard deviation of the random variable.

Skew: The third second-order central moment is a measure of asymmetry of the density function about the mean. Skewness can be considered as a measure of symmetry, or, in fact, the lack of symmetry. A distribution is said to be symmetric if, from a center point, the left side is identical to the right side. The skewness of the normal distribution is zero.

Asymmetry is called *skew* of the density function. It is given by

$$m_3^C = \text{Skew}(\mu_3)$$

$$= E[(X - \mu)^3]$$

$$= \int_{-\infty}^{\infty} (x - \mu)^3 f_X(x)\, dx. \tag{2.8}$$

Skewness: The normalized third-order central moment is called skewness of the density function. It is given by

$$\text{Skewness/coefficient of skewness} = \frac{\text{Skew}(\mu_3)}{\sigma_X^3}. \tag{2.9}$$

Skew is a word that relates to physical things. A painting hanging on the wall is skewed if it is leaning to one side or the other. One can claim that opinions are often skewed or biased. A movie can be said to be skewed toward one character more than the other.

In statistics, skewness represents an imbalance and asymmetry of the distribution graph from the mean of a distribution. Let us consider a normal data distribution using a bell curve, where the curve is perfectly symmetrical. Let us understand the meaning of a few basic terms—mean, median, and mode—and see how skewness affects them. Mean is the average of the numbers in the data distribution, median is the number that falls directly in the middle of the data distribution, and mode is the number that appears most frequently in the data distribution. Consider a normal data distribution. The mean is directly in the center of the bell curve. For a symmetric bell curve, the mean, median, and mode are all at the same position as shown in Figure 2.1. In a skewed distribution, the mean, median, and mode are located at different locations. A skewed data distribution may have a positive or negative skew. A skew is said to be positive when the mean is at a higher position than the median. The mean will be larger than the median in the skewed data set. A negative skew will lead to the opposite situation. Here, the mean is brought down and the median is larger than the mean.

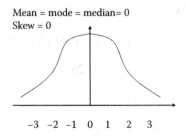

FIGURE 2.1
Normal distribution with mean, mode, and median at zero.

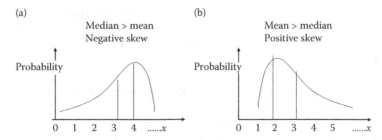

FIGURE 2.2
(a) Distribution with negative skew and (b) with positive skew.

A distribution is said to be skewed left or skewed right if the tail of the distribution is longer either on the right-hand side or on the left-hand side, respectively, as shown in Figure 2.2. The value of the skewness indicates the extent to which a distribution differs from a normal distribution.

$$\text{The formula to measure skew manually is } \frac{3(\text{Mean}-\text{Median})}{\text{Standard deviation}}. \tag{2.10}$$

In order to use this formula, we need to know the mean and median, of course. As we saw earlier, the mean is the average. It is the sum of the values in the data distribution divided by the number of values in the distribution. If the data distribution was arranged in numerical order, the median would be the value directly in the middle.

Kurtosis: Kurtosis is a measure relative to a normal distribution. It measures if the data are peaked or flat relative to a normal distribution. If the value of kurtosis is high, data sets tend to have a distinct peak near the mean; the data decline rapidly and will have heavy tails. On the contrary, the data sets with low kurtosis will have a flat top near the mean. A uniform distribution is thus the extreme case of low kurtosis.

Kurtosis is defined as the ratio of the fourth-order central moment and fourth power of standard deviation.

$$\text{Kurtosis} = \frac{(\mu_4)}{\sigma_X^4}. \tag{2.11}$$

The fourth-order central moment is defined as

$$m_4^C = (\mu_4) = E[(X-\mu)^4]$$

$$= \int_{-\infty}^{\infty} (x-\mu)^4 f_X(x)\, dx. \tag{2.12}$$

The kurtosis for a standard normal distribution can be shown to be equal to three. Hence, sometimes, the following definition of kurtosis is used.

$$\text{Kurtosis} = \frac{(\mu_4)}{\sigma_X^4 - 3}. \tag{2.13}$$

This is sometimes called *excess kurtosis* (it is also called the fourth cumulant). Using this definition, kurtosis for a normal distribution is zero. Using this second definition of kurtosis, positive kurtosis will indicate a "peaked" distribution and negative kurtosis will indicate a "flat" distribution.

Let us understand what kurtosis means. Kurtosis does not give the location of the peak. The location for the peak is, in fact, at the mode. Kurtosis is the standardized fourth moment. It is also known as peakedness. It measures how sharply curved the peak is for a given distribution. Kurtosis also describes heavy-tailedness. In reality, the actual fourth standardized moment measure does not quite measure either of those things!! Indeed, the first volume of Kendall and Stuart gives counter examples that show that a higher value of kurtosis is not necessarily associated with either higher peak or fatter tails. Kurtosis and skewness are strongly related. Interpretation of kurtosis is found to be somewhat easier when the distribution is nearly symmetric. Figure 2.3 shows what is peakedness and heavy-tailedness. If the distribution has higher standard deviation, it is said to have broad shoulders.

Let us write a MATLAB® program to measure skewness and kurtosis for a Rayleigh and a Gaussian distribution.

```
clear all;
clc;
x=(0:0.01:2);
p=raylpdf(x,0.5);
plot(x,p);
y=skewness(p);
s=3*(mean(p)-median(p))/std(p);
disp(s);
disp(y);
z=kurtosis(p);
disp(z);
sigma=2;
mean=1;
x1=(-3:0.01:5);
p1=normpdf(x1,1,2);
y1=skewness(p1);
disp(y1);
z1=kurtosis(p1);
disp(z1);
```

Skew Rayleigh = 0.7215

Skewness Rayleigh = 0.3572

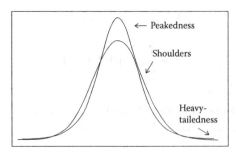

FIGURE 2.3
Peakedness and heavy tailedness.

Kurtosis Rayleigh = 1.5667

Skewness Gaussian = −0.0740

Kurtosis Gaussian = 1.5696

For a Rayleigh distribution, skew is positive as expected from the graph of the Rayleigh distribution pdf. Now, let us find the skew and kurtosis for a segment of the speech signal. A MATLAB program for this is given as follows.

```
clear all;
clc;
fp=fopen('watermark.wav', 'r');
fseek(fp,8000,-1);
a=fread(fp,2048);
subplot(2,1,1);
a=a-128;plot(a);
title('plot of speech segment');
xlabel('sample number');ylabel('amplitude');
p=hist(a,40);
subplot(2,1,2);bar(p);
title('histogram of speech');
xlabel('bin number');ylabel('frequency of occurrence');
s=3*(mean(p)-median(p))/std(p);disp(s);
z=kurtosis(p);
disp(z);
```

The speech segment and its histogram is plotted in Figure 2.4.

Skewness = −0.5816

Kurtosis = 2.9576

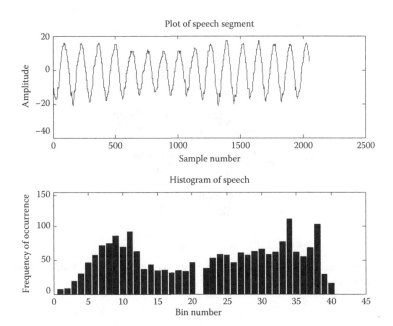

FIGURE 2.4
Plot of speech segment and its histogram.

Example 1

Find the mean value and variance of a uniform random variable varying between the values a and b.

Solution: Let us first evaluate the value of constant pdf. Let it be represented by K.

$$\int_a^b f_X(x)dx = \int_a^b K\,dx = 1.$$

$$[Kx]_a^b = 1. \tag{2.14}$$

$$K = \frac{1}{b-a}.$$

Let us find the mean value of x.

$$\mu = E[X] = \int_a^b xf_X(x)dx$$

$$= \int_a^b x \times K\,dx = K\left[\frac{x^2}{2}\right]_a^b$$

$$= \frac{(b^2 - a^2)}{2(b-a)} = \frac{b+a}{2}. \tag{2.15}$$

We will now find the variance of x.

$$\text{Variance} = E[X^2] - \mu^2 \left(\mu = \frac{a+b}{2}\right)$$

$$E[X^2] = \int_a^b [x^2 f_X(x)]dx$$

$$= \int_a^b Kx^2\,dx = K\left[\frac{x^3}{3}\right]_a^b$$

$$= \frac{1}{b-a} \times \frac{(b^3 - a^3)}{3} = \frac{b^2 + a^2 + ab}{3}.$$

$$\text{Variance} = \frac{b^2 + a^2 + ab}{3} - \frac{b^2 + a^2 + 2ab}{4}$$

$$= \frac{b^2 + a^2 - 2ab}{12} = \frac{(b-a)^2}{12}. \tag{2.16}$$

Example 2

Consider a random variable X with a pdf as shown in Figure 2.5. Find A, mean value of X, and variance of X.

Solution:

1. To find A: If the pdf is a valid distribution function, then the area under the curve, that is, the total probability, must be equal to 1.

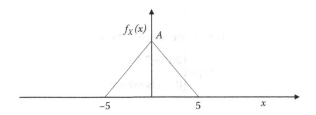

FIGURE 2.5
Pdf for a random variable x.

$$\frac{[5-(-5)]A}{2}=1.$$

$$\frac{10A}{2}=1$$

$$\Rightarrow A=\frac{1}{5}. \tag{2.17}$$

2. To find the mean value:

$$\mu = E[X] = \int_{-5}^{5} x f_X(x)\,dx$$

$$= \int_{-5}^{0} \frac{x(x-(-5))}{5dx} + \int_{0}^{5} \frac{x(5-x)}{5dx} = \frac{1}{5}\left[\int_{-5}^{0}(x^2+5x)\,dx + \int_{0}^{5}(5x-x^2)\,dx\right]$$

$$= \frac{1}{5}\left[\frac{x^3}{3}+\frac{5x^2}{2}\right]_{-5}^{0} + \frac{1}{5}\left[\frac{5x^2}{2}-\frac{x^3}{3}\right]_{0}^{5}$$

$$= \frac{1}{5}\left[-(-25)+\frac{125}{2}+\frac{125}{2}-25\right]$$

$$= \frac{1}{5}\left[25-25+\frac{125}{2}-\frac{125}{2}\right]=0. \tag{2.18}$$

3. To find the variance:

Variance $= E[X^2]-\mu^2$ (Note: $\mu = 0$)

$$= E[X^2] = \frac{1}{25}\left[\int_{-5}^{0}[x^3+5x^2]\,dx + \int_{0}^{5}[5x^2-x^3]\,dx\right.$$

$$= \frac{1}{25}\left\{\left[\frac{x^4}{4}+5\frac{x^3}{3}\right]\Big|_{-5}^{0} + \left[5\frac{x^3}{3}-\frac{x^4}{4}\right]\Big|_{0}^{5}\right\}$$

$$= \frac{1}{25}\left[\frac{-625}{4}+\frac{625}{3}+\frac{625}{3}-\frac{625}{4}\right]$$

$$= \frac{1}{25}[416.66-312.2]=4.1784. \tag{2.19}$$

Example 3

A random variable has a pdf given by the following equation

$$f(x) = \begin{cases} 0.1 & -3 \le x \le 7 \\ 0 & \text{elsewhere} \end{cases}.$$

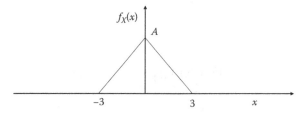

1. Find the mean value.
2. Find the mean square value.
3. Find the variance.

Solution: The mean value of the pdf is given as

$$\mu = E[X] = \int_{-3}^{7} 0.1x \, dx = 0.1 \frac{x^2}{2} \Big|_{-3}^{7} = 2. \tag{2.20}$$

The mean square value of the pdf is given as

$$E[X^2] = \int_{-3}^{7} 0.1x^2 \, dx = 0.1 \frac{x^3}{3} \Big|_{-3}^{7} = \frac{37}{3}. \tag{2.21}$$

The variance of the random variable X is

$$\sigma^2 = E[X^2] - \mu^2 = \frac{37}{3} - 4 = \frac{25}{3}. \tag{2.22}$$

Example 4

A random variable has an exponential pdf given by the following equation

$$f_X(x) = \begin{cases} \dfrac{1}{a_2} e^{-(x-a_1)/a_2} & x > a_1 \\[2mm] 0 & x < a_1 \end{cases}.$$

Find the mean value.

Solution: The mean value of the pdf is given as

$$\mu = E[X] = \frac{1}{a_2} \int_{a_1}^{\infty} x e^{-(x-a_1)/a_2} \, dx = \frac{e^{a_1/a_2}}{a_2} \int_{a_1}^{\infty} x e^{-x/a_2} \, dx.$$

Using the formula $\int x e^{ax} \, dx = e^{ax} \left[\dfrac{x}{a} - \dfrac{1}{a^2} \right]$ and putting $a = -\dfrac{1}{a_2}$,

$$E[X] = \frac{e^{a_1/a_2}}{a_2} \left\{ e^{-\frac{x}{a_2}} [-a_2 x + a_2^2] \right\} \Big|_{a_1}^{\infty} = a_1 + a_2. \tag{2.23}$$

Concept Check

- Name some statistical properties of a random variable.
- Define the mean or the expected value. Why is the mean value also called the expected value?
- Define the term variance for a random variable.
- What is skewness?
- Define the nth order moment for a random variable.
- Define the term kurtosis.
- If the value of kurtosis is high, what is your conclusion regarding the distribution of the random variable?
- If the value of kurtosis is low, what is your conclusion regarding the distribution of the random variable?
- What is the value of kurtosis for a normal distribution?
- What will be the value of kurtosis for a uniform distribution?

2.2 Functions for Finding Moments

There are two functions that allow calculation of moments. They are the characteristic function and the moment generating function. Let us understand the characteristic function first.

Characteristic Function: The characteristic function for any random variable X is defined as

$$\phi_X(\omega) = E[e^{j\omega X}] = \int_{-\infty}^{\infty} f_X(x)e^{j\omega X}\, dx. \tag{2.24}$$

Equation 2.24 can be easily recognized as the Fourier transform (FT) of $f_X(x)$ with the sign of ω reversed. Hence, the pdf ($f_X(x)$) can be obtained from the characteristic function by taking inverse FT. We have to reverse the sign of variable x. The advantage of using a characteristic function is that the nth moments can be easily obtained from the characteristic function given by

$$m_n = (-j)^n \frac{d^n(\phi_X(\omega))}{d\omega^n}\Big|_{\omega=0}. \tag{2.25}$$

Moment Generating Function: The moment generating function is defined as

$$M_X(\upsilon) = E(e^{\upsilon X}) = \int_{-\infty}^{\infty} f_X(x)e^{\upsilon X}\, dx. \tag{2.26}$$

Moments can be obtained from the moment generating function using the following equation.

$$m_x = \frac{d^n(M_X(\upsilon))}{d\upsilon^n} \Big\downarrow_{\upsilon=0} . \tag{2.27}$$

The moment generating function may not exist for all random variables and for all values of υ.

Example 5

The density function of the variable X is given by

$$f_x(x) = \begin{cases} \dfrac{1}{4} & -2 \le x \le 2 \\ 0 & \text{otherwise} \end{cases}.$$

Determine: (a) $P(X \le x)$, (b) $P(|x| \le 1)$, (c) mean and variance, and (d) the characteristic function.

Solution:

1. $P(X \le x) = \displaystyle\int_{-2}^{x} f_X(x)dx = \int_{-2}^{x} \frac{1}{4}dx = \frac{x}{4}\Big\downarrow_{-2}^{x} = \frac{x}{4} + \frac{1}{2}.$ \hfill (2.28)

2. $P(X \le 1) = \displaystyle\int_{-2}^{1} f_X(x)dx = \int_{-2}^{1} \frac{1}{4}dx = \frac{x}{4}\Big\downarrow_{-2}^{1} = \frac{3}{4}.$ \hfill (2.29)

3. The mean and the variance

$$\mu_x = E[X] = \int_{-2}^{2} \frac{x}{4}dx = \frac{x^2}{8}\Big|_{-2}^{2} = 0. \tag{2.30}$$

$$E[X^2] = \int_{-2}^{2} \frac{x^2}{4}dx = \frac{x^3}{12}\Big|_{-2}^{2} = \frac{4}{3}. \tag{2.31}$$

$$\sigma_x^2 = E[X^2] - \mu_x^2 = \frac{4}{3} - 0 = \frac{4}{3}. \tag{2.32}$$

4. The characteristic function for a continuous random variable is given as

$$\varphi_X(\omega) = \int_{-2}^{2} \frac{1}{4}e^{j\omega x} dx = \frac{1}{4j\omega}e^{j\omega x}\Big\downarrow_{-2}^{2} = \frac{e^{j2\omega} - e^{-j2\omega}}{4j\omega}$$

$$= \frac{2j\sin(2\omega)}{4j\omega} = \frac{\sin(2\omega)}{2\omega}. \tag{2.33}$$

Example 6

Find the characteristic function of the random variable X having density function

$$f_X(x) = e^{-\frac{1}{2}|x|} \text{ for all } x.$$

Solution: We will use the equation for the characteristic function.

$$\varphi_X(\omega) = \int_{-\infty}^{\infty} e^{-\frac{1}{2}|x|} e^{j\omega x}\, dx = \int_{-\infty}^{0} e^{\frac{1}{2}x} e^{j\omega x}\, dx + \int_{0}^{\infty} e^{-\frac{1}{2}x} e^{j\omega x}\, dx$$

$$= \frac{1}{(0.5)+j\omega} + \frac{1}{(0.5)-j\omega} = \frac{4}{1+4\omega^2}. \tag{2.34}$$

Example 7

Find the characteristic function and first moment of the random variable X having density function

$$f_X(x) = \begin{cases} \dfrac{1}{a_2} e^{-(x-a_1)/a_2} & x > a_1 \\ 0 & x < a_1 \end{cases}.$$

Solution:

$$\varphi_X(\omega) = \frac{1}{a_2}\int_{a_1}^{\infty} e^{-(x-a_1)/a_2} e^{j\omega x}\, dx = \frac{e^{a_1/a_2}}{a_2}\int_{a_1}^{\infty} e^{-\left(\frac{1}{a_2}-j\omega\right)x}\, dx$$

$$= \frac{e^{a_1/a_2}}{a_2}\left[\frac{e^{-\left(\frac{1}{a_2}-j\omega\right)x}}{-\left(\dfrac{1}{a_2}-j\omega\right)}\right]\Bigg\downarrow_{a_1}^{\infty} \tag{2.35}$$

$$= \frac{e^{j\omega a_1}}{(1-j\omega a_2)}.$$

$$m_1 = (-j)\frac{d}{d\omega}\varphi_X(\omega) = -j\left[e^{j\omega a_1}\left\{\frac{ja_1}{1-j\omega a_2} + \frac{ja_2}{(1-j\omega a_2)^2}\right\}\right]\Bigg\downarrow_{\omega=0}$$

$$= -j \times j(a_1+a_2) = a_1+a_2.$$

Example 8

Find the moment generating function and first moment of the random variable X having density function

$$f_X(x) = \begin{cases} \dfrac{1}{a_2} e^{-(x-a_1)/a_2} & x > a_1 \\ 0 & x < a_1 \end{cases}.$$

Solution:

$$M_X(\upsilon) = \frac{1}{a_2}\int_{a_1}^{\infty} e^{-(x-a_1)/a_2} e^{\upsilon x}\, dx = \frac{e^{a_1/a_2}}{a_2}\int_{a_1}^{\infty} e^{-\left(\frac{1}{a_2}-\upsilon\right)x}\, dx$$

$$= \frac{e^{a_1/a_2}}{a_2}\left[\frac{e^{-\left(\frac{1}{a_2}-\upsilon\right)x}}{-\left(\dfrac{1}{a_2}-j\omega\right)}\right]\Bigg\downarrow_{a_1}^{\infty} \tag{2.36}$$

$$= \frac{e^{\upsilon a_1}}{(1 - j\omega a_2)}.$$

$$m_1 = \frac{d}{d\upsilon} M_X(\upsilon) = \left[e^{\upsilon a_1} \left\{ \frac{a_1}{1 - \upsilon a_2} + \frac{a_2}{(1 - \upsilon a_2)^2} \right\} \right] \Bigg\downarrow_{\upsilon = 0}$$

$$= a_1 + a_2.$$

Concept Check

- Define the two moment generating functions.
- Define a characteristic function.
- How will you use a moment generating function for finding moments?
- How will you use a characteristic function for finding moments?

2.3 Transformations of a Random Variable

The transformation of a random variable refers to a change of one random variable into other random variable using linear or nonlinear or multiple-valued equations. Such change of a random variable into another random variable is required as the two random variables are often related to each other with some law or equation. This equation can be made use of to find the properties of one random variable if the properties of the other random variable are already known. The equation used for such a change of the random variable decides if it is a linear transformation or a nonlinear transformation. Consider a case wherein the user wants to transform or change one random variable x, say temperature in degree Celsius, to a new random variable y, say temperature in degree Kelvin. We can write the equation for this transformation as follows.

$$y \, (\text{degree Kelvin}) = x \, (\text{degree Celsius}) + 373. \tag{2.37}$$

This is a linear transformation. Let the transformation be denoted by T. We can thus represent the process of transformation as

$$y = T(x). \tag{2.38}$$

This equation will define the transfer characteristic represented as a graph of y vs. x. Here, the transformation T can be linear, nonlinear, segmented, multiple-valued, etc.

2.3.1 Monotonic Transformations of a Random Variable

Let us consider a simple transformation, say a linear transformation. We can represent this transformation as a linear system as shown in Figure 2.6. The designed system is studied

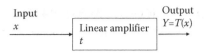

FIGURE 2.6
Linear transformation of a random variable.

by applying a random variable as input to the system. Let us consider just the amplification of the random variable. Let us find the distribution of the output random variable if the input random variable has a certain distribution. Refer to the transformation expressed in Equation 2.38. The transformation is said to be monotonic if for every value of X, there exists only one value of Y. The transformation can be monotonically increasing or monotonically decreasing. Let us assume that T is continuous and differentiable for all values of X. Let us find the density function of the transformed variable, namely, Y.

If T is a linear transformation, we can write

$$P(Y \leq y) = P(X \leq x).$$

$$F_Y(y) = F_X(x).$$

$$\int_{-\infty}^{y} f_Y(y)\,dy = \int_{-\infty}^{x} f_X(x)\,dx. \qquad (2.39)$$

Let us differentiate thiis equation with respect to y,

$$f_Y(y) = f_X(x)\frac{dx}{dy}.$$

To illustrate this concept, let us solve the following example. A linear amplifier is an example of a linear transformation. It transforms input voltage x into amplified output y. Here, $y = T(x) = \text{gain} \times x$.

Example 9

Let T be the linear transformation given by $y = T(x) = ax + b$, where a and b are real constants, respectively. Find the relation between the density functions of the two random variables.

Solution:

$$y = ax + b.$$

$$x = \frac{y - b}{a}. \qquad (2.40)$$

$$\frac{dx}{dy} = \frac{1}{a}, \quad f_Y(y) = f_X\left(\frac{y - b}{a}\right)\left|\frac{1}{a}\right|.$$

We will now consider an example where one discrete random variable is transformed into another discrete random variable.

Example 10

Consider a discrete random variable with the ensemble consisting of discrete values $x = \{-1, 0, 1, 2\}$, with probabilities given by the vector $P = \{0.1, 0.3, 0.4, 0.2\}$. The pdf can be written as $f_X(x)=0.1\times\delta(x+1)+0.3\times\delta(x)+0.4\times\delta(x-1)+0.2\times\delta(x-2)$. If x is transformed to a new random variable y given by $y=3-x^2+x^3/3$, find the density function for y.

Solution: Let us first map all x values in x vector to y values.

$$\text{For } x \;=\; \{-1,\,0,\,1,\,2\},\, y \;=\; \{5/3,\,3,\,7/3,\,5/3\}.$$

x values of –1 and 2 map onto $y=5/3$. The probabilities for $x=-1$ and $x=2$ will add up to give

$$f_y(y)=0.3\times\delta(y-5/3)+0.3\times\delta(y-7/3)+0.4\times\delta(y-3). \tag{2.41}$$

Example 11

Let x be the Gaussian distributed variable with mean equal to 1 and variance of 4. If it is exposed to a linear transformation given by $Y=2X+1$, find $f_Y(y)$.

Solution: It is given that x is a Gaussian distributed variable.

$$f_X(x)=\frac{1}{2\sqrt{2\pi}}e^{-(x-1)^2/8}.$$

$$f_Y(y)=f_X\left(\frac{y-1}{2}\right)\left|\frac{1}{2}\right|=\frac{1}{4\sqrt{2\pi}}e^{-\left(\frac{y-1}{2}-1\right)^2/8} \tag{2.42}$$

$$=\frac{1}{4\sqrt{2\pi}}e^{-(y-3)^2/32}.$$

Referring to the equation for Gaussian distribution, mean for y is 3 and standard deviation is 4.

Thus, it is found that the linear transformation does not affect the type of distribution. It is then possible to find the mean and standard deviation of the transformed variable in a simple way as follows.

Let us consider the same transformation used earlier, that is, $Y=2X+1$.

$$E[Y]=\mu_Y=E[2X+1]=2E[X]+1=2\times1+1=3.$$

$$\sigma_Y^2=E[Y^2]-\mu_Y^2.$$

$$E[Y^2]=E[(2X+1)^2]=E[4X^2+4X+1]$$
$$=4E[X^2]+4E[X]+1=4\times5+4\times1+1=25, \tag{2.43}$$

as $E[X^2]=\sigma_X^2+\mu_X^2=4+1=5$

$$\sigma_Y^2=25-9=16,\sigma_Y=4.$$

2.3.2 Multiple-Valued Transformations of a Random Variable

In general, the transformation may not be monotonic. Consider a two valued transformation as follows. The event $y\leq Y\leq y+dy$ corresponds to two equivalent events in the X domain, namely, $x_1\leq X\leq x_1+dx_1$ or $x_2\leq X\leq x_2+dx_2$. As the two events in the X domain are mutually exclusive, we may write

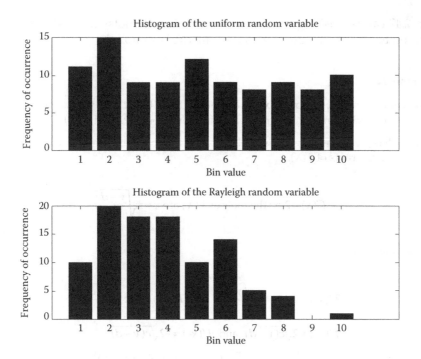

FIGURE 2.7

Histogram for a uniform and a Rayleigh variable with $N=100$.

$$P(y \leq Y \leq y+\mathrm{d}y) = P(x_1 \leq X \leq x_1+\mathrm{d}x_1) + P(x_2+\mathrm{d}x_2 \leq X \leq x_2).$$ (2.44)

In this case, $Y = T(X)$ has two roots.

$$\begin{aligned} f_Y(y)\mathrm{d}y &= P(y \leq Y \leq y+\mathrm{d}y). \\ f_X(x_1)\,|\,\mathrm{d}x_1\,| &= P(x_1 \leq X \leq x_1+\mathrm{d}x_1). \\ f_X(x_2)\,|\,\mathrm{d}x_2\,| &= P(x_2+\mathrm{d}x_2 \leq X \leq x_2). \\ f_Y(y)\mathrm{d}y &= f_X(x_1)\,|\,\mathrm{d}x_1\,| + f_X(x_2)\,|\,\mathrm{d}x_2\,|. \end{aligned}$$ (2.45)

As T is differentiable at all Xs, we know that

$$\begin{aligned} \mathrm{d}y &= T'(x_1)\mathrm{d}x_1. \\ \mathrm{d}y &= T'(x_2)\mathrm{d}x_2. \end{aligned}$$ (2.46)

T' represents the derivative.

$$\begin{aligned} f_Y(y)\mathrm{d}y &= f_X(x_1)\frac{\mathrm{d}y}{|T'(x_1)|} + f_X(x_2)\frac{\mathrm{d}y}{|T'(x_2)|}. \\ f_Y(y) &= \frac{f_X(x_1)}{|T'(x_1)|} + \frac{f_X(x_2)}{|T'(x_2)|}. \end{aligned}$$ (2.47)

Equation 2.47 can be easily generalized to n number of roots. To illustrate the concepts discussed up to now, let us consider some numerical examples.

Example 12

Consider the transformation $T(x) = cX^2 + d$ with $c > 0$. Let X be the Gaussian distributed variable with a mean of 1 and a standard deviation of 2. Find $f_Y(y)$.

Solution: It is given that X is Gaussian distributed variable. Let us first find the derivative of $T(X)$ and the roots of $T(X)$.

$$T'(x) = 2cx; \text{ the roots of } T(X) \text{ are } x = \pm\sqrt{\frac{y-d}{c}}. \tag{2.48}$$

These roots will exist only if $y > d$.

$$T'(x_1) = 2cx_1 = 2c\sqrt{\frac{y-d}{c}} \text{ and } T'(x_2) = -2c\sqrt{\frac{y-d}{c}}.$$

$$T'(x_1) = 2\sqrt{c(y-d)} \text{ and } T'(x_2) = -2\sqrt{c(y-d)}. \tag{2.49}$$

$$f_Y(y) = \frac{1}{2\sqrt{c(y-d)}}\left[f_X\left(\sqrt{\frac{y-d}{c}}\right) + f_X\left(-\sqrt{\frac{y-d}{c}}\right)\right].$$

Putting the values of mean and standard deviation in the equation, we get

$$f_X(x_1) = \frac{1}{2\sqrt{2\pi}}\exp\left[-\left(\sqrt{\frac{y-d}{c}}-1\right)^2/8\right] \text{ and}$$

$$f_X(x_2) = \frac{1}{2\sqrt{2\pi}}\exp\left[-\left(-\sqrt{\frac{y-d}{c}}-1\right)^2/8\right]. \tag{2.50}$$

$$f_Y(y) = \frac{1}{2\sqrt{c(y-d)}}\left[\begin{array}{l}\frac{1}{2\sqrt{2\pi}}\exp\left[-\left(\sqrt{\frac{y-d}{c}}-1\right)^2/8\right]\\+\frac{1}{2\sqrt{2\pi}}\exp\left[-\left(-\sqrt{\frac{y-d}{c}}-1\right)^2/8\right]\end{array}\right].$$

Example 13

A random variable X has a pdf given by $f_X(x) = \begin{cases}\frac{1}{2}\cos(x) & -\pi/2 < x < \pi/2\\0 & \text{elsewhere}\end{cases}$. Find the mean value of the function given by $Y = 4X^2$. Also find $f_Y(y)$.

Solution: Let us first find the derivative of $T(X)$ and the roots of $T(X)$.

$$T'(x) = 8x, \text{ the roots of } T(X) \text{ are } x = \pm\sqrt{y}\,/2. \tag{2.51}$$

$$T'(x_1) = 8x_1 = 4\sqrt{y} \text{ and } T'(x_2) = -4\sqrt{y}.$$

$$f_Y(y) = \frac{f_X(x_1)}{4\sqrt{y}} - \frac{f_X(x_2)}{4\sqrt{y}}. \tag{2.52}$$

$$f_Y(y) = \frac{\cos(\sqrt{y}\,/2)}{8\sqrt{y}} - \frac{\cos(-\sqrt{y}\,/2)}{8\sqrt{y}} = \frac{\cos(\sqrt{y}\,/2)}{4\sqrt{y}}.$$

Let us find the mean value of y.

$$E(Y) = 4E(X^2).$$

$$E(X^2) = \int_{-\pi/2}^{\pi/2} x^2 f_X(x)\,\mathrm{d}x = \frac{1}{2}\int_{-\pi/2}^{\pi/2} x^2 \cos(x)\,\mathrm{d}x$$

$$= \frac{1}{2}[2x\cos(x)\downarrow_{-\pi/2}^{\pi/2} + (x^2-2)\sin(x)\downarrow_{-\pi/2}^{\pi/2} \tag{2.53}$$

$$= \frac{1}{2}\left[\left(\frac{\pi^2}{4}-2\right)2\right] = \frac{\pi^2}{4} - 2.$$

$$E(Y) = \pi^2 - 8.$$

Example 14

A random variable X is uniformly distributed on the interval $(-5, 15)$. The random variable Y is formed as $Y = e^{-X/5}$. Find $E[Y]$.

Solution: The random variable X is uniformly distributed on the interval $(-5, 15)$. Hence, $f_X(x) = \frac{1}{20}$, $\mu_X = 5$, the transformation being a single-valued one.

$$f_Y(y) = \frac{f_X(x)}{\dfrac{\mathrm{d}T}{\mathrm{d}x}} = \frac{\frac{1}{20}}{-\frac{1}{5}e^{-X/5}} = -\frac{1}{4}e^{X/5}, \quad e^1 < y < e^{-3}. \tag{2.54}$$

$$E(Y) = -\frac{1}{4}\int_e^{e^{-3}} y f_Y(y)\,\mathrm{d}y = -\frac{1}{4}\int_e^{e^{-3}} e^{-X/5}e^{X/5}\,\mathrm{d}y = -\frac{y}{4}\downarrow_e^{e^{-3}} = -\frac{1}{4}[e^{-3}-e]$$

$$= \frac{1}{4}[e - e^{-3}]. \tag{2.55}$$

2.3.3 Computer Generation of a Transformed Variable

Let us consider the cumulative distribution function (CDF) equation for a Rayleigh random variable. Let $a = 0$.

$$F_X(x) = \begin{cases} 1 - e^{-(x-a)^2/2b^2} & x \geq a \\ 0 & x < a \end{cases}. \tag{2.56}$$

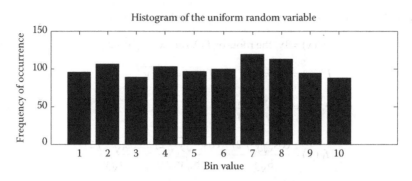

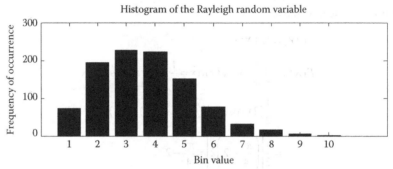

FIGURE 2.8

Histogram for a uniform and a Rayleigh variable with $N = 1000$

$$F_X(x) = \left\{ \begin{array}{ll} 1 - e^{-x^2/2b^2} & x \geq 0 \\ 0 & \text{elsewhere} \end{array} \right\}. \tag{2.57}$$

We will set
$$\begin{array}{ll} F_Y(Y) = 1 - e^{-y^2/2b^2} = x & \text{for } 0 < x < 1. \\ y = T(x) = \sqrt{-\sqrt{2}b \ln(1-x)} & \text{for } 0 < x < 1. \end{array} \tag{2.58}$$

We will now write a MATLAB program to generate a Rayleigh random variable.

```
clc;
clear all;
N=100;
x=rand(1,N);
b=1;
y=sqrt(-sqrt(2)*b*log(1-x));
z=hist(x);
w=hist(y);
subplot(2,1,1);
bar(z);
title('histogram of the uniform random variable');
xlabel('bin value');
ylabel('frequency of occurrence');
subplot(2,1,2);
bar(w);
```

```
title('histogram of the Rayleigh random variable');
xlabel('bin value');
ylabel('frequency of occurrence');
```

The histogram plots for the uniform random variable and Rayleigh random variable are shown in Figure 2.7. Let the value of the constant b be changed to 0.5. Let us generate 1000 random values. Now, we can see that the distribution closely approaches the Rayleigh distribution. The histogram for the Rayleigh variable is plotted in Figure 2.8.

Concept Check

- Define the transformation of a random variable.
- Why is such a transformation required?
- What is a linear transformation?
- How will you make use of this transformation to find properties of the transformed variable?
- What is a multiple-valued transformation?
- Define the equation to find the pdf of a transformed variable in case of multiple-valued transformations.
- Given a uniform variable, can you simulate another random variable using a transformation?
- Define the equation to simulate Rayleigh random variables.

2.4 Computation of the Energy Density Spectrum of a Deterministic Signal

Consider a deterministic signal that has finite energy and zero average power. This signal is then termed as an energy signal. We will estimate the energy density spectrum of an energy signal. Let us consider $x(t)$ as a finite duration signal, whose values are available with us. The problem is to estimate the true spectrum from this finite duration signal. If $x(t)$ is a finite energy signal, its energy E is finite and given by

$$E = \int_{-\infty}^{\infty} |x(t)|^2 \, dt < \infty. \tag{2.59}$$

Then, its FT exists and is given by

$$X(F) = \int_{-\infty}^{\infty} x(t)e^{-j2\pi Ft} \, dt \tag{2.60}$$

The energy of the time domain signal is the same as the energy of the Fourier domain signal. We can write

$$E = \int_{-\infty}^{\infty} |x(t)|^2 \, dt = \int_{-\infty}^{\infty} |X(F)|^2 \, dF. \tag{2.61}$$

$|X(F)|^2$ represents the distribution of signal energy with respect to frequency and is called the energy density spectrum. It is denoted as $S_{xx}(F)$. The total energy is simply the integration of $S_{xx}(F)$ over F. We will show that $S_{xx}(F)$ is a FT of an autocorrelation function.

The autocorrelation function of the finite energy signal is given by

$$R_{XX}(\tau) = \int_{-\infty}^{\infty} x^*(t) x(t+\tau) dt, \tag{2.62}$$

where $x^*(t)$ is the complex conjugate of $x(t)$. Let us calculate the FT of autocorrelation.

$$\int_{-\infty}^{\infty} R_{XX}(\tau) e^{-j2\pi F\tau} d\tau = \int_{-\infty}^{\infty} \int_{-\infty}^{\infty} x^*(t) x(t+\tau) e^{-j2\pi F\tau} \, dt \, d\tau. \tag{2.63}$$

Put $t' = t + \tau$, $dt' = d\tau$.

$$\int_{-\infty}^{\infty} x^*(t) e^{j2\pi Ft} \, dt \int_{-\infty}^{\infty} x(t') e^{-j2\pi Ft'} \, dt' = X^{*(F)} X(F) = |X(F)|^2 = S_{xx}(F). \tag{2.64}$$

Hence, $R_{XX}(\tau)$ and $S_{xx}(F)$ are FT pairs.

We can also prove a similar result for discrete time (DT) signals. Let $r_{xx}(k)$ and $S_{xx}(f)$ denote autocorrelation and energy density for a DT signal.

$$r_{XX}[k] = \sum_{n=-\infty}^{\infty} x^*[n] x[n+k]. \tag{2.65}$$

Put $n + k = n'$. FT of autocorrelation is given by

$$S_{xx}(f) = \sum_{k=-\infty}^{\infty} r_{xx}[k] e^{-j2\pi kf} = \sum_{n=-\infty}^{\infty} x^*[n] e^{j2\pi nf} \sum_{k=-\infty}^{\infty} x[n'] e^{-j2\pi n'f} = |X(f)|^2. \tag{2.66}$$

We can now list two different methods for computing the energy density spectrum, namely, direct method and indirect method.

Direct method: In the direct method, one computes the FT of the sequence $x(n)$ and then calculates the energy density spectrum as

$$S_{xx}(f) = |X(F)|^2 = \left| \sum_{n=-\infty}^{\infty} x[n] e^{-j2\pi fn} \right|^2. \tag{2.67}$$

Indirect method: In the indirect method, we have to first compute the autocorrelation and then find its FT to get the energy density spectrum.

$$S_{xx}(f) = \sum_{k=-\infty}^{\infty} \gamma_{xx}[k] e^{-j2\pi kf}. \tag{2.68}$$

There is a practical difficulty involved in computing energy density spectrum using Equations 2.67 and 2.68. The sequence is available only for a finite duration. Using a finite sequence is equivalent to multiplication of the sequence by a rectangular window. This will result in spectral leakage due to Gibbs phenomenon, as will be discussed in Chapter 5.

Autocorrelation indicates the similarity between a signal and its time-shifted version. The following are the properties of autocorrelation.

1. $R_{xx}(\tau)$ is an even function of τ.

$$R_{xx}(\tau) = \int_{-\infty}^{\infty} x^*(t)x(t+\tau)\,dt \text{ (put } t+\tau=t').$$

$$R_{xx}(-\tau) = \int_{-\infty}^{\infty} x(t'-\tau)x(t')\,dt' = R_{xx}(\tau). \tag{2.69}$$

2. $R_{xx}(0) = E = $ Energy of the signal.

$$R_{xx}(0) = \int_{-\infty}^{\infty} x^*(t)x(t)\,dt = \int_{-\infty}^{\infty} |x(t)|^2 \,dt = E. \tag{2.70}$$

3. $|R_{xx}(\tau)| \leq R_{xx}(0)$ for all τ. (2.71)
4. Autocorrelation and ESD are FT pairs. (Note: Already proved)

Following are the properties of ESD.

1. $S_{xx}(F) \geq 0$ for all F.
2. For any real valued signal $x(t)$, ESD is an even function.
3. The total area under the curve of ESD is the total energy of the signal.

Example 15

Find the autocorrelation of the function $x(t) = e^{-4t}u(t)$.

Solution: Let us find the autocorrelation for $\tau < 0$.

$$R_{xx}(\tau) = \int_0^{\infty} e^{-4t} e^{-4(t-\tau)} \,dt = e^{4\tau} \int_0^{\infty} e^{-8t} \,dt$$

$$= e^{4\tau} e^{-8t} / (-8) \Big\downarrow_0^{\infty} = \frac{e^{4\tau}}{8}. \tag{2.72}$$

Autocorrelation for $\tau > 0$,

$$R_{XX}(\tau) = \int_0^\infty e^{-4t}e^{-4(t+\tau)}\,dt = e^{-4\tau}\int_0^\infty e^{-8t}\,dt$$

$$= e^{-4\tau}e^{-8t}\,/(-8)\big\downarrow_0^\infty = \frac{e^{-4\tau}}{8}. \qquad (2.73)$$

$$R_{XX}(\tau) = \frac{e^{-4|\tau|}}{8}. \qquad (2.74)$$

Example 16

Find the ESD of $x(t) = e^{-4t}u(t)$ and verify that $R_{XX}(0) = E$.

Solution:

$$X(F) = \int_0^\infty e^{-4t}e^{-j\omega t}\,dt = \frac{e^{-(4+j\omega)t}}{-(4+j\omega)}\,\big\downarrow_0^\infty = \frac{1}{4+j\omega}. \qquad (2.75)$$

$$X(-F) = \int_{-\infty}^0 e^{4t}e^{-j\omega t}\,dt = \frac{e^{-(j\omega-4)t}}{-(j\omega-4)}\,\big\downarrow_{-\infty}^0 = \frac{1}{4-j\omega}. \qquad (2.76)$$

$$S_{XX}(F) = |X(F)|^2 = \frac{1}{4+j\omega}\times\frac{1}{4-j\omega} = \frac{1}{16+\omega^2}. \qquad (2.77)$$

We will find ESD as the FT of the autocorrelation already found.

$$S_{XX}(F) = \int_{-\infty}^\infty \frac{e^{-4|\tau|}}{8}e^{-j\omega\tau}\,d\tau = \frac{1}{8}\left[\int_0^\infty e^{-(4+j\omega)\tau}\,d\tau + \int_{-\infty}^0 e^{-(j\omega-4)\tau}\,d\tau\right]$$

$$= \frac{1}{8}\left[\frac{e^{-(4+j\omega)\tau}}{-(4+j\omega)}\,\big\downarrow_0^\infty + \frac{e^{-(j\omega-4)\tau}}{-(j\omega-4)}\,\big\downarrow_{-\infty}^0\right]$$

$$= \frac{1}{8}\left[\frac{1}{4+j\omega} + \frac{1}{4-j\omega}\right] = \frac{1}{8}\left[\frac{8}{16+\omega^2}\right] = \frac{1}{16+\omega^2}. \qquad (2.78)$$

Let us evaluate the energy and autocorrelation at zero.

$$R_{XX}(0) = \frac{e^{-4\times 0}}{8} = \frac{1}{8}.$$

$$E = \int_{-\infty}^\infty x^2(t)\,dt = \int_0^\infty e^{-8t}\,dt = -\frac{e^{-8t}}{8}\,\big\downarrow_0^\infty = \frac{1}{8}. \qquad (2.79)$$

Example 17

Find the cross-correlation of $x[n] = \{1, 2, 1\}$ and $y[n] = \{1, 4, 6, 4, 1\}$.

Solution: We have to skip the first step of convolution, that is, taking time reversal.

We will start from Step 2. We will take the sample by sample product.

$$Z[0] = \{1+8+6 = 15\}.$$

Then, we will shift $x[n]$ toward the right by one sample and again take sample by sample product.

$$Z[1] = \{4+12+4 = 20\}.$$

We will shift $x[n]$ toward the right again by one sample and take sample by sample product.

$$Z[2] = \{6+8+1 = 15\}, \; Z[3] = [4+2 = 6], \; Z[4] = [1].$$

We will now shift $x[n]$ toward the left by one sample and take sample by sample product.

$$Z[-1] = [4+2 = 6], \; Z[-2] = [1]; \text{ a total of seven values are nonzero.}$$

$$Z[n] = \{1, \, 6, \, 15, \, 20, \, 15, \, 6, \, 1\}.$$

(2.80)

Cross-correlation has maximum value at the first position.

Let us write a MATLAB program to find the cross-correlation and plot it. The plot is shown in Figure 2.9.

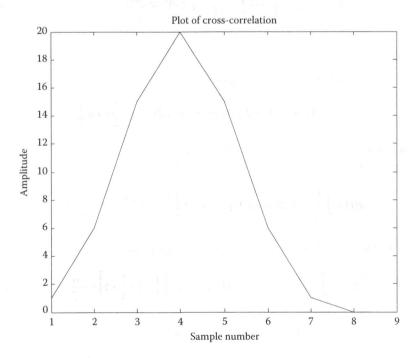

FIGURE 2.9
Cross-correlation of two signals.

```
clear all;
clc;x= [1 2 1 0 0];y= [1 4 6 4 1];
z=xcorr(x,y);
plot(z);
z= Columns 1 through 7
      1.0000    6.0000    15.0000    20.0000    15.0000    6.0000    1.0000
```

Example 18

Find autocorrelation and ESD for $x(t) = A \times \text{rect}(t)$ for $-\frac{1}{2} \leq t \leq \frac{1}{2}$.

Solution: For $\tau < -1$, there is no overlap, so the output is zero.

$$\text{For } -1 < \tau < 0, R_{XX}(\tau) = \int_{-\frac{1}{2}-\tau}^{\frac{1}{2}} A^2 \, dt = A^2 \left[\frac{1}{2} + \frac{1}{2} + \tau \right] = A^2 [1+\tau]. \tag{2.81}$$

$$\text{For } 0 < \tau < 1, R_{XX}(\tau) = \int_{-\frac{1}{2}}^{\frac{1}{2}-\tau} A^2 \, dt = A^2 \left[\frac{1}{2} - \tau + \frac{1}{2} \right] = A^2 [1-\tau]. \tag{2.82}$$

To find ESD, we will find the FT of the signal and its energy.

$$E = |X(f)|^2 = \left| \int_{-\frac{1}{2}}^{\frac{1}{2}} A e^{-j\omega t} \, dt \right|^2 = \left| A \frac{e^{-j\omega t}}{-j\omega} \Big|_{-\frac{1}{2}}^{\frac{1}{2}} \right|^2 = A^2 \left| \frac{2j \sin(\omega/2)}{-j\omega} \right|^2$$

$$= A^2 \left[\frac{\sin(\omega/2)}{\omega/2} \right]^2 = A^2 \left[\frac{\sin(\pi f)}{\pi f} \right]^2 = A^2 \sin c^2(f). \tag{2.83}$$

Example 19

Find the cross-correlation of the two signals

$$x(t) = A \times \text{rect}(t) \text{ and } y(t) = -A \times \text{rect}(2t) \text{ for } -\frac{1}{2} \leq t \leq \frac{1}{2}.$$

Solution: There is no overlap when $\tau < -3/4$, so output is zero.
Consider the interval $-3/4 < \tau < -1/4$; output $z(t)$ is given by

$$z(t) = \int_{-\frac{1}{2}}^{\frac{1}{4}-\tau} -A^2 \, dt = -A^2 [t] \Big|_{-\frac{1}{2}}^{\frac{1}{4}-\tau} = -A^2 \left[\frac{1}{4} - \tau + \frac{1}{2} \right] = -A^2 \left[\frac{3}{4} - \tau \right]. \tag{2.84}$$

Consider the interval $-1/4 < \tau < 1/4$, output $z(t)$ is given by

$$z(t) = \int_{-\frac{1}{4}-\tau}^{\frac{1}{4}-\tau} -A^2 \, dt = -A^2 [t] \Big|_{-\frac{1}{4}-\tau}^{\frac{1}{4}-\tau} = -A^2 \left[\frac{1}{4} - \tau + \frac{1}{4} + \tau \right] = -\frac{A^2}{2}. \tag{2.85}$$

Consider the interval $1/4 < \tau < 3/4$, output $z(t)$ is given by

$$z(t) = \int_{-\frac{1}{4}-\tau}^{\frac{1}{2}} -A^2 \, dt = -A^2 [t] \Big|_{-\frac{1}{4}-\tau}^{\frac{1}{2}} = -A^2 \left[\frac{1}{2} + \tau + \frac{1}{4} \right] = -A^2 \left[\frac{3}{4} + \tau \right]. \tag{2.86}$$

There is no overlap when $\tau > 3/4$, so output is zero.

Correlogram: A correlogram is a plot of one signal against another signal. It tells us about the similarity between the two signals. If the correlogram is a straight line graph, the two signals are highly correlated. If the slope of the graph is positive, they are positively correlated. If the slope is negative, they are negatively correlated. If the graph is random, the signals are not correlated at all.

Example 20

Find the correlogram of the sine and cosine functions $\sin(\omega t)$ and $\cos(\omega t)$.

Solution: We will find the values of the functions from 0 to 2π radians and plot one function against another using the following MATLAB program. We can also do it using manual calculations.

```
clear all;
clc;x=0:0.1:2*pi;
y=sin(x);
z=cos(x);
plot(y,z);
title('correlogram of the two signals');xlabel('sin
value');ylabel('cosine value');
```

The correlogram is plotted in Figure 2.10. We can see that it follows a definite pattern. The signals are related in some sense.

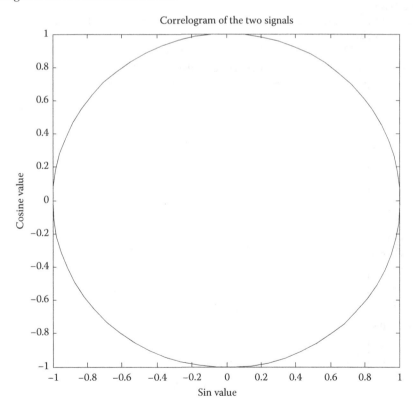

FIGURE 2.10
Correlogram of the sine and cosine function.

Example 21

Find the correlogram of the exponential functions e^{at} and e^{-at}.

Solution: Following is a MATLAB program to find the correlogram. The plot of the correlogram is shown in Figure 2.11. The first 16 values calculated for the two signals are listed as follows.

$$e^{at} = [1.0000 \quad 1.0202 \quad 1.0408 \quad 1.0618 \quad 1.0833 \quad 1.1052$$

$$1.1275 \quad 1.1503 \quad 1.1735 \quad 1.1972 \quad 1.2214 \quad 1.2461 \quad 1.2712$$

$$1.2969 \quad 1.3231 \quad 1.3499] \tag{2.87}$$

$$e^{-at} = [1.0000 \quad 0.9802 \quad 0.9608 \quad 0.9418 \quad 0.9231 \quad 0.9048$$

$$0.8869 \quad 0.8694 \quad 0.8521 \quad 0.8353 \quad 0.8187 \quad 0.8025 \quad 0.7866$$

$$0.7711 \quad 0.7558 \quad 0.7408] \tag{2.88}$$

```
clear all;
clc;
a=0.2;
t=0:0.1:2*pi;
y=exp(a*t)
z=exp(-a*t)
plot(y,z); title('correlogram of the two signals');xlabel('e^at
value');ylabel('e^-at value');
```

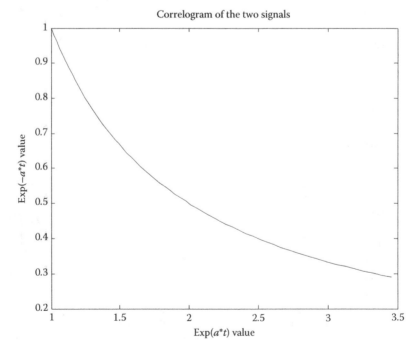

FIGURE 2.11

Correlogram of the exponential functions e^{at} and e^{-at}.

2.4.1 Estimation of Power Density Spectrum of Random Signal

There is a class of signals called stationary random signals that are of infinite duration and have infinite energy. The average power is, however, finite. Such signals are classified as power signals. We need to estimate the power density spectrum for such signals. For a random process, the autocorrelation is given by

$$\gamma_{XX}(\tau) = E[x^*(t)x(t+\tau)],\tag{2.89}$$

where E denotes the expected value. This cannot be calculated if the pdf is not known. For a random process, we do not know the pdf. We need to only estimate the value of autocorrelation as

1. We have only finite duration record.
2. The pdf is not known.

We assume that the process is wide-sense stationary and ergodic. We calculate time average autocorrelation as it is the same as ensemble autocorrelation in case of the ergodic process. Hence, we calculate

$$\gamma_{XX}(\tau) = \lim_{\tau \to \infty}[R_{XX}(\tau)].\tag{2.90}$$

The estimate $P_{XX}(F)$ of the power density spectrum is calculated as the FT of autocorrelation, and the actual value of the power density spectrum is the expected value of $P_{XX}(F)$ in the limit as $T_0 \to \infty$.

$$\begin{aligned}
P_{XX}(F) &= \int_{-T_0}^{T_0} R_{XX}(\tau)e^{-j2\pi F\tau}\,d\tau \\
&= \frac{1}{2T_0}\int_{-T_0}^{T_0}\left[\int_{-T_0}^{T_0}x^*(t)x(t+\tau)dt\right]e^{-j2\pi F\tau}\,d\tau \\
&= \frac{1}{2T_0}\left|\int_{-T_0}^{T_0}x(t)e^{-j2\pi Ft}\,dt\right|^2.
\end{aligned}\tag{2.91}$$

The actual power density spectrum is obtained as

$$\Gamma_{XX}(F) = \lim_{T_0 \to \infty}E(P_{XX}(F))\tag{2.92}$$

There are two approaches for computing the power density spectrum, namely, the direct method or the indirect method, by first calculating the autocorrelation and then finding the FT of it to get the power density spectrum.

In practice, we have only finite samples of the DT signals. The time average autocorrelation is computed using the equation,

$$\gamma_{XX}[m] = \frac{1}{n-|m|}\sum_{n=|m|}^{N-1}x^*[n]x[n+m],\quad m=-1,-2,\ldots,1-N.\tag{2.93}$$

The normalization factor results in a mean value of autocorrelation computed in the same manner as the actual autocorrelation. We then find the FT of the mean value of autocorrelation to get an estimate of the power spectrum. The estimate of the power spectrum can also be written as

$$P_{XX}(f) = \frac{1}{N}\left|\sum_{n=0}^{N-1} x[n]e^{-j2\pi fn}\right|^2 = \frac{1}{N}|X(f)|^2. \tag{2.94}$$

Equation 2.94 represents the well-known form of the power density spectrum estimate and is called a periodogram. To find the pdf, one draws the histogram for a signal with finite duration and uses curve fitting tests such as chi-square test or Kolmogorov–Smirnov test (K–S test) to fit a standard probability distribution for a given signal. The mathematical equation for that standard pdf can now be used in the calculations.

Properties of PSD can be listed as follows.

1. $P_{XX}(f) \geq 0$ for all f.

2. $P_{XX}(f) = \lim_{T \to \infty} \frac{1}{2T}|X(f)|^2$.

3. PSD of a real valued signal $x(t)$ is an even function of f. $P_{XX}(f) = P_{XX}(-f)$.

4. The total area under the curve of the PSD of the power signal equals the average signal power.

5. $S_{YY}(f) = |H(f)|^2 S_{XX}(f)$, where $H(f)$ stands for the transfer function of the linear time invariant system (LTI) system.

Example 22

Find the autocorrelation, spectral density, and power of the power signal given by $x(t) = A\sin(\omega_c t)$.

Solution:

$$R_{XX}(\tau) = \lim_{T \to \infty} \frac{1}{2T}\int_{-T}^{T} x(t)x(t-\tau)\,dt = \lim_{T \to \infty} \frac{1}{4T}\int_{-T}^{T} A\sin(\omega_c t)\sin(\omega_c t - \omega_c \tau)\,dt$$

$$= \lim_{T \to \infty} \frac{A^2}{4T}\left[\int_{-T}^{T}\cos(\omega_c \tau)\,dt - \int_{-T}^{T}\cos(2\omega_c t - \omega_c \tau)\,dt\right]$$

$$= \frac{A^2}{2}\cos(\omega_c \tau). \tag{2.95}$$

$$R_{XX}(0) = \frac{A^2}{2} = \text{Total power}.$$

PSD is the FT of autocorrelation.

$$PSD = FT\left[\frac{A^2}{2}\cos(\omega_c \tau)\right] = \frac{A^2}{4}FT\,[e^{j\omega_c \tau} + e^{-j\omega_c \tau}]$$

$$= \frac{A^2}{4}[\delta(\omega - \omega_c) + \delta(\omega - \omega_c)]. \tag{2.96}$$

Note that we have used the shifting property of FT.

2.4.2 Use of Discrete Fourier Transform (DFT) for Power Spectrum Estimation

In this section, we will discuss the basic method of estimating power spectrum, namely, finding the spectrum using DFT. Consider N samples of data. Following is the algorithm to find the samples of the periodogram.

1. Read N data values.
2. Find the N point DFT.
3. Computation of DFT gives N points of the periodogram.

$$P_{XX}\left(\frac{k}{N}\right) = \frac{1}{N}\left|\sum_{n=0}^{N-1} x(n)e^{-j2\pi nk/N}\right|^2 \quad \text{for } k = 0, 1, \dots, N-1. \tag{2.97}$$

4. Increase the length of the sequence by padding extra zeros to interpolate the spectrum at more frequencies.

The block schematic for using DFT to find the periodogram is shown in Figure 2.12.

Let us take speech samples (N), calculate DFT, and find N points of the periodogram. A MATLAB program to find the periodogram is given here. Figure 2.13 shows the plot of speech signal samples and the plot of its periodogram.

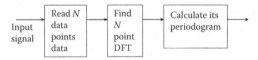

FIGURE 2.12
Block schematic for finding periodogram of a signal.

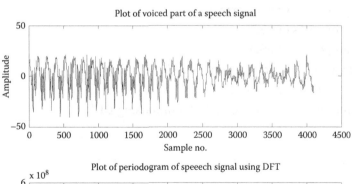

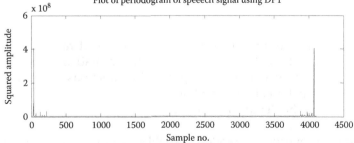

FIGURE 2.13
Plot of voiced part of speech signal and its periodogram.

```
%a program to find periodogram using DFT
clear all;
fp=fopen('watermark.wav');
fseek(fp,224500,-1);
a=fread(fp,4096);
a=a-128;
subplot(2,1,1);plot(a);title('plot of voiced part of a signal');
xlabel('sample no.');
ylabel('amplitude');
b=fft(a);
b1=abs(b.*b);
subplot(2,1,2);plot(b1);
title('plot of periodogram using DFT');
xlabel('sample no.');ylabel('squared amplitude');
```

Concept Check

- Define energy signal.
- Define power signal.
- Define autocorrelation and cross-correlation.
- What is the relation between autocorrelation and ESD?
- What is the relation between correlation and PSD?
- What is a correlogram?
- How will you find a correlogram?
- State the properties of autocorrelation.
- State the properties of ESD.
- State the properties of PSD.
- How will you find the total energy of a signal?
- How will you find total power?
- Write a procedure for plotting periodogram using DFT.

Summary

The chapter began with an introduction to the properties of random variables such as mean and standard deviation. We have described different statistical properties of random variables, characteristic functions, moment generating functions, transformation of a random variable, and estimation of ESD and PSD.

The highlights of the chapter are as follows.

1. The statistical properties of random variables are discussed. Statistical properties include mean, variance, skew, kurtosis, and higher order moments. All these properties are defined. The physical meaning of skew and kurtosis is defined and

explained using graphs. Several numerical examples are solved to calculate different statistical properties. The calculation of skew and kurtosis for speech signals
and Rayleigh distribution is described using the MATLAB program.

2. The moment generating function and characteristic function are defined. The use
of these functions for finding the moments of the random variable is illustrated
using numerical examples. Transformation of a random variable is discussed.
Linear as well as nonlinear and multiple-valued transformations are described.
The calculation of the pdf of the transformed variable in each case is explained.
The calculation of the mean value of the transformed variable is illustrated using
simple examples. The computer generation of a transformed variable is illustrated
using MATLAB program. Rayleigh variable is simulated using a transformation
of a uniformly distributed random variable.

3. Energy and power signals are defined. Autocorrelation and cross-correlation
for the signals are also defined. The relation between autocorrelation and ESD
is explained. They are FT pairs. Similarly, for power signals, autocorrelation and
PSD are FT pairs. The problems in the estimation of ESD and PSD are discussed.
Simple examples are solved to illustrate the concepts. The correlogram is defined.
MATLAB programs are written to draw correlograms for signals. If the graph is a
straight line, the signals are highly correlated. If the graph has some regularity, the
signals are related. If the graph is of random nature, the signals are not correlated.
Properties of autocorrelation are listed and proved. Properties of ESD and PSD
are listed. Use of DFT for calculation of periodogram is explained using MATLAB
program.

Key Terms

Estimation of random variables
Mean
Standard deviation
Variance
Higher order moments
Central moment
Skew
Kurtosis
Moment generating function
Characteristic function
Calculation of moments
Positive skew
Negative skew
Peakedness
Tailedness
Transformation of a random variable
Linear transformation
Nonlinear transformation
Multiple-valued transformation

Computer generation of a random variable
Rayleigh variable
Energy signal
Power signal
Energy spectral density (ESD)
Power spectral density (PSD)
Correlogram
Autocorrelation
Cross-correlation
Discrete Fourier transform (DFT)

Multiple-Choice Questions

1. The mean for a random variable is to estimated because
   a.   The frequency of occurrence cannot be calculated.
   b.   The pdf for the signal is not known.
   c.   The mean varies from segment to segment.
   d.   The signal is available for infinite duration.

2. Standard deviation is the
   a.   Square root of the second-order central moment.
   b.   Square of the second-order central moment.
   c.   Second-order central moment.
   d.   Second-order moment

3. The skewness is a measure of
   a.   Symmetry in the distribution.
   b.   Similarity in the distribution.
   c.   Asymmetry in the distribution.
   d.   Variance of the distribution.

4. Kurtosis measures the
   a.   Tailedness of the distribution.
   b.   Peakedness of the distribution.
   c.   Asymmetry of the distribution.
   d.   Peakedness and tailedness of the distribution.

5. The kurtosis for a standard normal distribution is
   a.   Equal to two.
   b.   Equal to three.
   c.   Equal to zero.
   d.   Equal to four

6. The mean value of a uniform random variable between −3 and 7 is
   a.   0.25
   b.   0.2
   c.   0.1
   d.   0.3

7. The characteristic function can be easily recognized as
   a.   The FT of the function with the sign of ω reversed.
   b.   The FT of the function.
   c.   Inverse FT of the function.
   d.   Inverse FT of the function with sign of ω reversed.

8. The moment generating function is used for
   a. Generating the moments.
   b. Evaluation of moments.
   c. Evaluation of skew.
   d. Evaluation of kurtosis.

9. If $T(x)$ is a linear transformation given by $y = T(x) = ax + b$, then
   a. $f_Y(y) = f_X\left(\dfrac{y-b}{a}\right)\left|\dfrac{1}{a}\right|$.
   b. $f_Y(y) = f_X\left(\dfrac{y-b}{a}\right)|a|$.
   c. $f_Y(y) = f_X(y-b)|a|$.
   d. $f_Y(y) = f_X(y-b)/|a|$.

10. Energy signal is one for which
    a. Energy is infinite and power is zero.
    b. Total energy is finite and average power is zero.
    c. Energy is finite and power is finite.
    d. Energy is infinite but power is finite.

11. Power signal is one for which
    a. Energy is finite and power is finite.
    b. Energy is infinite and power is infinite.
    c. Total energy is infinite but average power is finite.
    d. Total energy is finite and average power is finite.

12. Autocorrelation and ESD are
    a. Related to each other.
    b. Not related.
    c. DCT pairs.
    d. FT pairs.

13. The total energy under the curve of the PSD of power signal equals
    a. Average signal power.
    b. Total signal power.
    c. Total energy of the signal.
    d. Mean power of the signal.

14. Periodogram is the method for estimation of
    a. PSD.
    b. ESD.
    c. Average power.
    d. Average energy.

15. Correlogram is a plot of one signal against another signal. It tells us about
    a. The dissimilarity between the two signals.
    b. The similarity between the signals.
    c. The degree of matching of the two signals.
    d. The exact relation between the two signals.

Review Questions

1. Name and explain different statistical properties of any random variable.
2. Define mean or expected value of a random variable. Why is it termed as the expected value?
3. How will you fit a standard distribution to an unknown random variable?
4. Define standard deviation and variance of a random variable. What is the relation between the standard deviation and variance?
5. State the relation between second-order moment and the variance.
6. Define higher order moments and higher-order central moment.
7. Define skew. What is the physical significance of a skew?
8. Define fourth-order central moment. How is it related to the shape of the distribution?
9. Record a speech file in your own voice and find the histogram. Find the skew and kurtosis for the signal using a MATLAB program.
10. Define the characteristic function and a moment generating function.
11. Explain the use of the characteristic function and the moment generating function for finding moments.
12. Define autocorrelation. When will you say that the signals are correlated?
13. Define ESD and PSD. What are the difficulties in estimation of PSD?
14. State the properties of autocorrelation and prove them.
15. State the properties of ESD and PSD.
16. Define correlogram. How will you plot a correlogram of two signals?
17. How will you find PSD using DFT. Use the periodogram method to find the PSD?
18. State a procedure to evaluate the ESD.

Problems

1. Write a MATLAB program to measure the skewness and kurtosis for an exponential distribution.

2. Find the mean value and variance of a uniform random variable varying between the values –7 and 9.

3. Consider a random variable X with pdf shown in the following figure. Find A, mean value of X, and variance of X.

4. A random variable has pdf given by the following equation

$$f(x)=\begin{cases} 0.125 & -4\leq x\leq 4 \\ 0 & \text{elsewhere} \end{cases}$$

   a. Find the mean value.
   b. Find the mean square value.
   c. Find the variance.

5. A random variable has exponential pdf given by the following equation

$$f_X(x)=\begin{cases} \dfrac{1}{5}e^{-(x-2)/5} & x>2 \\ 0 & x<2 \end{cases}.$$

Find the mean value.

6. The density function of the variable X is given by

$$f_x(x)=\begin{cases} \dfrac{1}{3} & -3\leq x\leq 3 \\ 0 & \text{otherwise} \end{cases}.$$

Determine: (a) $P(X\leq x)$, (b) $P(|x|\leq 1)$ (c) mean and variance, and (d) The characteristic function.

7. Find the characteristic function of the random variable X having density function

$f_X(x)=e^{-\frac{1}{7}x}$ for all $x>0$.

8. Find the characteristic function and first moment of the random variable X having density function

$$f_X(x)=\begin{cases} \dfrac{1}{2}e^{-(x-1)/2} & x>1 \\ 0 & x<1 \end{cases}.$$

9. Find the moment generating function and first moment of the random variable X having density function

$$f_X(x)=\begin{cases} \dfrac{1}{3}e^{-(x-1)/3} & x>1 \\ 0 & x<1 \end{cases}.$$

10. Let T be the linear transformation given by $y = 2x+3$. Find the relation between the density functions of the two random variables x and y.

11. Consider a discrete random variable with ensemble consisting of discrete values $x=\{-1, 0, 1, 2\}$ with probabilities given by the vector $P=\{0.3, 0.1, 0.4, 0.2\}$. The pdf function can be written as
$f_X(x)=0.3\times\delta(x+1)+0.1\times\delta(x)+0.4\times\delta(x-1)+0.2\times\delta(x-2)$. If x is transformed to a new random variable y given by $y = 1-2x^2+x^3$, find the density function for y.

12. Let x be the Gaussian distributed variable with mean equal to 3 and variance of 9. If it is exposed to a linear transformation given by $Y=X+1$, find $f_Y(y)$.

13. Consider the transformation $T(x)=2X^2+3$. Let X be the Gaussian distributed variable with mean of 1 and standard deviation of 2. Find $f_Y(y)$.

14. A random variable X has a pdf given by
$$f_X(x)=\begin{cases}\cos(x) & -\pi/2<x<\pi/2\\0 & \text{elsewhere}\end{cases}$$. Find the mean value of the function given by
$Y = X^2$. Find $f_Y(y)$.

15. A random variable X is uniformly distributed on the interval (–2, 4). The random variable Y is formed as $Y = e^{-X}$. Find $E[Y]$.

16. Generate a Rayleigh variable with $a = 1$ using a transformation of a variable.

17. Find the autocorrelation of the function $x(t)=e^{-7t}u(t)$.

18. Find the ESD of $x(t)=e^{-2|t|}u(t)$ and verify that $R_{XX}(0)=E$.

19. Find the cross-correlation of $x[n]=\{1, 1, 1\}$ and $y[n]=\{1, 1, 1, 1, 1\}$.

20. Find the autocorrelation and ESD for $x(t)=2\times\text{rect}(t)$ for $-\frac{1}{4}\le t\le\frac{1}{4}$.

21. Find the correlogram of sine and cosine functions $\sin(2\omega t)$ and $\cos(3\omega t)$.

22. Find the correlogram of the exponential functions e^{2t} and e^{-2t}.

23. Find the autocorrelation, spectral density, and power of the power signal given by $x(t)=A\cos(\omega_c t)$.

Answers

Multiple-Choice Questions

1. (b)
2. (a)
3. (c)
4. (d)
5. (b)
6. (c)

7. (a)
8. (b)
9. (a)
10. (b)
11. (c)
12. (d)
13. (a)
14. (a)
15. (b)

Problems

1. Skew = 1.5301
   Kurtosis = 4.3831

2. $k = \dfrac{1}{16}$, mean = 1, variance = 21.33

3. $A = \dfrac{1}{3}$, mean = 0, variance = 1.61

4. Mean = 0, mean square value = 8.53, variance = 8.53

5. Mean = 7

6. $P(X \le x) = (x/3) + 1$, $P(X \le 1) = \dfrac{4}{3}$, mean = 0, variance = 6, $\varphi_X(\omega) = \dfrac{\sin(3\omega)}{3\omega}$

7. $\varphi_X(\omega) = \dfrac{1}{\left(\dfrac{1}{7}\right) - j\omega}$

8. First moment = 3

9. First moment = 4

10. $f_Y(y) = f_X\left(\dfrac{y-3}{2}\right)\left|\dfrac{1}{2}\right|$

11. $f_Y(y) = 0.3 \times \delta(y-2) + 0.4 \times \delta(y) + 0.3 \times \delta(y-1)$

12. $f_Y(y) = \dfrac{1}{3\sqrt{2\pi}} e^{-(y-4)^2/18}$, $\mu_Y = 4$, $\sigma_Y^2 = 9$

13. $f_Y(y) = \dfrac{1}{2\sqrt{2(y-3)}}\left[\dfrac{1}{2\sqrt{2\pi}}\exp\left[-\left(\sqrt{\dfrac{y-3}{2}}-1\right)^2/8\right] + \dfrac{1}{2\sqrt{2\pi}}\exp\left[-\left(-\sqrt{\dfrac{y-3}{2}}-1\right)^2/8\right]\right]$

14. Mean = $(\pi2/2) - 2$

15. $E(Y) = \dfrac{1}{6}[e^2 - e^{-4}]$

16. Refer to the MATLAB program in the text.

$$F_Y(Y) = 1 - e^{-(y-1)^2/2b^2} = x \qquad \text{for } 0 < x < 1$$

$$y = T(x) = 1 + \sqrt{-\sqrt{2}\,b\ln(1-x)} \qquad \text{for } 0 < x < 1$$

17. $R_{XX}(\tau) = \dfrac{e^{-7\tau}}{14}$

18. $S_{XX}(F) = |X(F)|^2 = \dfrac{1}{2+j\omega} \times \dfrac{1}{2-j\omega} = \dfrac{1}{4+\omega^2}$, $R_{XX}(0) = \dfrac{1}{4} = E$

19. Cross-correlation $= [1\ 2\ 3\ 3\ 3\ 2\ 1\ 0\ 0]$

20. For $-1 < \tau < 0$, $R_{XX}(\tau) = \displaystyle\int_{-\frac{1}{4}-\tau}^{\frac{1}{4}} 4\ dt = 4\left[\dfrac{1}{4} + \dfrac{1}{4} + \tau\right] = 4[1/2 + \tau]$

  For $0 < \tau < 1$, $R_{XX}(\tau) = \displaystyle\int_{-\frac{1}{4}}^{\frac{1}{4}-\tau} 4\,dt = 4\left[\dfrac{1}{4} - \tau + \dfrac{1}{4}\right] = 4[1/2 - \tau]$

  $$E = |X(f)|^2 = 2\left[\dfrac{\sin(\omega/4)}{\omega/4}\right]^2$$

21. Cross-correlation is plotted in the following figure.

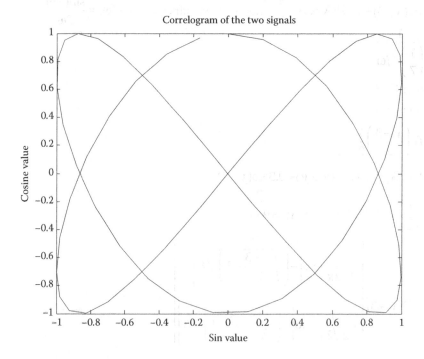

Correlogram of the two signals

22. Correlation is plotted in the following figure.

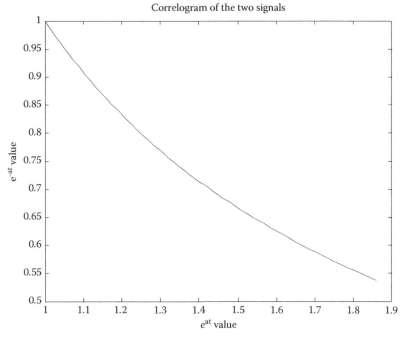

Correlogram of the two signals

23. $R_{XX}(\tau) = \dfrac{A^2}{2}\sin(\omega_c \tau)$, PSD $= \dfrac{A^2}{4j}\left[\delta(\omega - \omega_c) - \delta(\omega - \omega_c)\right]$

$R_{XX}(0) = 0$

3

Multiple Random Variables and Random Process

LEARNING OBJECTIVES

- Multiple random variables
- Properties of multiple random variables
- Marginal density function evaluation
- Joint probability density function (pdf) and cumulative distribution function (CDF)
- Statistical independence
- Orthogonality of two random variables
- Operations on multiple random variables
- Correlation and covariance
- Linear combination of random variables
- Modeling a random signal
- Autoregressive (AR), moving average (MA), and autoregressive moving average (ARMA) modeling
- Random process
- Properties of random process
- Wide sense stationary
- Ergodicity

The random variable is a useful concept for evaluating the parameters for variables such as mean and standard deviation. In some situations, it is necessary to extend the concept to more than one random variable. In engineering, many interesting applications, such as autocorrelation, cross-correlation, and covariance, can be handled by the theory of two random variables. Hence, we will focus on two random variables.

3.1 Multiple Random Variables

The random variable is introduced in Chapter 1, and the properties of random variable are discussed in Chapter 2. Consider a process consisting of a number of random variables such as temperature, pressure, and humidity. We are normally interested in the relationship between two or more random variables. We will generalize the concept termed multiple random variables to N random variables. Let us first discuss this concept of continuous random variables. Later on, this concept can be easily extended to discrete random variables.

Let us start with two random variables X and Y. We will define the joint pdf or simply the joint density function of X and Y as

$$f_{XY}(x, y) \geq 0$$

and   (3.1)

$$\int_{-\infty}^{\infty} \int_{-\infty}^{\infty} f_{XY}(x, y) \, dx \, dy = 1$$

The simple pdf for a single random variable spans a two-dimensional space. The joint pdf spans a three-dimensional space and will represent a surface. The total volume that is bounded by this surface and xy plane is unity. The probability that x lies between x_1 and x_2 and y lies between y_1 and y_2 is given by

$$P(x_1 < X < x_2, y_1 < Y < y_2) = \int_{-y_1}^{y_2} \int_{-x_1}^{x_2} f_{XY}(x, y) \, dx \, dy$$ (3.2)

The joint CDF $F_{XY}(x, y)$ is the probability that the joint events $\{X \leq x, Y \leq y\}$ hold good. The joint CDF can be written as

$$F_{XY}(x, y) = P(X \leq x, Y \leq y) = \int_{-\infty}^{y} \int_{-\infty}^{x} f_{XY}(x, y) \, dx \, dy$$ (3.3)

The properties of joint CDF can be listed as follows:

1. $0 \leq F_{XY}(x, y) \leq 1$
2. $F_{XY}(\infty, \infty) = 1$
3. Joint CDF is a non decreasing function of both x and y
4. $F_{XY}(-\infty, -\infty) = F_{XY}(x, -\infty) = F_{XY}(-\infty, x) = 0$
5. $P(x_1 < X < x_2, Y \leq y) = F_{XY}(x_2, y) - F_{XY}(x_1, y) \geq 0$
6. $P(X \leq x, y_1 < Y < y_2) = F_{XY}(x, y_2) - F_{XY}(x, y_1) \geq 0$
7. $P(x_1 < X < x_2, y_1 < Y < y_2) = F_{XY}(x_2, y_2) - F_{XY}(x_1, y_2) - F_{XY}(x_2, y_2) + F_{XY}(x_1, y_1)$
8. $F_{XY}(x, \infty) = F_X(x) \, \& \, F_{XY}(\infty, y) = F_Y(y)$

The joint pdf can be obtained from the joint distribution function by taking the derivative with respect to x and y.

$$f_{xy}(x, y) = \frac{d^2}{dx dy}(F_{XY}(x, y))$$ (3.4)

The properties of joint density functions are as follows:

1. $0 \leq f_{XY}(x, y) \leq 1$

2. $\int_{-\infty}^{\infty} \int_{-\infty}^{\infty} f_{XY}(x, y) = 1$

3. $F_{XY}(x, y) = \int_{-\infty}^{x} \int_{-\infty}^{y} f_{XY}(x_1, y_1) dx_1 dy_1$

4. $F_X(x) = \int_{-\infty}^{x} \int_{-\infty}^{\infty} f_{XY}(x_1, y_1) dx_1 dy_1$, $F_Y(y) = \int_{-\infty}^{\infty} \int_{-\infty}^{y} f_{XY}(x_1, y_1) dx_1 dy_1$

5. $P(x_1 < X < x_2, y_1 < Y < y_2) = \int_{x_1}^{x_2} \int_{y_1}^{y_2} f_{XY}(x, y) dx dy$

6. $f_X(x) = \int_{-\infty}^{\infty} f_{XY}(x, y) dy$, $f_Y(y) = \int_{-\infty}^{\infty} f_{XY}(x, y) dx$

If we want to test that the given joint density function is a valid density function, conditions 1 and 2 stated earlier can be used as sufficient conditions. To illustrate this concept, we will solve one numerical example.

Example 1

Find the value of constant "a" for which the function $f_{XY}(x, y) = ae^{-2x} \cos(y)$ $0 \leq x \leq 2, 0 \leq y \leq \pi/2$ is a valid density function.

Solution: Let us check conditions 1 and 2.

$$\int_0^2 \int_0^{\pi/2} ae^{-2x} \cos(y) dx dy = a \int_0^2 e^{-2x} dx \int_0^{\pi/2} \cos(y) dy = -\frac{a}{2} e^{-2x} \Big|_0^2 \sin(y) \Big|_0^{\pi/2}$$

$$= \frac{a}{2}(1 - e^{-4}) = 1$$

$$a = \frac{2}{1 - e^{-4}} \tag{3.5}$$

3.1.1 Marginal Density Functions and Statistical Independence

The distribution function for one random variable, namely, $F_X(x)$ and $F_Y(y)$, can be obtained using property 8 stated earlier for joint distribution function, i.e., by setting the value of the other variable to infinity in the joint distribution function. The functions $f_X(x)$ and $f_Y(y)$ are called marginal density functions.

Let us recall that the two events A and B are statistically independent, if

$$P(A \cap B) = P(A)(B) \tag{3.6}$$

Let us apply this condition to two random variables X and Y by defining two events A and B. We can write that the two events are statistically independent if

$$P(X \leq x, Y \leq y) = P(X \leq x)P(Y \leq y)$$

by definition of distribution function

$$F_{XY}(x, y) = F_X(x)F_Y(y) \tag{3.7}$$

by differentiation

$$f_{XY}(x, y) = f_X(x)f_Y(y)$$

These conditions serve as the conditions for sufficiency or conditions for test of independency of two random variables.

Let us now derive the condition for the conditional distribution and conditional density.

$$F_X(x/Y \le y) = \frac{P(X \le x, Y \le y)}{P(Y \le y)} = \frac{F_{XY}(x, y)}{F_Y(y)} = F_X(x) \qquad (3.8)$$

Similarly, $F_X(y/X \le x) = \dfrac{P(X \le x, Y \le y)}{P(X \le x)} = \dfrac{F_{XY}(x, y)}{F_X(x)} = F_Y(y)$ \qquad (3.9)

Conditional density function can also be found by differentiating the distribution functions.

$$f_X(x/Y \le y) = \frac{f_{XY}(x, y)}{f_Y(y)} = f_X(x) \qquad (3.10)$$

$$f_Y(y/X \le x) = \frac{f_{XY}(x, y)}{f_X(x)} = f_Y(y) \qquad (3.11)$$

Any of the conditions given earlier can be used for testing if the two random variables are statistically independent. We will illustrate the concepts by solving some numerical examples.

3.1.2 Operations on Multiple Random Variables

We will go through the operations on multiple random variables such as expected value, joint moments about the origin, joint central moments, and joint characteristic functions. We will define correlation, covariance, and the correlation coefficient for two random variables.

Expected Value: When multiple random variables are involved, expectation value must be calculated with respect to all the variables involved. Let us consider a function of two random variables, say $g(x, y)$. The expected value of the function is defined as

$$E[g(x, y)] = \int_{-\infty}^{\infty} \int_{-\infty}^{\infty} g(x, y) f_{XY}(x, y) \, dx \, dy \qquad (3.12)$$

Example 2

Let the function be given by $g(x, y) = aX + bY$. Find the mean value of the function in terms of the mean of the individual random variables.

Solution:

$$E[aX + bY] = \int_{-\infty}^{\infty} \int_{-\infty}^{\infty} (aX + bY) f_{XY}(x, y) \, dx \, dy$$

$$= a \int_{-\infty}^{\infty} X \left[\int_{-\infty}^{\infty} f_{XY}(x, y) \, dy \right] dx + b \int_{-\infty}^{\infty} Y \left[\int_{-\infty}^{\infty} f_{XY}(x, y) \, dx \right] dy$$

$$= a \int_{-\infty}^{\infty} X f_X(x) \, dx + b \int_{-\infty}^{\infty} Y f_Y(y) \, dy$$

$$= aE[X] + bE[Y] \qquad (3.13)$$

Thus, we can conclude that the mean of the weighted sum of random variables is equal to the weighted sum of the mean values.

Joint Moments about the Origin

The joint moments about the origin can be defined as

$$m_{nk} = E[X^n Y^k] = \int_{-\infty}^{\infty} \int_{-\infty}^{\infty} X^n Y^k f_{XY}(x, y) \, dx \, dy \tag{3.14}$$

The order of the moment is $n+k$. The first-order moments, also known as the center of gravity of the joint density function, are defined as

$$m_{01} = E[Y], \; m_{10} = E[X] \tag{3.15}$$

The second-order moments are

$$m_{02} = E[Y^2], \; m_{20} = E[X^2], \; m_{11} = E[XY] \tag{3.16}$$

The second-order moment $m_{11} = E[XY]$ is called as correlation of X and Y. If $m_{11} = R_{XY} = E[XY] = E[X]E[Y]$, X and Y are said to be uncorrelated. The converse, however, is not true. If $R_{XY} = 0$, then the random variables X and Y are said to be orthogonal.

Joint Central Moments

We now define joint central moments as

$$mc_{nk} = E[(X - \mu_X)^n (Y - \mu_Y)^k]$$

$$= \int_{-\infty}^{\infty} \int_{-\infty}^{\infty} (X - \mu_X)^n (Y - \mu_Y)^k f_{XY}(x, y) \, dx \, dy \tag{3.17}$$

The second-order center moments given below are called as variances of X and Y, respectively.

$$mc_{20} = E[(X - \mu_X)^2] = \sigma_X^2, \; mc_{02} = E[(Y - \mu_Y)^2] = \sigma_Y^2 \tag{3.18}$$

The second-order joint moment is of great importance. It is called as covariance. It is defined as

$$mc_{11} = C_{XY} = E[(X - \mu_X)(Y - \mu_Y)]$$

$$= E[XY - \mu_X Y - \mu_Y X + \mu_X \mu_Y]$$

$$= E[XY] - \mu_X E[Y] - \mu_Y E[X] + \mu_X \mu_Y$$

$$(E[X] = \mu_X, \; E[Y] = \mu_Y)$$

$$= E[XY] - \mu_X \mu_Y \tag{3.19}$$

If the two variables are independent or uncorrelated, the covariance is zero. If X and Y are orthogonal,

$$C_{XY} = -\mu_X\mu_Y, \quad E[XY] = 0 \tag{3.20}$$

The normalized second-order moment is known as correlation coefficient of X and Y and is defined as

$$\rho = E\left[\frac{(X-\mu_X)(Y-\mu_Y)}{\sigma_X\sigma_Y}\right], \quad -1 \leq \rho \leq 1 \tag{3.21}$$

Example 3

Given the joint pdf for two random variables X and Y

$$f(x,y) = \begin{cases} 9\exp(-3x)\exp(-3y) & x,y \geq, 0 \\ 0 & \text{elsewhere} \end{cases}$$

Check if it is a valid density function. Find the probability that X and Y lie between the limits 0 and 1. Find the marginal density functions. Check if the variables X and Y are statistically independent.

Solution: In order to check if it is a valid density function, we have to integrate the function over the complete range and check if the total probability is 1.

$$\int_0^\infty \int_0^\infty f(x,y)\,\mathrm{d}x\,\mathrm{d}y = 1 \tag{3.22}$$

$$\int_0^\infty 9e^{-3x}\,\mathrm{d}x \int_0^\infty e^{-3y}\,\mathrm{d}y = 9\frac{e^{-3x}}{-3}\Big|_0^\infty \times \frac{e^{-3y}}{-3}\Big|_0^\infty$$

$$= 9 \times \frac{1}{3} \times \frac{1}{3} = 1 \tag{3.23}$$

Hence, it is a valid density function.

1. To find probability that the variables x and y lie from 0 to 1 can be found by integrating the function between the limit 0 and 1.

$$\int_0^1 \int_0^1 9 \times e^{-3x}e^{3y}\,\mathrm{d}x\,\mathrm{d}y = 9 \times \frac{e^{-3x}}{-3}\Big|_0^1 \times \frac{e^{-3y}}{-3}\Big|_0^1 = 9 \times \frac{e^{-3}-1}{-3} \times \frac{e^{-3}-1}{-3}$$

$$= (0.0497 - 1)^2 = (-0.9502)^2 = 0.902 \tag{3.24}$$

2. Find the marginal density functions

$$f_Y(y) = \int_0^\infty f(x,y)\,\mathrm{d}x$$

$$f_Y(y) = 9\int_0^\infty e^{-3x} \times e^{-3y}\,\mathrm{d}x \tag{3.25}$$

$$f_Y(y) = -3e^{-3y}$$

$$f_X(x) = \int_0^\infty f(x,y)\,dy$$

$$f_X(x) = 9\int_0^\infty e^{-3x} \times e^{-3y}\,dy \tag{3.26}$$

$$f_X(x) = -3e^{-3x}$$

To check if the variables are statistically independent, we have to check if $f_{XY}(x,y) = f_X(x)f_Y(y)$, which is indeed holding good. Hence, the two variables are statistically independent.

Example 4

Given the joint pdf for two random variables X and Y

$$f(x,y) = \begin{cases} \dfrac{4}{225}xy & 0 \le x \le 3, 0 \le y \le 5 \\ 0 & \text{elsewhere} \end{cases}$$

Check if it is a valid density function. Find the probability that X is lying between 0 and 2 and Y is between 0 and 4. Find the marginal density functions. Check if the variables X and Y are statistically independent.

Solution: In order to check if it is a valid density function, we have to integrate the function over the complete range and check if the total probability is 1.

$$\int_0^3\int_0^5 f(x,y)\,dx\,dy = 1 \tag{3.27}$$

$$\frac{4}{225}\int_0^3 x\,dx \int_0^5 y\,dy = \frac{4}{225}\times\frac{x^2}{2}\Big\downarrow_0^3 \times \frac{y^2}{2}\Big\downarrow_0^5 = 1$$

Hence, it is a valid density function.

1. What is the joint distribution function?

   To find the probability that the variables x and y lie from 0 to 2 can be found by integrating the function between the limit 0 and 4.

$$\int_0^2\int_0^4 \frac{4}{225}\times xy\,dx\,dy = \frac{4}{225}\times\frac{x^2}{2}\Big\downarrow_0^2\times\frac{y^2}{2}\Big\downarrow_0^4 = \frac{4}{225}\times\frac{4}{2}\times\frac{16}{2} = \frac{64}{225} = 0.284 \tag{3.28}$$

2. Find the marginal density functions.

$$f_Y(y) = \int_0^3 f(x,y)\,dx$$

$$f_Y(y) = \int_0^\infty \frac{4}{225}\times xy\,dx \tag{3.29}$$

$$f_Y(y) = \frac{4}{225}\times y \times \frac{x^2}{2}\Big\downarrow_0^3 = \frac{2y}{25}$$

$$f_X(x) = \int_0^5 f(x,y)\,dy$$

$$f_X(x) = \frac{4x}{225}\int_0^5 y\,dy \qquad (3.30)$$

$$f_X(x) = \frac{4x}{225}\times\frac{y^2}{2}\downarrow_0^5 = \frac{2x}{9}$$

To check if the variables are statistically independent, we have to check if $f_{XY}(x,y) = f_X(x)f_Y(y)$, which is indeed holding good. Hence, the two variables are statistically independent.

Example 5

Given the joint pdf for two random variables X and Y

$$f(x,y) = \begin{cases} kx(1+y) & 0 < x \le 2, 0 < y \le 1 \\ 0 & \text{elsewhere} \end{cases}$$

Find the value of k. Find joint distribution function and marginal density function. What is the probability that $y > 0$?

Solution:

1. What is the value of k?
   In order to find k, the integral over the joint probability density equal to 1, that is

$$\int_{-\infty}^{\infty}\int_{-\infty}^{\infty} f(x,y)\,dx\,dy = 1 \qquad (3.31)$$

$$\int_0^2\int_0^1 kx(1+y)\,dx\,dy = 1 \qquad (3.32)$$

$$k\int_0^2 x\,dx\int_0^1 (1+y)\,dy = 1 \qquad (3.33)$$

$$k.2.\frac{3}{2} = 1 \quad \therefore k = \frac{1}{3} \qquad (3.34)$$

$$\therefore f(x,y) = \begin{cases} \dfrac{1}{3}x(1+y) & 0 < x \le 2, 0 < y \le 1 \\ 0 & \text{elsewhere} \end{cases} \qquad (3.35)$$

2. What is the joint distribution function?
   The joint distribution function is defined by the following equation:

$$F_{XY}(x,y) = P(X \le x, Y \le y) \qquad (3.36)$$

$$F_{XY}(x,y) = P(X \le 2, Y \le 1)$$

3. Find the marginal density function $f_Y(y)$

$$f_Y(y) = \int_0^2 f(x, y) \mathrm{d}x$$

$$f_Y(y) = \int_0^2 \frac{1}{3} x(1 + y) \mathrm{d}x \tag{3.37}$$

$$f_Y(y) = \frac{(1 + y)}{3} \int_0^2 x \mathrm{d}x$$

$$\therefore f_Y(y) = \frac{2}{3}(1 + y) \quad 0 < y \le 1 \tag{3.38}$$

$$f_X(x) = \int_0^1 f(x, y) \mathrm{d}y$$

$$f_X(x) = \int_0^1 \frac{1}{3} x(1 + y) \mathrm{d}y \tag{3.39}$$

$$f_X(x) = \frac{x(y + y^2 / 2)}{3} \downarrow_0^1 = \frac{x}{2}$$

4. What is the probability that $y > 0$?
   Now, as per property,

$$F_{XY}(x, 0) = 0 \tag{3.40}$$

$$\therefore P(Y \le 0) = 0 \tag{3.41}$$

$$\therefore P(Y > 0) = 1$$

Example 6

Two random variables X and Y have joint pdf of the form

$$f(x, y) = \begin{cases} \dfrac{3}{2}(x^2 + y^2) & 0 \le x \le 1, 0 \le y \le 1 \\ 0 & \text{elsewhere} \end{cases}$$

Check if it is a valid density function. Find the probability that x and y both are greater than 0.5. Find marginal density functions, mean, and standard deviation of x.

Solution: First, we will confirm that the given pdfs $f(x, y)$ is a valid density function. This can be done by testing property

$$\int_{-\infty}^{\infty} \int_{-\infty}^{\infty} f(x, y) \mathrm{d}x \mathrm{d}y = 1 \tag{3.42}$$

$$= \int_0^1 \int_0^1 \frac{3}{2}(x^2 + y^2)\,dx\,dy$$

$$= \frac{3}{2} \int_0^1 \int_0^1 (x^2 + y^2)\,dx\,dy$$

$$= \frac{3}{2} \int_0^1 \left(\frac{x^3}{3} + xy^2 \right)_0^1 dx\,dy$$

$$= \frac{3}{2} \int_0^1 \left(\frac{1}{3} + y^2 \right) dy$$

$$= \frac{3}{2} \left(\frac{y}{3} + \frac{y^3}{3} \right)_0^1$$

$$= \frac{3}{2} \left(\frac{1}{3} + \frac{1}{3} \right)$$

$$= 1 \tag{3.43}$$

Hence, $f(x, y)$ is a valid density function.

Now, probability that both x and Y are larger than 0.5 is given by

$$= \int_{0.5}^1 \int_{0.5}^1 \frac{3}{2}(x^2 + y^2)\,dx\,dy \tag{3.44}$$

$$= \frac{3}{2} \int_{0.5}^1 \left[\left(\frac{x^3}{3} + xy^2 \right) \downarrow_{0.5}^1 \right] dy$$

$$= \frac{3}{2} \int_{0.5}^1 \left(\frac{7}{24} + \frac{y^2}{2} \right) dy \tag{3.45}$$

$$= \frac{3}{2} \left[\frac{7}{24} y + \frac{y^3}{6} \right] \downarrow_{0.5}^1 = \frac{7}{16}$$

1. Find the mean value of X.

We first determine marginal density function $f_X(x)$

$$f_X(x) = \int_{-\infty}^{\infty} f_{XY}(x, y)dy$$

$$f_X(x) = \int_0^1 \frac{3}{2}(x^2 + y^2)dy \tag{3.46}$$

$$f_X(x) = \frac{3}{2} \left(x^2 y + \frac{y^3}{3} \right)_0^1$$

$$f_X(x) = \frac{3}{2} \left(x^2 + \frac{1}{3} \right)$$

$$f_Y(y) = \int_{-\infty}^{\infty} f_{XY}(x,y)dx$$

$$f_Y(y) = \int_0^1 \frac{3}{2}(x^2 + y^2)dx$$

$$f_Y(y) = \frac{3}{2}\left(y^2 x + \frac{x^3}{3}\right)\Big|_0^1 \tag{3.47}$$

$$f_Y(y) = \frac{3}{2}\left(y^2 + \frac{1}{3}\right)$$

Now,

$$E[X] = \int_0^1 x f_X(x)dx$$

$$= \frac{3}{2}\int_0^1 x\left(x^2 + \frac{1}{3}\right)dx = \frac{3}{2}\left(\frac{x^4}{4} + \frac{x^2}{6}\right)\Big|_0^1 = \frac{5}{8} = \mu_X \tag{3.48}$$

2. Find the standard deviation σ_x.
   We know that

$$\sigma_x^2 = E[X^2] - \mu_x^2$$

$$E[X^2] = \int_0^1 x^2 f_X(x)dx = \int_0^1 x^2 \times \frac{3}{2}\left(x^2 + \frac{1}{3}\right)dx$$

$$= \frac{3}{2}\left(\frac{x^5}{5} + \frac{x^3}{9}\right)\Big|_0^1 = \frac{7}{15} \tag{3.49}$$

$$\sigma_x^2 = \frac{7}{15} - \left(\frac{5}{8}\right)^2 = 0.076$$

$$\sigma_x = 0.2756$$

Example 7

Two random variables x and y have a joint pdf of the form

$$f(x,y) = \begin{cases} ke^{-(2x+3y)} & x > 0, \quad y > 0 \\ 0 & \text{elsewhere} \end{cases}$$

What is the value of k? Are X and Y statistically independent? Why? Find the conditional pdf $f(x/y)$

Solution: In order to find k, the integral over joint pdf must be evaluated. We have

$$\int_0^\infty \int_0^\infty k e^{-(2x+3y)} \, dx \, dy = 1$$

$$k \int_0^\infty e^{-2x} \, dx \int_0^\infty e^{-3y} \, dy = 1 \qquad (3.50)$$

$$k \frac{(-1)}{-2} \times \frac{(-1)}{-3} = 1$$

$$k = 6$$

1. For x and y to be statistically independent, we have to prove that

$$f_{XY}(x, y) = f_X(x) f_Y(y) \qquad (3.51)$$

Hence,

$$f_X(x) = \int_{-\infty}^\infty f_{XY}(x, y) dy$$

$$= 6 \int_0^\infty e^{-2x} e^{-3y} \, dy = 6e^{-2x} \frac{(-1)}{(-3)} \qquad (3.52)$$

$$f_X(x) = 2e^{-2x} \quad \text{for } x > 0$$

$$f_Y(y) = \int_{-\infty}^\infty f_{XY}(x, y) dx$$

$$= 6e^{-3y} \int_0^\infty e^{-2x} \, dx = 6e^{-3y} \frac{(-1)}{(-2)} \qquad (3.53)$$

$$f_Y(y) = 3e^{-3y} \quad \text{for } y > 0$$

Therefore,

$$f_{XY}(x, y) = f_X(x) f_Y(y) = 2e^{-2x} \times 3e^{-3y} = 6e^{-2x} e^{-3y}$$

Hence, X and Y are statistically independent.

2. Find the conditional pdf $f(x/y)$

We know that X and Y are statistically independent. We can write

$$f(x/y) = \frac{f_{XY}(x, y)}{f_Y(y)} = \frac{6e^{-2x} e^{-3y}}{3e^{-3y}} = 2e^{-2x} = f_X(x) \qquad (3.54)$$

Example 8

Given two random variables X and Y, each uniformly distributed between -1 and $+1$ with a joint pdf

$$f(x,y)=\begin{cases}0.25 & -1\le x\le 1,-1\le y\le 1\\ 0 & \text{elsewhere}\end{cases}$$

1. Are X and Y statistically independent? Why? Find the marginal density functions $f_X(x)$ and $f_Y(y)$
2. Find the mean values, mean square values, and variances of X and Y. Find also the correlation coefficient between X and Y and state if the variables are uncorrelated.
3. Find the mean value of the random variable Z, its mean square value, and its variance.

Solution: Let us find the marginal density functions $f_X(x)$ and $f_Y(y)$ as

$$f_X(x)= \int_{-\infty}^{\infty} f_{XY}(x,y)\mathrm{d}y = \int_{-1}^{1}0.25\,\mathrm{d}y = 0.25(y)\big\vert_{-1}^{1}=0.5 \tag{3.55}$$

Similarly, the reader can verify that

$$f_Y(y)=0.25(x)\big\vert_{-1}^{1} = 0.5 \tag{3.56}$$

The reader can also verify that

$$f_{XY}(x,y)= f_X(x)f_Y(y), \tag{3.57}$$

Hence, X and Y are statistically independent.

1. Find mean values, mean square values, and variances of X and Y. Find also the correlation coefficient between X and Y.

$$E[X]=\mu_x = \int_{-1}^{1} xf_X(x)\mathrm{d}x = \int_{-1}^{1}\frac{x}{2}\mathrm{d}x = \left(\frac{x^2}{4}\right)\Bigg\vert_{-1}^{1} =0 \tag{3.58}$$

$$E[X^2]= \int_{-1}^{1} x^2 f_X(x)\mathrm{d}x = \int_{-1}^{1}\frac{x^2}{2}\mathrm{d}x = \frac{x^3}{6}\Bigg\vert_{-1}^{1} =\frac{1}{3} \tag{3.59}$$

$$\sigma_x^2 = E[X^2]-\mu_x^2 =\frac{1}{3}-0=\frac{1}{3} \tag{3.60}$$

On similar lines, the reader can verify that

$$E[Y]=\mu_y = 0, \quad E[Y^2]=\frac{1}{3}, \quad \sigma_y^2 =\frac{1}{3} \tag{3.61}$$

The correlation coefficient between two random variables is defined as

$$\rho_{XY} = \frac{\text{cov}(X,Y)}{\sigma_x \sigma_y} = \frac{E[XY]-\mu_x\mu_y}{\sigma_x\sigma_y} \tag{3.62}$$

Now,

$$E[XY]= \int_{-1}^{1} \int_{-1}^{1} xy f_{XY}(x,y)dxdy$$

$$= \int_{-1}^{1}\int_{-1}^{1} 0.25xy\,dx\,dy = 0.25 \int_{-1}^{1} x\,dx\left[\frac{y^2}{2}\Big|_{-1}^{1}\right] \tag{3.63}$$

$$=0.25\left[\frac{x^2}{2}\right]_{-1}^{1} \times 0 = 0\times 0 = 0$$

$$\rho_{XY} = \frac{\text{cov}(x,y)}{\sigma_x\sigma_Y}=0$$

Hence, the two variables X and Y are uncorrelated.

2. Find the mean value, mean square value, and variance of the random variable Z.

$$E[Z]= E[X]+E[Y]=\mu_x +\mu_y = 0$$

$$E[Z^2]= E[(X+Y)^2]= E[X^2]+E[Y^2]+2E[XY]$$

$$=\frac{1}{3}+0+\frac{1}{3}=\frac{2}{3} \tag{3.64}$$

$$\sigma_Z^2 =\sigma_X^2 +\sigma_Y^2 =\frac{1}{3}+\frac{1}{3}=\frac{2}{3}$$

As the random variables x and y are uncorrelated.

Example 9

Let $x(t)=2$ for $1\le t \le 2$, $h(t)=1$ *for* $0\le t \le 3$, if $z = x+h$. Find and sketch the pdf $f_Z(z)$. Find $x(t)*h(t)$.

Solution: Because the two random variables are statistically independent, the pdf of their sum Z is found by obtaining the convolution of the respective pdfs. The convolution is solved as follows.
    We start with step 1, i.e., drawing $x(\tau)$ and $h(-\tau)$.

Step 1: Let us draw both these waveforms. Figure 3.1 shows plots of $x(\tau)$ and $h(t-\tau)$ for different intervals.

Step 2: Start with time shift t, which is large and negative. Let t vary from minus infinity to 1. We find that until t crosses 1, there is no overlap between the two signals. So, the convolution integral has a value of zero from minus infinity to zero. At $t = 1$, the right edge of $h(\tau)$ touches the left edge of $x(t)$.

Step 3: Consider the second interval between $t = 1$ and 2. $x(\tau)h(t-\tau) = 2$.

The overlapping interval will be between 0 and t. The output can be calculated as

$$y(t) = \int_1^t 2 \times 1 \, d\tau = 2\tau \downarrow_1^t = (2t-2) \tag{3.65}$$

Step 4: Consider the second interval between $t = 2$ and 4. $x(\tau)h(t-\tau) = 2$.

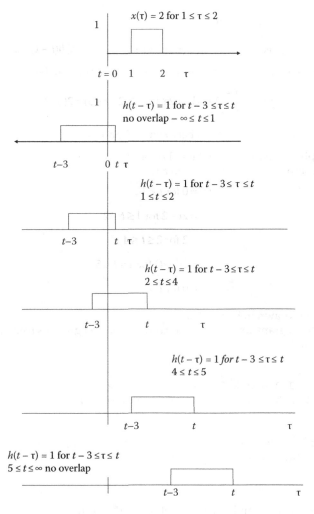

FIGURE 3.1
Plots of $x(\tau)$ and $h(t-\tau)$ for different intervals.

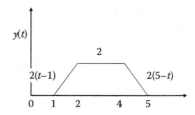

FIGURE 3.2
Plot of the output of the system.

The overlapping interval will be from 1 to 2. The output can be calculated as

$$y(t) = \int_1^2 2 \times 1 \, d\tau = 2\tau \Big|_1^2 = 2 \tag{3.66}$$

The output is constantly equal to 2.

Step 5: Consider the second interval between $t = 4$ and 5. $x(\tau)h(t-\tau) = 2$.

The overlapping interval will be from $t–3$ to 2. The output can be calculated as

$$y(t) = \int_{t-3}^2 2 \times 1 \, d\tau = 2\tau \Big|_{t-3}^2 = 2(2-t+3) = 2(5-t) \tag{3.67}$$

Step 6: Consider the second interval between $t = 5$ and infinity.

No overlapping interval will be there. The output is zero.
The overall output can be summarized as

$$y(t) = 0 \text{ for } t < 1$$

$$= 2t - 2 \text{ for } 1 \le t \le 2$$

$$= 2 \text{ for } 2 \le t \le 4 \tag{3.68}$$

$$= 2(5-t) \text{ for } 4 \le t \le 5$$

$$= 0 \text{ for } t > 5$$

The output $y(t)$ is drawn in Figure 3.2.
A MATLAB program for finding the convolution integral is given. The output is shown in Figure 3.3.

```
clear all;
x=[0 0 0 0 0 0 0 0 0 2 2 2 2 2 2 2 2 2 2 2 0];
y=[1 1 1 1 1 1 1 1 1 1 1 1 1 1 1 1 1 1 1 1 1 1 1 1 1 1 1 1 1 1 1 0];
z=conv(x,y);
n=1:1:53;
plot(n,z);title('convolved output for example 8');
xlabel('time');ylabel('amplitude');
```

Example 10

Two random variables X and Y have a joint pdf of the form

$$f_{XY}(x,y) = \begin{cases} A(x+y) & 0 \le x \le 1, 0 \le y \le 1 \\ 0 & \text{elsewhere} \end{cases}$$

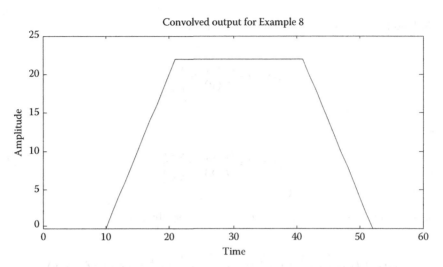

FIGURE 3.3
Convolved output for Example 8.

1. Find the value of A. Find the marginal density functions and conditional density functions.
   We find the value of A by solving the following equations:

$$
\int_0^1 \int_0^1 f_{XY}(x,y)\,dx\,dy = \int_0^1 \int_0^1 A(x+y)\,dx\,dy
$$

$$
= \int_0^1 \int_0^1 Ax\,dx\,dy + \int_0^1 \int_0^1 Ay\,dx \tag{3.69}
$$

$$
= A[y]\Big|_0^1 \frac{x^2}{2}\Big|_0^1 + A[x]\Big|_0^1 \frac{y^2}{2}\Big|_0^1
$$

$$
= A
$$

hence, $A = 1$

2. Find the marginal pdfs $f_X(x)$ and $f_Y(y)$.

$$
f_X(x) = \int_{-\infty}^{\infty} f_{XY}(x,y)\,dy = \int_0^1 (x+y)\,dy = \begin{cases} x + \dfrac{1}{2} & 0 \le x \le 1 \\[2mm] 0 & \text{elsewhere} \end{cases} \tag{3.70}
$$

Likewise,

$$
f_Y(y) = \begin{cases} y + \dfrac{1}{2} & 0 \le y \le 1 \\[2mm] 0 & \text{elsewhere} \end{cases} \tag{3.71}
$$

3. Find the conditional pdfs $f(x/y)$ and $f(y/x)$:

$$f_X(x/y) = \frac{f_{XY(x,y)}}{f_Y(y)} = \frac{x+y}{y+\dfrac{1}{2}}$$ (3.72)

and

$$f_y(y/x) = \frac{f_{XY(x,y)}}{f_x(x)} = \frac{x+y}{x+\dfrac{1}{2}}$$ (3.73)

4. Are X and Y statistically independent? Why or why not?

$$f_X(x).f_Y(y) = \left(x+\frac{1}{2}\right)\left(y+\frac{1}{2}\right) = xy + \frac{1}{2}x + \frac{1}{2}y + \frac{1}{4} \neq (x+y) = f_{XY}(x,y)$$ (3.74)

Hence, X and Y are not statistically independent.

Example 11

Two random variables X and Y are related by $Y = 2X - 1$, and X is uniformly distributed from 0 to 1. Find the correlation coefficient for X and Y.

Solution: X is uniformly distributed from 0 to 1.
  Hence,

$$f_X(x) = 1 \text{ for } 0 \leq x \leq 1$$ (3.75)

Therefore,

$$E[X] = \int_0^1 x f_X(x) dx = \int_0^1 x\, dx = \frac{1}{2},$$ (3.76)

$$E[X^2] = \int_0^1 x^2 f_X(x) dx = \int_0^1 x^2\, dx = \frac{1}{3}$$ (3.77)

$$E[X] = \int_0^1 x f_X(x) dx = \int_0^1 x\, dx = \frac{1}{2}$$ (3.78)

Therefore,

$$E[Y] = E[2X - 1] = 2E[X] - 1 = 0,$$ (3.79)

$$E[XY] = E[X(2X - 1)] = 2E[X^2] - E[X] = \frac{2}{3} - \frac{1}{2} = \frac{1}{6}$$ (3.80)

$$E[Y^2] = E[(2X-1)(2X-1)] = 4E[X^2] - 4E[X] + 1 = \frac{4}{3} - \frac{4}{2} + 1 = \frac{1}{3} \qquad (3.81)$$

$$\therefore \rho_{XY} = \frac{\text{cov}(X,Y)}{\sigma_x \sigma_y} = \frac{E[XY] - \mu_x \mu_y}{\sigma_x \sigma_y}$$

$$\rho_{XY} = \frac{\frac{1}{6} - 0}{\frac{1}{3} \cdot \frac{1}{12}} = 6 \qquad (3.82)$$

Example 12

A random variable X has a mean value of 10 and a variance of 36. Another random variable Y has a mean value of -5 and a variance of 64. The correlation coefficient for X and Y is -0.25. Find the variance of the random variable $Z = X + Y$.

Solution: The correlation coefficient between two random variables is defined as

$$\rho_{XY} = \frac{\text{cov}(X,Y)}{\sigma_x \sigma_y} = \frac{E[XY] - \mu_x \mu_y}{\sigma_x \sigma_y} \qquad (3.83)$$

$$-0.25 = \frac{\text{cov}(X,Y)}{(6).(8)}$$

$$\text{cov}(X,Y) = -12 \qquad (3.84)$$

The variance of sum of two random variables is given by

$$\sigma_z^2 = \sigma_x^2 + \sigma_y^2 + 2\text{cov}(X,Y) \qquad (3.85)$$

$$\sigma_z^2 = 36 + 64 - 24 = 76$$

Concept Check

- State the application where you need to consider two or more random variables?
- Define a joint CDF for two random variables.
- What is the relation between joint CDF and joint pdf?
- State the properties of joint CDF.
- State the properties of joint pdf.
- What are marginal density functions?

- State sufficient conditions for checking the statistical independence of two random variables.
- Define correlation and covariance.
- Define joint moments and joint central moments.
- Define joint characteristic function.

3.2 Modeling a Random Signal

The commonly used models for signals can be classified as parametric or nonparametric models. The methods that do not make any assumption about how the data are generated in a signal are called nonparametric methods. Nonparametric models have two basic limitations. First, the autocorrelation estimate is assumed to be zero outside the signal. Second, the inherent limitation is that the signal is assumed to be periodic with a period of N samples. Both these assumptions are nonrealistic. Nonparametric models use statistical parameters for signal modeling. Parametric methods do not use such assumptions. Parametric methods extrapolate the autocorrelation. Extrapolation is possible due to some predictability.

Parametric methods rely on the process of signal generation. Consider a speech signal. We know that it has high correlation between successive signal samples. We can predict further samples if the previous samples are available. This indicates that the limitation of finite data record will not be there for a parametric approach. We will be in a position to predict further samples when we have already calculated the predictor coefficients. The parametric methods include AR model parameters, MA model parameters, and ARMA model parameters.

3.2.1 AR, MA, and ARMA Modeling

The AR model is commonly used model for signals like speech. AR systems have only poles, and zeros exist only at zero. It is termed as an all-pole model. The vocal tract is modeled as all-pole system in the case of speech signal. Parametric methods are based on the modeling of data sequence as the output of a linear system characterized by the transfer function with poles and zeros. When the system has poles and zeros, it is termed pole-zero model or ARMA model. It is of the form

$$H(Z) = \frac{\sum_{k=0}^{M} b_k Z^{-k}}{1 + \sum_{k=1}^{N} a_k Z^{-k}} \tag{3.86}$$

If the system has only poles and zeros are only at $Z = 0$, the system is called all-pole or AR model. The transfer function of the system in Z domain is written as

$$H(Z) = \frac{1}{1 + \displaystyle\sum_{k=1}^{N} a_k Z^{-k}} \tag{3.87}$$

If the system has only zeros and poles exist only at Z equal to infinity, then the system is called as all-zero or MA model. The transfer function of MA system is written as

$$H(Z) = \sum_{k=0}^{M} b_k Z^{-k} \tag{3.88}$$

When the system has poles, the stability of the system must be confirmed before its actual implementation. The system stability can be easily checked in Laplace or Z domain. If the poles lie on the left-hand side of S domain, the system is stable. Similarly, if the poles lie within the unit circle, the system is stable. The system is said to be causal, and its current output depends on present or past inputs and past outputs. The normal procedure for using these models for random signals is to use a small segment of the signal over which the signal parameters can be assumed to be constant. Now, evaluate the parameters and use the model for estimation of power spectral density, autocorrelation, covariance, etc. These models, namely, MA, AR, and ARMA, are used for finite impulse response (FIR) and infinite impulse response (IIR) filter implementation, respectively. We will make use of these methods and models for spectral estimation, and it is discussed in detail in Chapter 5.

Concept Check

- What are nonparametric models?
- What are the basic limitations of nonparametric models?
- What are parametric models?
- Define AR models.
- Define MA model.
- Define ARMA mode.
- State applications of parametric models.

3.3 Random Processes

We now extend the concept of random variable to include time. A random variable is a function of the outcomes of the random experiment called as ensemble. If we allow the random variable to be a function of time too denoted by $X(s,t)$, then a family of such functions is called a random process. Figure 3.4 shows a random process. If the value of *s* is fixed, we get a random variable that is a function of time. If the value of t is fixed, we get a random variable that is a function of *s*, i.e., an ensemble function. Consider an example of temperature measurement using a number of transducers. If we measure the value of temperature for different transducers at the same time *t*, we get the random variable called ensemble function. The mean value of this random variable is called ensemble average. If we measure the value of temperature using the same transducer at different time, we get a random variable that is a function of time. The mean value of this random variable is called time average. If we can predict the future values of any sample function from its past values, we call it deterministic process. If the future values of a sample function cannot be predicted, then it is called nondeterministic process.

3.3.1 Stationary and Nonstationary Process

We can also classify the random processes as stationary or nonstationary random processes. If all the statistical properties, namely, the mean, moments, and variance, for the random variables of the random process are constant with respect to time, the process is a stationary random process. The two processes are statistically independent if random variable groups of one process are independent of random variable groups of other process. We define different levels of stationarity for a process, namely, first-order stationarity and second-order stationarity.

3.3.2 Stationary Processes

We will define first-order stationary process. A random process is said to *stationary to order one* if its density function does not shift with the shift in time origin and the mean value is constant.

$$f_X(x_1 : t_1) = f_X(x_1 : t_1 + \delta) \tag{3.89}$$

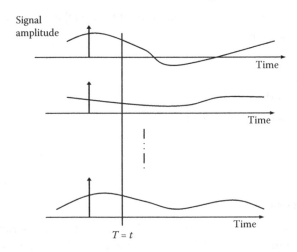

FIGURE 3.4
A continuous random process.

A random process is said to be *stationary to order two*, if its second-order density function is invariant to a time difference and if its autocorrelation function is a function of time difference and not the absolute time.

$$f_X(x_1, x_2 : t_1, t_2) = f_X(x_1, x_2 : t_1 + \delta, t_2 + \delta) \quad (3.90)$$

If we put $\delta = -t_1$, we can conclude that the second-order density function is a function of time difference.

A random process is said to be *wide sense stationary* if the two conditions are met: first, if the mean is constant; second, if the autocorrelation of a process is a function of time difference.

3.3.3 N^{th} Order or Strict Sense Stationary Process

A random process is stationary to order N if its nth order density function is invariant to a time origin shift. Stationarity of order N implies stationarity to all orders $< N$. A process stationary to all orders is called strict sense stationary process. Figure 3.4 shows a continuous random process.

We will illustrate the meaning of *wide sense stationary* with the help of a simple example.

Example 13

Consider a random process given by $x(t) = A\cos(\omega_0 t + \theta)$. Prove that the process is wide sense stationary if values of A and ω_0 are constants and ϑ is a uniformly distributed random variable on the interval 0 to 2π.

Solution: Let us find the mean value of a process.

$$E[x(t)] = \frac{1}{2\pi} \int_0^{2\pi} A\cos(\omega_0 t + \theta) \, d\theta = 0 \quad (3.91)$$

Let us find the autocorrelation function.

$$R_{xx}(\tau) = E[x(t)x(t+\tau)]$$

$$= E[A\cos(\omega_0 t + \theta) A\cos(\omega_0 t + \omega_0 \tau + \theta)]$$

$$= \frac{A^2}{2}[\cos(\omega_0 \tau) + \cos(2\omega_0 t + \omega_0 \tau + 2\theta)]$$

$$= \frac{A^2}{2}\cos(\omega_0 \tau)$$

(3.92)

The second term in the equation integrates to zero. We find that the autocorrelation is a function of time difference. The mean value is constant. These two conditions confirm that the process is a wide sense stationary process.

Ergodic Process: The process is ergodic if the time average is equal to ensemble average and time average autocorrelation is equal to the ensemble autocorrelation. Equation 1.76 specifies the ensemble autocorrelation. The condition of ergodicity requires that

$$\text{Time average autocorrelation} = \lim_{T \to \infty} \frac{1}{2T} \int_{-T}^{T} x(t)x(t+\tau) \, dt$$

(3.93)

$$\text{Ensemble autocorrelation} = E[x(t)x(t+\tau)]$$

Ergodicity is a restrictive form of stationarity. It is normally very difficult to prove that the given process is ergodic. For simplicity, we assume the given process to be ergodic so that we can find the ensemble average from the time average. We can find ensemble autocorrelation from the time average autocorrelation. The computations are simplified. The argument may be made that due to assumption of ergodicity, our analysis is on shaky grounds. The fact is that the theory only suggests a model for real situation. The problem is it is solved as an approximation problem. Hence, it makes no difference if we assume that the process is also ergodic.

Concept Check

- Define a random process.
- Define a deterministic process.
- Define a stationary process.
- Define wide sense stationary process.
- Define the ergodic process.
- What is time average autocorrelation?
- Define ensemble autocorrelation.

Summary

We started with the introduction of multiple random variables, their properties, correlation, and covariance. We have described the transformations of a random variable such as linear transformation, nonlinear transformation, and multiple-valued transformations. We have dealt with the random process and properties of random processes such as ergodicity and wide sense stationarity.

The highlights for the chapter are as follows:

1. The theory of multiple random variables is introduced. The terms like joint CDF and joint pdf are defined with their properties. The joint pdf is the derivative of joint CDF. The marginal density functions are defined. The marginal density functions are obtained by integrating over the other variable. The condition for statistical independence of two random variables is stated in terms of marginal density functions. The related conditions are also developed for statistical independence based on other properties such as probability, joint pdfs, and joint CDFs. The statistical properties of the joint variables are defined, namely, the expected value, joint center moments, joint moments, correlation, cross-correlation, covariance, and correlation coefficient.

2. The random signal models are discussed in the next section. The models are divided in two categories, parametric and nonparametric models. Nonparametric models use statistical properties. Parametric models are based on poles and zeros

of the system generating these signals. The three basic models, namely, AR, MA, and ARMA models, are discussed. AR models are also called all pole models. These systems have only poles. The zeros, if any, will exist at origin. MA models are all-zero models. They have no poles except at infinity. ARMA models are pole-zero models. We use a small segment of the signal over which the signal parameters can be assumed to be constant and evaluate the parameters. The model is used for estimation of power spectral density, autocorrelation, covariance, etc. These models, namely, MA, AR, and ARMA, are used for FIR and IIR filter implementation, respectively.

3. The third section deals with the random processes. A random variable is a function of the outcomes of the random experiment called ensemble. If we allow the random variable to be a function of time too, denoted by $X(s, t)$, then a family of such functions is called a random process. The classification of the random process such as deterministic/nondeterministic and stationary/nonstationary is described. A random process is said to be *wide sense stationary* if the mean is constant and the autocorrelation of a process is a function of time difference. The process is said to be ergodic process if the time average is equal to ensemble average and time average autocorrelation is equal to the ensemble autocorrelation.

Key Terms

Joint random variables

Joint CDF

Joint pdf

Statistical independence

Marginal density functions

Statistical properties of joint variables

Joint moments

Joint center moments

Correlation

Cross-correlation

Correlation coefficient

Autoregressive model

Moving average model

Autoregressive moving average model

Poles and zeros

Finite impulse response filter

Infinite impulse response filter

Parametric models

Nonparametric models

Random process

Stationary process

Strict sense stationary process

Wide sense stationary process

Ergodic process

Multiple-Choice Questions

1. A joint density function spans a
   a. One-dimensional space
   b. Two-dimensional space
   c. Three-dimensional space
   d. Four-dimensional space

2. The joint density function is a valid density function if

   a. $\displaystyle\int_{-\infty}^{\infty}\int_{-\infty}^{\infty} f_{XY}(x,y)\,\mathrm{d}x\,\mathrm{d}y = 1$

   b. $\displaystyle\int_{-\infty}^{\infty}\int_{-\infty}^{\infty} f_{XY}(x,y)\,\mathrm{d}x\,\mathrm{d}y = \infty$

   c. $\displaystyle\int_{-\infty}^{\infty} f_{XY}(x,y)\,\mathrm{d}x = 1$

   d. $\displaystyle\int_{-\infty}^{\infty}\int_{-\infty}^{\infty} f_{XY}(x,y)\,\mathrm{d}x\,\mathrm{d}y = 0$

3. The marginal CDF $F_X(x)$ can be found from joint CDF as

   a. $F_X(x) = F_{XY}(x, \infty)$

   b. $F_X(x) = F_{XY}(\infty, y)$

   c. $F_X(x) = F_{XY}(x, 0)$

   d. $F_X(x) = F_{XY}(0, y)$

4. The marginal pdf can be found from joint pdf using

   a. $\displaystyle f_X(x) = \int_{-\infty}^{\infty} f_{XY}(x, y)\,\mathrm{d}x$

   b. $\displaystyle f_X(x) = \int_{-\infty}^{\infty} f_{XY}(x, y)\,\mathrm{d}y$

   c. $\displaystyle f_X(x) = \int_{-\infty}^{\infty} f_{XY}(x, 0)\,\mathrm{d}x$

   d. $\displaystyle f_X(x) = \int_{-\infty}^{\infty} f_{XY}(0, y)\,\mathrm{d}y$

5. Two events are statistically independent if

   a. $P(A \cap B) = P(A) + P(B)$

   b. $P(A \cap B) = P(A) * P(B)$

   c. $P(A \cap B) = P(A) - P(B)$

   d. $P(A \cap B) = P(A)P(B)$

6. If the two events A and B are statistically independent, then

   a. $\displaystyle f_X(x / Y < y) = \frac{f_{XY}(x, y)}{f_Y(y)}$

   b. $f_X(x / Y < y) = f_X(x)$

   c. $\displaystyle f_X(x / Y < y) = \frac{f_{XY}(x, y)}{f_X(x)}$

   d. $\displaystyle f_X(x / Y < y) = \frac{f_X(x)}{f_Y(y)}$

7. The correlation $E[XY]$ is the

   a. Joint moment m_{11}

   b. Joint moment m_{10}

   c. Joint moment m_{01}

   d. Joint moment m_{00}

8. The joint center moment mc_{11} is given by

   a. $E[XY] - \mu_{XY}$

   b. $E[XY] - \mu_{XY}^2$

   c. $E[XY] - \mu_X \mu_Y$

   d. $E[XY] - \mu_X^2 \mu_Y^2$

9. The correlation coefficient is given by

   a. $\rho = \dfrac{E(x - \mu_X)(Y - \mu_Y)}{\sigma_X \sigma_Y}$

   b. $\rho = \dfrac{E(x - \mu_X)(Y - \mu_Y)}{\sigma_X^2 \sigma_Y^2}$

   c. $\rho = \dfrac{E(x - \mu_X^2)(Y - \mu_Y^2)}{\sigma_X \sigma_Y}$

   d. $\rho = E\left[\dfrac{(x - \mu_X)(Y - \mu_Y)}{\sigma_X \sigma_Y} \right]$

10. AR model has

   a. All poles

   b. All zeros

   c. Poles and zeros

   d. Only zeros

11. ARMA model has

   a. All poles

   b. All zeros

   c. Poles and zeros

   d. Only zeros

12. A nonparametric model for a signal uses

   a. Pole locations

   b. Mean, standard deviation, etc.

   c. All except statistical measures

   d. Zero locations

13. The process is said to be ergodic process if the time average is equal to ensemble average

   a. And time average autocorrelation is constant

   b. Or time average autocorrelation is equal to the ensemble autocorrelation

   c. And time average autocorrelation is equal to the ensemble autocorrelation

   d. And the ensemble autocorrelation is constant

14. If we allow the random variable to be a function of time too, then a family of such functions is called
    a. A stationary process
    b. A random process
    c. An ergodic process
    d. A random variable

15. A random process is said to be wide sense stationary
    a. If the mean is constant and the autocorrelation of a process is a function of time difference
    b. If the mean is constant
    c. If the autocorrelation of a process is a function of time difference
    d. If the mean is constant or the autocorrelation of a process is a function of time difference

Review Questions

1. Define joint CDF and joint pdf for joint random variables. What is the relation between joint CDF and joint pdf?
2. State the properties of joint CDF.
3. State the properties of joint pdf.
4. Define joint moments and joint center moments.
5. Define correlation, covariance, and correlation coefficient.
6. State the different sufficient conditions for statistical independence of two random variables.
7. Define AR and MA model for a signal. State the properties of AR and MA models.
8. How will you check the stability of the AR and ARMA models?
9. What are the applications of these models?
10. Define a random process. What is ensemble average and what is a time average? Define time average autocorrelation and ensemble autocorrelation.
11. Define a wide sense stationary process. When will you classify the process as ergodic?
12. How will you classify the random processes? Define different degrees of stationarity for a random process.
13. Define ergodic process. Why is an unknown process assumed to be ergodic?

Problems

1. Find the value of constant "b" for which the function
   $f_{XY}(x, y) = be^{-x} \cos(2y) \quad 0 \le x \le 2, 0 \le y \le \pi/4$ is a valid density function.

2. Given the joint pdf for two random variables X and Y

$$f(x,y) = \begin{cases} k\exp(-2x)\exp(-2y) & x, y \geq 0 \\ 0 & \text{elsewhere} \end{cases}$$

Find k. Check if it is a valid density function. Find the probability that X and Y lie between the limits 0 and 1. Find the marginal density functions. Check if the variables X and Y are statistically independent.

3. Given the joint pdf for two random variables X and Y

$$f(x,y) = \begin{cases} kxy & 0 \leq x \leq 2, 0 \leq y \leq 2 \\ 0 & \text{elsewhere} \end{cases}$$

Find k for which it is a valid density function. Find the probability that X is between 0 and 1 and Y is between 0 and 1. Find the marginal density functions. Check if the variables X and Y are statistically independent.

4. Give the joint pdf for two random variables X and Y

$$f(x,y) = \begin{cases} k(x+1)(y+2) & 0 < x \leq 1, 0 < y \leq 1 \\ 0 & \text{elsewhere} \end{cases}$$

Find the value of k. Find joint distribution function and marginal density function. What is the probability that $y > 0$?

5. Two random variables X and Y have a joint pdf of the form

$$f(x,y) = \begin{cases} \dfrac{3}{10}(x^2 + y^2) & 0 \leq x \leq 1, 0 \leq y \leq 2 \\ 0 & \text{elsewhere} \end{cases}$$

Check if it is a valid density function. Find the probability that x and y both are greater than 0.5. Find marginal density functions, mean, and standard deviation of x.

6. Two random variables x and y have a joint pdf of the form

$$f(x,y) = \begin{cases} ke^{-(x+y/2)} & x > 0, \ y > 0 \\ 0 & \text{elsewhere} \end{cases}$$

What is the value of k? Are X and Y statistically independent? Why? Find the conditional pdf $f(x/y)$.

7. Given two random variables X and Y, each is uniformly distributed between –2 and +2 with a joint pdf

$$f(x,y) = \begin{cases} 0.2 & -2 \leq x \leq 2, -2 \leq y \leq 2 \\ 0 & \text{elsewhere} \end{cases}$$

1. Are X and Y statistically independent? Why? Find the marginal density functions $f_X(x)$ and $f_Y(y)$.

2. Find the mean values, mean square values, and variances of X and Y. Find also the correlation coefficient between X and Y and state if the variables are uncorrelated.

3. Find the mean value of the random variable Z, its mean square value, and its variance if $Z = X+Y$.

8. Let $x(t)=2$ for $0 \le t \le 2$, $h(t)=t$ for $0 \le t \le 3$, if $z = x+y$. Find $f_Z(z)$, find $x(t)*h(t)$.

9. Two random variables X and Y have a joint pdf of the form

$$f_{XY}(x,y) = \begin{cases} A(2x+3y) & 0 \le x \le 1, 0 \le y \le 1 \\ 0 & \text{elsewhere} \end{cases}$$

1. Find the value of A. Find the marginal density functions and conditional density functions.

10. Two random variables X and Y are related by $Y = 4X - 3$ and X is uniformly distributed from 0 to 1. Find the correlation coefficient for X and Y.

11. A random variable X has a mean value of 5 and a variance of 25. Another random variable Y has a mean value of 1 and a variance of 4. The correlation coefficient for X and Y is –0.25. Find the variance of the random variable $Z = X+Y$.

12. Consider a random process given by $x(t)=10\sin(200\pi t + \vartheta)$. Prove that the process is *wide sense stationary* if ϑ is a uniformly distributed random variable on the interval 0 to 2π.

Answers

Multiple-Choice Questions

1. (c)
2. (a)
3. (a)
4. (b)
5. (d)
6. (b)
7. (a)
8. (c)
9. (d)
10. (a)
11. (c)
12. (b)

13. (c)

14. (b)

15. (a)

Problems

1. $b = \dfrac{2}{1-e^{-2}}$

2. $k = 4$, $P(X \le 1, Y \le 1) = 4(e^{-2} - 1)^2$, $f_X(x) = -2e^{-2x}$, $f_Y(y) = -2e^{-2y}$, the two variables are statistically independent.

3. $k = 1/4$, $P(X \le 1, Y \le 1) = \dfrac{1}{16}$, $f_X(x) = x/2$, $f_Y(y) = y/2$, the two variables are statistically independent.

4. $k = 4/15$, $f_X(x) = 2(x+1)/3$, $f_Y(y) = 2(y+2)/5$, $P(Y > 0) = 1$

5. It is a valid density function, $P(X, Y > 0.5) = 21/40 = 0.525$,

   $$f_X(x) = \frac{3}{5}\left(x^2 + \frac{4}{3}\right),\ f_Y(y) = \frac{3}{10}\left(y^2 + \frac{1}{3}\right),\ E[X] = 0.55,\ \sigma_x^2 = 0.08416,\ \sigma_x = 0.2901.$$

6. $k = \tfrac{1}{2}$, $f_X(x) = e^{-x}$, $f_Y(y) = \dfrac{1}{2}e^{-y/2}$, the two variables are statistically independent, $f(x/y) = f_X(x)$.

7. The two variables are statistically independent as $f_{XY}(x, y) = f_X(x)f_Y(y)$. $E[X] = 0$,

   $E[Y] = 0$, $E[XY] = 0$, covariance $= 0$, $\sigma_x^2 = \dfrac{4}{3}$, $\sigma_Y^2 = \dfrac{4}{3}$

   $E[Z] = E[X] + E[Y] = \mu_x + \mu_y = 0$, $E[Z^2] = \dfrac{8}{3}$, $\sigma_Z^2 = \dfrac{8}{3}$.

8. $z(t) = 0$ for $t < 0$

   $= t^2$ for $0 \le t \le 2$

   $= 4t - 4$ for $2 \le t \le 3$

   $= 4t - t^2 + 5$ for $3 \le t \le 5$

   $= 0$ for $t > 5$

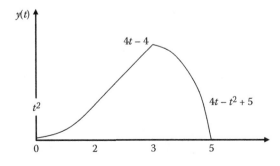

9. $A = 2/5$, $f_X(x) = \begin{cases} \dfrac{4}{5}x + \dfrac{6}{5} & \text{for } 0 \le x \le 1 \\ 0 \end{cases}$, $f_Y(y) = \begin{cases} \dfrac{6}{5}y + \dfrac{4}{5} & \text{for } 0 \le y \le 1 \\ 0 \end{cases}$

The two variables are not statistically independent.

10. $E[X] = \frac{1}{2}$, $E[Y] = -1$, $E[XY] = -1/6$, $\rho = \dfrac{3}{7}$

11. $\sigma_z^2 = 24$

12. The process is *wide sense stationary* as the mean is constant and autocorrelation is a function of time difference.

4

Detection and Estimation

This chapter describes the main components of a system, namely, the transmitter, channel, and receiver. The first section introduces the basics of communication theory and deals with signal transmission through channels, distortionless transmission, bandwidth of the signal, relation between bandwidth and rise time, the Paley–Wiener theorem, ideal low-pass filters (LPF), high-pass filters (HPF), bandpass filters (BPF), and so on. The next section discusses optimum detection methods. These methods include weighted probability tests, maximum likelihood (ML) criterion, Bayes criteria, minimax criteria, Neyman–Pearson criteria, and receiver operating characteristics (ROC). The estimation theory will be described in a further section. Finally, three types of estimators, namely, the Bayesian estimator, maximum likelihood estimator (MLE), and minimum mean-square estimation (MMSE), are discussed. The Cramer–Rao inequality will also be explained.

4.1 Basics of Communication Theory

As mentioned earlier, a digital communication system consists of a transmitter, a channel, and a receiver, as shown in Figure 4.1.

4.1.1 Transmitter

The transmitter consists of a sampler, a quantizer, and a coder. The raw signal like speech or a video is always in analog form. The analog signal is first sampled using the sampling frequency decided by the sampling theorem. The sampling theorem can be stated as follows. If the analog signal is band limited with a bandwidth of W, then if it is sampled with a sampling frequency greater than or equal to $2W$, the signal can be recovered from the signal samples using a sinc function as the interpolation function. The sampled signal is called a discrete time (DT) signal. This is then quantized and converted into a digital signal.

4.1.1.1 Justification for Sampling and Quantization of Analog Signals

The use of digital computers for signal processing has reduced the number of computations as less data are required. Converting the analog signal to its DT version is justifiable according to the sampling theorem. The DT signal is further quantized to get the digital signal. This digitization is also justifiable because of the nature of the human ear and eye. Human ear cannot discriminate if the signal amplitude is changed by a small amount. Thus, if there is small quantization noise, we may not be able to detect it. However, we have to select the step size for quantization so that it fits in the range where humans will not detect the difference. Similarly, very small changes in image intensity are not detected by the human eye.

The digital signal is further coded using lossless or lossy coding techniques. For further details on coding techniques, the reader can refer to any textbook on digital communication.

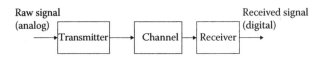

FIGURE 4.1
Basic communication system.

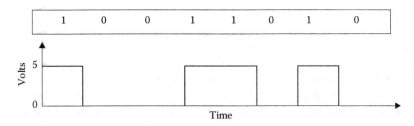

FIGURE 4.2
PCM-coded waveform for an 8 bit binary code—10011010.

Speech coders producing a reconstructed signal that converges toward the original signal with decreasing quantization noise are called waveform coders. Waveform coders minimize the error between the synthesized and the original speech waveforms.

A pulse code modulation (PCM) system is a basic coding scheme used for telephone speech signal. The signal is first appropriately sampled. The samples are then quantized by either a mid-riser or a mid-thread quantizer using a basic quantization technique. Let us say we are using 8 bit quantization; then, every sample will be encoded using an 8 bit binary code and transmitted in the pulse-coded form. Pulse coding using return to zero (RZ coding) is done as follows. A one is represented as a high voltage level, say +5 volts, and a zero is represented as a low voltage level, say zero volts, as shown in Figure 4.2.

The quantized and coded signal is now in the form of voltage levels, like in the case of PCM.

4.1.2 Channel

The transmitted signal then passes over the channel. The bandwidth of the channel is always limited. Hence, the channel functions as a LPF. The signal gets distorted when it passes over the channel. External noise also gets added in the signal. Channel equalizers are used to overcome the problem of band limitation of the channel. Equalizers are used at the receiver end. The channel coding will include techniques like error correcting codes such as inclusion of cyclic redundancy check (CRC) characters. When the PCM-coded coded signal gets transmitted over the channel, it will get distorted due to limited channel band width and the square waveform will be received as a smooth sine like waveform, as shown in Figure 4.3.

4.1.3 Receiver

The receiver has a channel equalizer at the start to correct for channel errors. For a digital system, the receiver will include a CRC to correct the received bits. The received signal contains quantization noise. This noise is removed using a LPF. The output of the LPF is a waveform, as shown in Figure 4.3. The receiver samples the analog voltage values and takes a decision whether the detected signal is a 1 or a 0. The threshold

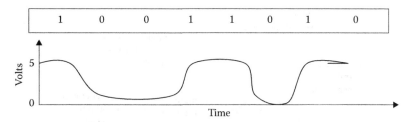

FIGURE 4.3
Waveform of a received signal that has passed the LPF in a receiver.

is selected at the receiver end. If the received signal level is greater than the threshold, it is received as a "1" and if it is less than the threshold, it is received as a "0," For further details, the reader can refer to any textbook on digital communication. The next section deals with the optimum detection of "1" and "0" at the receiver end.

Concept Check

- How will you convert an analog signal into a digital signal?
- Justify sampling and quantization.
- Explain PCM waveform coding technique for coding a quantized signal.
- Explain the behavior of a channel.
- How will you detect the signal at the receiver end?
- Draw the nature of the received signal.

4.2 Linear but Time-Varying Systems

The characteristic properties of linear systems do not always remain constant. The properties change with time. Consider the example of a speech signal. This is a naturally occurring signal. It has some random components. There is the voice box called the larynx that has a pair of vocal folds. These vocal folds generate excitation in the form of a train of impulses that passes via a linear system called the vocal tract. The train of impulses convolves with the vocal tract response to generate speech. Speech properties obviously change with time as we utter different words. The properties of vocal tract, that is, its impulse response, vary with time. The linear system of the vocal tract will then be a linear but time-varying (LTV) system. When we speak, we utter different words. This is possible because we can change the resonant modes of the vocal cavity and we can stretch the vocal cords, to some extent, to modify the pitch period for different vowels. Thus, the speech signal can be described by a time-varying model (LTV system). The impulse response is time varying, as shown in Figure 4.4.

4.2.1 Filter Characteristics of Linear Systems

A filter is a frequency selective device/system that is normally used to limit the spectrum of the signal to only some bands of frequencies. The linear system behaves as a

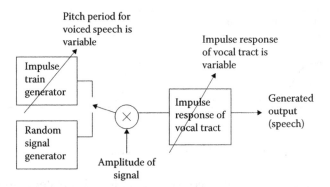

FIGURE 4.4
LTV model for speech signal.

filter because all frequencies that pass via a system do not have the same gain. Let us understand the concept of gain. The frequency response to a system can be plotted by sending different frequencies with the same amplitude as input to the system. The output is plotted with respect to the frequency of the signal. We find that the response is not flat, that is, all frequencies do not have the same gain as they pass via a linear system. The response is not flat means the gain, i.e., the ratio of output voltage to input voltage is not constant for all frequencies (refer to Figure 4.5). Hence, we say that the linear system behaves as a filter, filtering out or attenuating certain frequency signals. Frequency response measures the output spectrum of a system in response to any stimulus. It is used to characterize the dynamic behavior of the system. It is plotted as magnitude and phase of the output as a function of input frequency. Let us put it in more simple terms. If a sine wave of a certain frequency is applied as input to a system, a linear system will respond at that frequency with a certain magnitude and a certain phase angle relative to the input. For a linear system, if the amplitude is doubled, the amplitude of the output will also be doubled. In addition, if the system is linear time invariant (LTI), then the frequency response also will not vary with time. Let us consider a typical frequency response of a resistance–capacitance (RC) coupled amplifier as shown in Figure 4.5. It is seen that the amplifier attenuates very low and very high frequencies. Generally, the lower and upper cut-off frequencies are determined to detect the bandwidth of the filter response. The cut-off frequencies are evaluated as the frequencies at which the gain falls to $1/\sqrt{2}$ of its maximum value or the power reduces to half its maximum value. This is because the human ear cannot hear the sound when the power drops to a value below the half power. The cut-off frequencies F_1 and F_2 are shown in Figure 4.5.

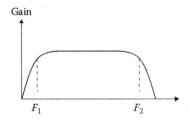

FIGURE 4.5
Frequency response of an RC coupled amplifier.

4.2.2 Distortionless Transmission through a System

Let us first understand the meaning of distortionless transmission. Consider a signal that passes over a communication channel, which can be treated as a linear system functioning as a filter. If the signal received at the receiver end is the replica of the input signal transmitted from the transmitter end except for the change in amplitude and a constant phase delay, the channel is said to be a distortionless channel. To put it mathematically, we can write

$$y(t) = Kx(t - t_0). \tag{4.1}$$

K stands for gain, which refers to change in amplitude and t_0 accounts for delay in the transmission channel. Let us transform this condition into the frequency domain. We will use the Fourier transform (FT) of Equation 4.1.

$$Y(f) = KX(f)\exp(-j2\pi f t_0). \tag{4.2}$$

In this equation, we have used the time shifting property of FT. The reader can learn more about these properties in Chapter 6. Hence, the transfer function of the channel can be written in general as

$$H(f) = \frac{Y(f)}{X(f)} = K\exp(-j2\pi f t_0 \pm n\pi). \tag{4.3}$$

In order to have distortionless transmission, the transfer function of the channel must satisfy the following conditions.

1. The amplitude response $|H(f)|$ should be constant for all frequencies.
2. The phase must vary linearly with frequency f, passing through the origin or a multiple of π at zero frequency.
3. If the spectrum of the signal is limited to a band, Conditions 1 and 2 must be satisfied only for that band of frequencies, that is, over the passband only. Pass band is the frequency band that is passed by the filter with gain equal to one.

In general, we can say that for any audio system, the LTI filter has to have a flat magnitude response and a linear phase response. There can be two kinds of distortions taking place.

1. *Amplitude distortion*: This occurs when the amplitude response is not constant over a passband.
2. *Phase distortion*: This occurs when the phase response is not linear. When the phase is linear as shown in Figure 4.6, the group delay, defined as the rate of variation of phase with frequency, is constant. The group delay is a time delay introduced for each frequency. When the group delay is constant, all frequencies are delayed by the same amount when they pass via a filter or a channel. This means there is no phase distortion. If there is phase distortion, it is easily noticeable by the human ear. Linear phase is a desirable feature when signals such as speech are processed. The group delay is mathematically defined as

$$\text{Group delay}\,(\tau) = \frac{d(\text{phase})}{df}. \tag{4.4}$$

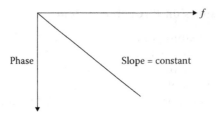

FIGURE 4.6
Linear phase characteristic.

4.2.2.1 Signal Bandwidth

Let us define the signal bandwidth. The input signal contains different frequencies. The signal may be a low-pass signal or a bandpass signal. Consider a low-pass or a baseband signal. Low-pass filter is one that passes the low frequencies with gain equal to one. The baseband signal is the message signal frequency band. The bandpass filter is a filter that passes the frequency band with gain equal to one. Let the highest frequency in the signal be W. We then define the band width of the signal as W. The signal spectrum is represented in the form of a triangle, as shown in Figure 4.7. It does not mean that the different frequencies shown have the magnitude given by the triangle magnitude. It just says that the signal has frequencies up to W. The signal can also be a bandpass signal and its spectrum will be represented as shown in Figure 4.8.

4.2.3 Ideal Low-Pass Filter, High-Pass Filter, and Bandpass Filter Characteristics

Some LTI systems behave as low-pass filters (LPFs), some as high-pass filters (HPFs), and some as bandpass filters (BPFs) or band reject filters (BRFs). Let us study the ideal characteristics of such filters. We will consider an LPF first. The ideal LPF is also called a brick wall filter because its walls look like a brick wall as shown in Figure 4.9.

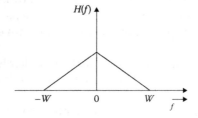

FIGURE 4.7
Signal spectrum with bandwidth W.

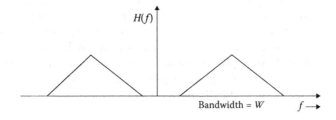

FIGURE 4.8
Bandpass signal.

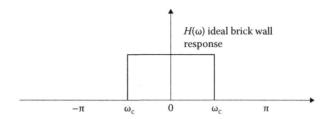

FIGURE 4.9
Ideal brick wall response of LPF filter characteristics.

We know that the frequency domain specifications for an ideal brick wall LPF can be given as

$$H(\omega) = \begin{cases} 1 & \text{for } |\omega| \le \omega_c \\ 0 & \text{for } \omega_c < \omega < \pi \end{cases}, \tag{4.5}$$

where ω_c is called the cut-off angular frequency of the filter and the sampling frequency is denoted by f_s. The angular frequency is plotted as a normalized angular frequency given by

$$\text{Normalized angular frequency} = \frac{2\pi f}{f_s}. \tag{4.6}$$

The maximum value of normalized frequency will be $\dfrac{2\pi f_N}{f_s} = \pi$.

$$f_N = \frac{f_s}{2} = \text{Nyquist frequency}. \tag{4.7}$$

The phase response is linear in the passband. We know that when a continuous time signal is appropriately sampled by multiplying it with a train of impulse, we get a replicated spectrum in the transformed domain, namely, the Fourier domain called a discrete time Fourier transform (DTFT). This transform domain signal is periodic with a period equal to 2π. The filter specification given by Equation 4.5 represents such a periodic signal in the Fourier domain. Hence, when we revert to the time domain, we must get a sampled signal in the time domain, which is the impulse response of the filter. This impulse response can be obtained by taking the inverse FT of the DTFT. That is,

$$h_n = \frac{1}{2\pi} \int_{-\omega_c}^{\omega_c} e^{j\omega n} \, d\omega. \tag{4.8}$$

$$h_n = \frac{1}{2\pi} \left[\frac{1}{jn} e^{j\omega n} \right]_{-\omega_c}^{\omega_c}. \tag{4.9}$$

$$h_n = \begin{cases} \dfrac{1}{2\pi} \left[\dfrac{1}{jn} (e^{j\omega_c n} - e^{-j\omega_c n}) \right] = \dfrac{\omega_c}{\pi} \dfrac{\sin(\omega_c n)}{\omega_c n} & \text{for } n \ne 0 \\[4mm] \dfrac{\omega_c}{\pi} \sin(\omega_c n) & \text{for } n = 0. \end{cases} \tag{4.10}$$

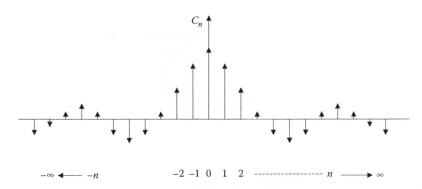

FIGURE 4.10
Infinite impulse response for the ideal LPF in time domain.

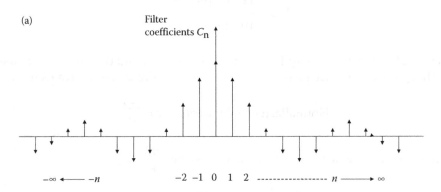

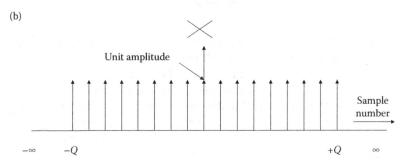

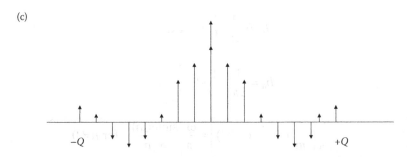

FIGURE 4.11
(a) Infinite unit impulse response of ideal low pass filter, (b) rectangular DT window function of unity amplitude from -Q to +Q, and (c) windowed impulse response for ideal brick wall filter.

The impulse response exists for all n from $-\infty$ to $+\infty$ as shown in Figure 4.10.

It is not possible to implement the LPF using these infinite coefficients. Hence, the response is truncated between some $-Q$ and $+Q$. Figure 4.11 shows the truncated response using a rectangular window. Figures 4.11a, 4.11b and 4.11c show infinite unit impulse response of ideal low-pass filter, rectangular DT window function of unity amplitude from $-Q$ to $+Q$, and windowed impulse response for ideal brick wall filter, respectively. The truncated response is still difficult to use for practical implementation as it is not causal. To make it realizable, the response is shifted along the time axis by Q samples. We say that we introduce a delay of Q samples. The introduction of the delay of Q samples makes the filter act as a linear phase filter.

This suggests that to implement the ideal filter, the system response must include finite coefficients. Second, the response must exist only for positive values of n so that it will be causal, i.e., the current output of the system depends on the current and past inputs and past outputs. We can thus conclude that the design of an ideal brick wall filter is not possible.

Similarly, the ideal HPF and BPF frequency response is shown in Figures 4.12 and 4.13, respectively. The same concepts apply for HPF and BPF. Hence, it is not repeated here. The impulse response in the time domain will again be the sinc function with a different width of the main lobe.

4.2.4 Causality and Paley–Wiener Criteria

We can see that if we can make the number of coefficients in the impulse response finite and causal, then it is possible to design finite impulse response (FIR) filters that closely approximate the frequency response specifications. This is exactly the meaning of the well known Paley–Wiener theorem. A typical magnitude and phase response of the realizable filter is shown in Figure 4.14. The figure shows that the response of the filter is constant over the passband and oscillates in the stopband. The gain is very less or the attenuation is high only at some discrete frequencies as shown in Figure 4.14. The phase response is seen to be linear in the passband.

Let us consider it more in detail. Consider a causal system. The causality condition in the time domain means that the impulse response is zero for negative time. If this is true, let us transform this condition into the frequency domain. If the impulse response of a system

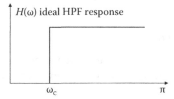

FIGURE 4.12
Ideal brick wall response for HPF.

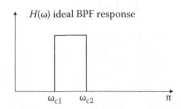

FIGURE 4.13
Ideal brick wall response for BPF.

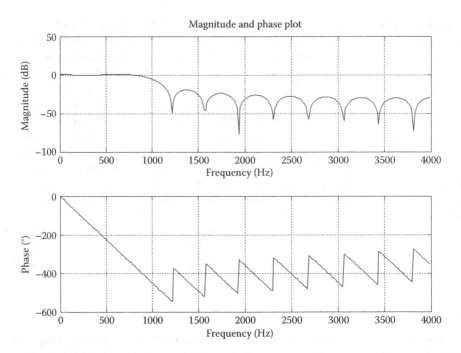

FIGURE 4.14
Typical magnitude and phase response of a realizable LPF.

is zero for all negative time, the frequency response will not be a brick wall response with sharp cut off and a zero gain or infinite attenuation over the entire stopband. The frequency response will have oscillations near the band edge of the filter and the infinite attenuation will only be obtained for discrete frequencies decided by the stopband oscillatory response as shown in Figure 4.14. This is the frequency domain condition of causality that is otherwise known as the Paley–Wiener criteria.

We will now state the Paley–Wiener criteria.

4.2.4.1 Statement of the Theorem

The Paley–Wiener theorem states that a necessary and sufficient condition for any gain function to be the gain of a causal filter is that the integral given in the following equation must be convergent.

$$\int_{-\infty}^{\infty} \frac{|\alpha(t)|}{1+f^2}\, \mathrm{d}f < \infty \tag{4.11}$$

If the condition stated in Equation 4.11 is satisfied, then the phase may be associated with this gain function such that the resulting filter has a causal impulse response that is zero for negative time. We can say that the Paley–Wiener criterion is the frequency equivalent of the causality requirement. The realizable gain characteristic may have infinite attenuation for a discrete set of frequencies, but cannot have infinite attenuation over a band of frequencies. In order that these conditions hold good, the amplitude response must be square integrable, that is, FT must exist.

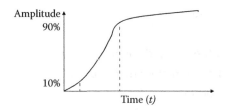

FIGURE 4.15
The plot of the rise time calculation.

$$\int_{-\infty}^{\infty} |H(f)|^2 \, df < \infty \tag{4.12}$$

4.2.5 Relation between Bandwidth and Rise Time

The transmission channel is always band limited. The linear system of the channel has a finite bandwidth. This means that the frequencies outside the band of the channel will be attenuated and the magnitude response will not be constant for all frequencies. The response of the system can be tested by applying the square wave as the input to a system. This is usually referred to as square wave testing of the system. The square wave has a sharp rising edge. When a square wave passes via a band limited system, it gets distorted. The sharp rising edge indicates that the input is changing at a fast rate or there is a sudden change in the input. The sharp edge is representative of all frequencies from zero to infinity. This is because the FT of a square wave is a sinc function as already studied in Section 4.2.3.

This sinc function extends up to infinity. When the square wave is applied as input to a linear system, the high frequencies above the cut-off frequency are all attenuated and the sharp rising edge in the input appears as a slow rising waveform at the output. We can measure the rise time of the output wave to find the relation between the rise time and the higher cut-off frequency, namely, the band width. This is also referred to as a slewing rate in reference to the response of the operational amplifier. The rising edge and measurement of rise time is shown in Figure 4.15. The time required for the amplitude at the output to rise from 10% to 90% of the maximum output is measured and is called the rise time. The upper cut-off frequency, namely, the bandwidth is given by

$$\text{Bandwidth} = \frac{0.35}{\text{Rise time}} \tag{4.13}$$

Concept Check

- What is a filter?
- What is an LTV system?
- What is a distortionless transmission?
- State the conditions for the transfer function to get a distortionless transmission.
- What is a phase distortion?
- What is amplitude distortion?

- Define a linear phase system.
- Define a group delay.
- State the condition for the group delay to get no phase distortion.
- Define a brick wall filter characteristic for an LPF.
- What are ideal HPFs and BPFs?
- What does causality mean in time domain?
- State the Paley–Wiener theorem.
- What is the relation between bandwidth of the filter and the rise time of the output waveform?

4.3 Optimum Detection

The aim of any communication, control, or radar system is to design a receiver that will maximize the information of the detected signal. The received signal is mixed with noise. The decision taken at the receiver end is based on the statistics of the noisy signal. The theory of optimum receivers is based on the concepts of information theory and detection theory. The receivers are designed to maximize signal-to-noise ratio (SNR) or minimize the error rate. We will go through six different approaches one by one for optimum detection.

1. *Maximum average SNR criterion*: This approach is used for amplitude modulation (AM), frequency modulation (FM), phase modulation (PM), and pulse position modulation (PPM) systems. The gain in the output SNR is measured as compared to the carrier-to-noise ratio (CNR) at the receiver.
2. *Maximum peak SNR criterion*: This approach is mostly used for radar signal detection. Maximum peak SNR measurement gives maximum matched filter output indicating maximum visibility of radar echoes in the background noise.
3. *Inverse probability criterion*: The received signal contains maximum information if the inverse probability $P(x/y)$ is maximized. $P(x/y)$ denotes the probability of transmission of x given that y is received at the receiver end. Hence, it is also called the maximum *a posteriori* criterion (MAP).
4. *Maximum likelihood (ML) criterion*: This approach is used for parameter estimation. Here, the likelihood function, namely, $P(y/x)$ is maximized for optimum solution.
5. *Maximum information transmission criterion*: This is equivalent to the minimum error rate criterion. The receiver uses optimum value of the threshold to detect the signal.
6. *Mean square error criterion (MSE)*: The receiver takes the decision based on minimization of mean square error.

4.3.1 Weighted Probabilities and Hypothesis Testing

The channel is noisy and can be modeled as a LPF. The received signal is distorted and the receiver has to take a decision regarding if one is transmitted or a zero is transmitted in the presence of noise. The observable signal at the receiver end is a random signal and detecting the actual transmitted signal becomes a statistical decision problem. The received signal is $y(t) = x(t) + n(t)$, where $y(t)$ is the received signal, $x(t)$ is the transmitted signal, and $n(t)$ is noise. We will define two hypotheses.

Hypothesis H_0: Null hypothesis, that is, no signal is present.

Hypothesis H_1: Signal is present.

The decision rule will be based on the probabilistic criterion. Let us consider the decision rule as follows.

$$\text{Decide on } H_1 \text{ if } p(H_1/y) > p(H_0/y), \tag{4.14}$$

$$\text{Otherwise decide on } H_0.$$

This decision rule is called the MAP probability criterion. The probabilities $p(H_1/y)$ and (H_0/y) are posteriori probabilities. Let us apply Bayes' theorem. We can write

$$p(H_1/y) = \frac{P_1 p(y/H_1)}{p(y)}, \quad P_1 = p(1) = p(H_1)$$

$$\text{and } p(H_0/y) = \frac{P_0 p(y/H_0)}{p(y)}, \quad P_0 = p(0) = p(H_0). \tag{4.15}$$

$$P_1 + P_0 = 1.$$

We say restate the decision rule as
Decide on H_1 if

$$p(H_1/y) > p(H_0/y),$$

$$\text{i.e., } \frac{P_1 p(y/H_1)}{p(y)} > \frac{P_0 p(y/H_0)}{p(y)}, \tag{4.16}$$

$$\text{i.e., } \lambda_1 = \frac{p(y/H_1)}{p(y/H_0)} > \frac{P_0}{P_1}.$$

Decide on H_0 if

$$p(H_1/y) < p(H_0/y),$$

$$\text{i.e., } \frac{P_1 p(y/H_1)}{p(y)} < \frac{P_0 p(y/H_0)}{p(y)}, \tag{4.17}$$

$$\text{i.e., } \lambda_0 = \frac{p(y/H_0)}{p(y/H_1)} > \frac{P_1}{P_0}.$$

The ratios λ_0 and λ_1 are called likelihood ratios. The test is called the likelihood ratio test. An alternative decision rule can be stated by selecting a threshold y_0 as follows.

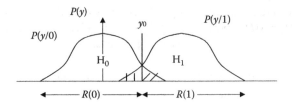

FIGURE 4.16
Conditional probabilities of two events and decision regions.

$$\text{If } y > y_0 \quad H_1 \text{ is true}$$

$$\text{and} \qquad\qquad\qquad\qquad\qquad\qquad (4.18)$$

$$\text{if } y < y_0 \quad H_0 \text{ is true.}$$

The threshold y_0 is so selected such that the overall error probability is minimized. Consider the conditional probabilities $p(y/0)$ and $p(y/1)$ as shown in Figure 4.16. The threshold y_0 divides the range of y into two regions, namely, $R(0)$ and $R(1)$.

Let us define the false alarm probability and probability of miss as follows.

$$P_f = P_0 \int\limits_{R(1)} P(y/0)\,\mathrm{d}y = P_0 \int\limits_{y_0}^{\infty} P(y/0)\,\mathrm{d}y. \qquad (4.19)$$

$$P_f = P_1 \int\limits_{R(0)} P(y/1)\,\mathrm{d}y = P_1 \int\limits_{-\infty}^{y_0} P(y/1)\,\mathrm{d}y. \qquad (4.20)$$

The total error probability can be written as $P_e = P_f + P_m$. The total probability can be minimized by selecting y_0 at the proper position. To minimize the error probability, we set $\partial P_e / \partial y_0 = 0$. This makes

$$P_e = P_f + P_m.$$

$$\frac{\partial P_e}{\partial y_0} = 0 \Rightarrow P_0 P(y/0) = P_1 P(y/1).$$

$$\qquad\qquad\qquad\qquad\qquad\qquad (4.21)$$

$$\frac{P(y_0/1)}{P(y_0/0)} = \frac{P_0}{P_1} = \lambda_t.$$

The decision rule is H_1 if $\lambda_1 > \lambda_t$ and H_0 if $\lambda_0 > 1/\lambda_t$.

The decision rule given in Equation 4.21 is called the ideal observer test. Comparing Equations 4.21 and 4.17, we see that this ideal observer test also leads to the likelihood ratio test.

4.3.2 Bayes Criterion

The ideal observer criterion is based on unweighted probabilities. Often, different decisions have different levels of risk. Based on the cost of the risk, the weights for the probabilities are decided. In the case of radar, the cost of missing a target is definitely greater than the cost of hitting it. The cost of taking a correct decision is considered as zero. The cost matrix can be written as

$$[C] = \begin{bmatrix} 0 & C_m \\ C_f & 0 \end{bmatrix}.$$ (4.22)

The probability matrix can be written as

$$[P] = \begin{bmatrix} 0 & P_m \\ P_f & 0 \end{bmatrix}.$$ (4.23)

We can now write the average risk per decision as

$$\bar{C}(y_0) = P_0 C_f \int_{y_0}^{\infty} p(y/0) \, dy + P_1 C_m \int_{-\infty}^{y_0} p(y/1) \, dy.$$ (4.24)

The criterion for taking a decision is to minimize the average risk per decision. This is done by selecting a proper value of threshold y_0 obtained by making the derivative of the cost zero with respect to the threshold. We set

$$\frac{\partial C(\bar{y}_0)}{\partial y_0} = 0 \text{ and calculate the value of } y_0.$$

$$P_0 C_f p(y/0) = P_1 C_m p(y/1) \Rightarrow \lambda_t = P_0 C_f / P_1 C_m.$$ (4.25)

This equation is same as that obtained for the ideal observer.

4.3.3 Minimax Criterion

Often, *a priori* probabilities, namely, P_0 and P_1, are not known. The Bayes risk cannot be calculated without knowing *a priori* probability. In this situation, one finds Bayes minimum risks for different values of *a priori* probability. The graph of minimum risk is plotted against *a priori* probability as shown in Figure 4.17. The maximum value of this curve is the worst case risk value. The value of the decision threshold is found out for this worst case risk value.

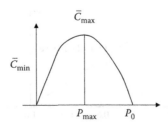

FIGURE 4.17
Minimax risk (cost function) plotted with respect to probability.

4.3.4 Neyman–Pearson Criterion

In case of applications like radar, both *a priori* probability and cost matrix are not known. The false alarm probability tolerated by the application is assigned. The receiver observes that this false alarm probability is not exceeded. The probability of detection $P_d = 1 - P_m$ is also maximized. This is known as the Neyman–Pearson criterion. Such a problem is called an optimization problem and is solved using Lagrangian multipliers. The optimization problem optimizes values of Q given by the following equation for a given value of P_f.

$$Q = P_m + \mu P_f.$$

(4.26)

μ : Lagrangian multiplier.

In mathematics, optimization problems are solved using the method of Lagrange multipliers. It is a strategy for finding the local maxima and minima of a function and was invented by Joseph Louis Lagrange. If we substitute $P_1 C_m = 1$ and $P_0 C_f = \mu$ in Equation 4.24, we get

$$C(y_0) = \mu \int_{y_0}^{\infty} p(y/0)\mathrm{d}y + \int_{-\infty}^{y_0} p(y/1)\mathrm{d}y$$

(4.27)

$$= \mu P_f + P_m.$$

Comparing Equations 4.26 and 4.27, we see that the Neyman–Pearson criterion is a special case of the Bayes criterion. The threshold value obtained by the Bayes criterion is given by Equation 4.25. Putting $P_1 C_m = 1$ and $P_0 C_f = \mu$ in Equation 4.25, we get

$$\lambda_t = P_0 C_f / P_1 C_m = \mu.$$

(4.28)

4.3.5 Receiver Operating Characteristics

We observe that the Neyman–Pearson criterion contains all the information required to use Bayes criterion and the minimax criterion. In the theory of hypothesis testing, the false alarm probability P_f is termed as the size of the test and the detection probability P_d is called the power of the test. The power of the test is plotted against the size of the test and the curve is called the ROC or the operating characteristic of the test. The nature of ROC

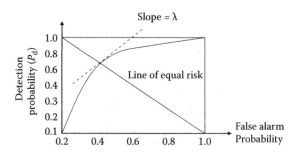

FIGURE 4.18
Receiver operating characteristic.

is as shown in Figure 4.18. It will depend on the probabilities and not on the cost or *a priori* probabilities. The slope of ROC at any point gives the value of the threshold. ROC may be used to determine Bayes minimum risk.

Concept Check

- How will you design the optimum receiver?
- Name six different approaches for optimum detection.
- Define two hypotheses for detecting a signal at the receiver.
- State the decision rule for the maximum *a posteriori* approach.
- How will you decide a threshold for minimizing error probability?
- Define a rule for ideal observer.
- Define the cost matrix for Bayes criterion.
- What is the method of Lagrange multiplier?
- Define ROC.

4.4 Estimation Theory

Estimation theory is a branch of statistics and signal processing dealing with the estimation of the values of parameters based on a measured data set. The estimated parameters will describe an underlying distribution of the measured data. An estimator must approximate the unknown parameters using the data set. We will study three types of estimators, namely, the Bayesian estimator, the MLE, and the MMSE. The finite number of data values that we work with comes from some random distribution. The random distribution has to be identified. Once the distribution is identified, we can estimate the parameters such as mean, variance, and moments using the knowledge of the distribution. The first few moments of the random data set can be assumed to be the moments of the identified distribution. The values of further moments can be easily estimated. This method is called the method of moment's estimation.

1. *Maximum likelihood estimator*: It is based on a simple idea as follows. Suppose, we have different populations of data sets available with us. Consider a single sample. It is more likely to come from a particular population. We have to identify the population based on ML. Consider a data vector $u = [u_1, u_2,..., u_n]$. We have to estimate the parameter vector $\theta = [\theta_1, \theta_2,..., \theta_n]$. Given that the data vector u has occurred, we have to maximize the conditional joint probability density function by properly selecting the parameter vector. Here, the data vector is fixed and the parameter vector is variable. The likelihood function is defined as the conditional probability density function viewed as the function of the parameter vector.

$$l(\theta) = f(u \mid \theta). \tag{4.29}$$

This function is differentiated with respect to the parameter vector and equated to zero. The value of the parameter vector estimated is the solution of the differential equation. In many cases, it is more convenient to work with the logarithm of the likelihood function rather than with the likelihood function. We say that the estimate is unbiased if the estimated parameters match the measured parameters from the data set. ML estimates compare with the lower bound and are found to be consistent.

2. *Cramer–Rao inequality*: Consider a data vector $u = [u_1, u_2, ..., u_n]$. We have to estimate the parameter vector $\theta = [\theta_1, \theta_2, ..., \theta_n]$. The conditional joint probability density function is $f(u|\theta)$. Using log likelihood function, we can write $I(\theta) = \ln[f(u|\theta)]$. We will form the K-by-K matrix as follows (each element in the matrix is the double derivative of the information I with respect to the parameter vector elements).

$$J = \begin{bmatrix} E\left[\dfrac{\partial^2 I}{\partial \theta_1^2}\right] & E\left[\dfrac{\partial^2 I}{\partial \theta_1 \partial \theta_2}\right] & \cdots & E\left[\dfrac{\partial^2 I}{\partial \theta_1 \partial \theta_K}\right] \\ E\left[\dfrac{\partial^2 I}{\partial \theta_2 \partial \theta_1}\right] & E\left[\dfrac{\partial^2 I}{\partial \theta_2^2}\right] & \cdots & E\left[\dfrac{\partial^2 I}{\partial \theta_2 \partial \theta_K}\right] \\ \vdots & \vdots & \cdots & \vdots \\ E\left[\dfrac{\partial^2 I}{\partial \theta_K \partial \theta_1}\right] & E\left[\dfrac{\partial^2 I}{\partial \theta_K \partial \theta_2}\right] & \cdots & E\left[\dfrac{\partial^2 I}{\partial \theta_K^2}\right] \end{bmatrix}. \tag{4.30}$$

The matrix J is called Fisher's information matrix. Let I denote the inverse of J, then I_{ii} will denote the ith diagonal element. Let $\hat{\theta}_i$ be an estimate of θ_i. The Cramer–Rao inequality says that the variance of the estimate is always greater than or equal to the ith diagonal element, that is,

$$\text{var}[\hat{\theta}_i] \geq I_{ii} \quad \text{for } i = 1, 2, ..., K. \tag{4.31}$$

This inequality allows us to specify the lower bound for the variance; this lower bound is called the Cramer–Rao lower bound (CRLB). If the estimator has a variance equal to CRLB, then the estimator is said to be efficient. MLEs are asymptotically efficient, that is,

$$\lim_{M \to \infty} \left[\frac{\text{var}[\theta_{i,ml} - \theta_i]}{I_{ii}}\right] = 1 \quad \text{for } i = 1, 2, ..., K. \tag{4.32}$$

$\theta_{i,ml}$ is the ML estimate of θ_i.

Asymptotically means for a large sample size of greater than 50 samples. Similarly, it is observed that MLEs are asymptotically Gaussian.

3. *Bayes estimator (conditional mean estimator/MMSE estimator)*: We will derive the formula for conditional mean estimates from first principles and then show that this estimate is the same as a MMSE estimate. Let the observation be denoted by y.

Consider a random variable x that is a function of the observation y. We can consider x as the parameter extracted from the data value y. Let us denote the estimate of x as $\bar{x}(y)$. Let us now define a cost function for Bayes estimator as $C(x, \bar{x}(y))$. Using the Bayes estimation theory, we will write the Bayes risk.

$$R = E[C(x, \bar{x}(y))]$$

$$= \int_{-\infty}^{\infty} dx \int_{-\infty}^{\infty} C(x, \bar{x}(y)) f_{X,Y}(x, y) \, dy,$$

(4.33)

where $f_{x,y}(x,y)$ is the joint probability density function of x and y.

The Bayes estimate of $\bar{x}(y)$ is the estimate that minimizes the risk. The cost function of interest is the mean-squared error. Let us define the estimation error as $\varepsilon = x - \bar{x}(y)$. The cost function becomes $C = \varepsilon^2 = (x - \bar{x}(y))^2$. We will put this value of cost function in Equation 4.32 to find the risk.

$$R = E[C(x, \bar{x}(y))]$$

$$= \int_{-\infty}^{\infty} dx \int_{-\infty}^{\infty} (x - \bar{x}(y))^2 f_{X,Y}(x, y) dy,$$

(4.34)

where $f_{x,y}(x,y) = f_x(x/y) f_y(y)$ using Bayers rule.

$$R = \int_{-\infty}^{\infty} dy \, f_Y(y) \int_{-\infty}^{\infty} (x - \bar{x}(y))^2 f_X(x/y) dx.$$

(4.35)

We can see that the inner integral and $f_Y(y)$ both are positive quantities. Hence, the risk can be minimized by minimizing the inner integral. We will differentiate the inner integral and equate it to zero.

$$\frac{dI}{d\bar{x}} = -2 \int_{-\infty}^{\infty} x f_X(x/y) dx + 2\bar{x}(y) \int_{-\infty}^{\infty} f_X(x/y) dx.$$

(4.36)

$$\frac{dI}{d\bar{x}} = -2 \int_{-\infty}^{\infty} x f_X(x/y) dx + 2\bar{x}(y) \text{ as } \int_{-\infty}^{\infty} f_X(x/y) dx = 1.$$

$\bar{x}_{ms}(y)$: Mean square estimate of x.

(4.37)

$$\bar{x}_{ms}(y) = \int_{-\infty}^{\infty} x f_X(x/y) dx.$$

The plot of cost function as the mean-squared error is shown in Figure 4.19. The estimate is the MMSE estimate. From Equation 4.37, we can recognize the estimate

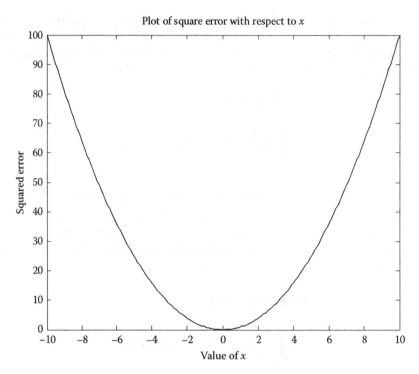

FIGURE 4.19
Plot of cost function as a mean-squared error with respect to x.

as the conditional mean of parameter x. Hence, the estimator is called conditional mean estimator and the minimum value of risk is the inner integral, which is just the conditional variance (refer to Equation 4.34).

Concept Check

- What is estimation theory?
- What is the method of moments estimator?
- What is the idea behind MLE?
- What is the Cramer–Rao inequality?
- What is the meaning of asymptotically efficient?
- Define the cost function for the Bayes estimator.

Summary

The chapter began with an introduction to basic communication theory. We discussed the detection methods and the estimation theory. The following concepts are highlighted:

1. The basic blocks of the communication system, mainly the transmitter, channel, and receiver, are described. The PCM-coded signal is explained. The channel has a limited bandwidth. The signal gets distorted when it passes the channel. The received signal is then used to take a decision if a "1" or a "0" is transmitted.

2. The characteristics of the LTV system are defined. The channel functions as an LPF. The evaluation of cut-off frequency is explained. The input signal passing through the channel gets distorted. The signal transmission through the channel is discussed. Distortionless transmission is defined and the conditions for amplitude and phase response to get distortionless transmission are specified. The linear phase characteristic is defined and the relation between group delay and phase characteristic is explained. The bandwidth of the signal is defined. The ideal characteristics for LPF, HPF, and BPF are discussed. To get the ideal brick wall response, the impulse response of the filter in time domain must be a sinc function varying from minus infinity to infinity. This response is required to be truncated to get the FIR so that it becomes realizable. The system will further become causal when a delay of Q samples is introduced. This introduction of delay makes the phase characteristic linear. The causality in frequency domain is explained and the Paley–Wiener criterion is stated. Finally, the relation between the rise time of the output waveform and the bandwidth of the system is explained.

3. Detection theory was introduced. Six approaches for optimum detection were described. We defined the hypotheses for taking the decision at the receiver. MAP and likelihood ratio test were discussed. Bayes criterion uses the minimization of risk approach using a cost function. We then described the minimax criterion, Neyman–Pearson criterion, and ROC in detail.

4. Estimation theory was then introduced. We discussed the MLE and the Cramer–Rao inequality. We then described the Bayes estimator. The cost function was defined as the squared error and it was shown that the Bayes estimator is MMSE. The estimate is recognized as the conditional mean. Hence, the estimator is called the conditional mean estimator and the minimum value of risk is just the conditional variance.

Keywords

Basic digital communication system
Digital transmitter
Channel
Receiver
Distortionless channel
LTI system
LTV system
Linear phase system
Group delay
Phase distortion

Bandwidth
Rise time
Paley–Wiener criterion
Detection
Pulse code modulation (PCM)
Bayes criterion
Optimum receiver
Neyman–Pearson criterion
Minimax criterion
Maximum a posteriori criterion

Likelihood ratio test   Minimum mean-squared error (MMSE)
Receiver operating characteristics   estimator
Maximum likelihood estimator   Cramer–Rao inequality
Bayes estimator   Asymptotically efficient

Multiple-Choice Questions

1. A digital transmitter consists of
    a. A sampler.
    b. A sampler and a quantizer.
    c. A sampler and a coder.
    d. A sampler, quantizer, and a coder.

2. Converting analog signal to a DT signal with proper sampling frequency is justifiable because of the
    a. Sampling theorem in the time domain.
    b. Sampling theorem in the frequency domain.
    c. Sampling theorem with sampling frequency < signal frequency.
    d. None of the above.

3. The conversion of DT signal to a digital signal is justifiable because
    a. Small quantization noise is tolerable to humans.
    b. Small quantization noise is tolerable to the human eye and ear.
    c. Small quantization noise is annoying.
    d. Quantization noise is usually large.

4. Waveform coders minimize the error between
    a. The synthesized and the modified speech waveforms.
    b. The synthesized and the main speech waveforms.
    c. The synthesized and the generated speech waveforms.
    d. The synthesized and the original speech waveforms.

5. The 8 bit PCM converts each speech sample into
    a. 8 bit code.
    b. 16 bit code.
    c. 1 bit code.
    d. 32 bit code.

6. The channel distorts the signal as the channel behaves as a
    a. High-pass filter.
    b. Low-pass filter.

    c. Bandpass filter.

    d. Band reject filter.

7. The receiver uses a comparator with a proper threshold selected.

    a. If the received signal is > threshold, it receives a "1."

    b. If the received signal is < threshold, it receives a "1."

    c. If the received signal is > threshold, it receives a "1" or else receives a "0."

    d. If the received signal is < threshold, it receives a "1" or else receives a "0."

8. The distortionless transmission requires a constant passband amplitude response and

    a. Constant passband phase response.

    b. Linear passband phase response.

    c. Piecewise linear phase response.

    d. Constant phase response over all bands.

9. The Paley–Wiener criterion is related to

    a. Causality condition in the time domain.

    b. Non-causal systems.

    c. Causality condition in the frequency domain.

    d. Distortionless systems

10. Bandwidth of a system and the rise time of the output waveform are related by the equation

    a. BW = 1/rise time.

    b. BW = 0.35/rise time.

    c. BW = 0.35∗rise time.

    d. BW = 1/(0.25∗rise time).

11. Ideal LPF response in the frequency domain cannot be obtained.

    a. The required impulse response is of infinite duration.

    b. The required impulse response extends from minus infinity to infinity and is non-causal.

    c. The required impulse response cannot be truncated.

    d. The required impulse response can never be made causal.

12. The maximum peak SNR criteria is one of the approaches for

    a. Optimum detection at the receiver end.

    b. Optimum decision at the transmitter end.

    c. Maximum detection.

    d. Maximizing the error.

13. Hypothesis H_0 means

    a. Taking a decision for "0".

    b. Taking a decision for "1".

   c. Taking a decision for not a "0".

   d. Taking a decision for not a "1".

14. Likelihood ratio test says that taking a decision for H_1 if auto regressive moving average (ARMA) model has

   a. $\lambda_1 = \dfrac{p(y/H_1)}{p(y/H_0)} > \dfrac{P_0}{P_1}$.

   b. $\lambda_1 = \dfrac{p(y/H_1)}{p(y/H_0)} < \dfrac{P_0}{P_1}$.

   c. $\lambda_1 = \dfrac{p(y/H_1)}{p(y/H_0)} = \dfrac{P_0}{P_1}$.

   d. $\lambda_1 = \dfrac{p(y/H_1)}{p(y/H_0)} > \dfrac{P_1}{P_0}$.

15. The threshold for detection of "0" and "1" is so selected that

   a. The overall error probability is maximized.

   b. The overall error probability is minimized.

   c. The overall error probability is optimized for "0".

   d. The overall error probability is zero.

16. In case of Bayes criteria

   a. The weights for the probabilities are decided based on the cost of risk,.

   b. The weights are decided based on Bayes theorem.

   c. Risk is decided based on probabilities.

   d. Risk is based on probability.

17. The worst case risk value is calculated based on

   a. Maximum risk values for different probabilities.

   b. Maximum of the minimum risks for all probabilities.

   c. Maximum possible risk.

   d. Minimum risk values for different probabilities.

18. In case of Neyman–Pearson criteria

   a. The detection probability is minimized.

   b. The detection probability is maximized.

   c. The miss probability is maximized.

   d. The total probability is maximized.

19. The receiver operating characteristic plots

   a. The power of the test against the size of the test.

   b. The size of the test against the probability.

   c. The power of the test against the probability.

   d. The size of the test against the power of the test.

20. MMSE criteria is used for
   a. Estimation of parameters.
   b. Estimation of test.
   c. Estimation based on minimum mean square error.
   d. Estimation based on maximum mean square error.
21. The Cramer–Rao inequality says that the variance of the estimate is
   a. Always less than or equal to the *i*th diagonal element.
   b. Always equal to the *i*th diagonal element or time average autocorrelation is equal to the ensemble autocorrelation.
   c. Always smaller than or equal to the *i*th diagonal element.
   d. Always greater than or equal to the *i*th diagonal element.
22. The estimator is said to efficient if
   a. If the estimator has a variance greater than or equal to CRLB.
   b. If the estimator has a variance equal to CRLB.
   c. If the estimator has a variance less than CRLB.
   d. If the estimator has a variance greater than or less than CRLB.

Review Questions

1. Draw the basic block diagram of a digital communication system and explain it.
2. Justify sampling and quantization of analog signal.
3. Explain why the signal gets distorted when it passes over the channel?
4. What is a linear time-varying system? Explain the LTV model for the speech signal.
5. Explain the conditions required for a channel to behave as a distortionless channel.
6. Explain phase distortion. What is the meaning of linear phase characteristic? Can it achieve no phase distortion?
7. Define the Paley–Wiener criterion for causality. Is it related to the causality condition in the frequency domain?
8. Define characteristics for ideal LPF, HPF, and BPF.
9. Explain the relation between the bandwidth of the channel and the rise time of the output waveform.
10. Explain how a decision is taken at the receiver end for detecting a "1" or a "0".
11. Write six different approaches for optimum detection.
12. Explain maximum a priori probability criterion.
13. Explain likelihood ratio test for optimum detection.
14. Explain selection of threshold for optimum detection.

15. Define a cost function. Explain Bayes criterion.

16. Explain the minimax criteria.

17. Explain the Neyman–Pearson criterion.

18. What is the power of test and size of test? What is ROC?

19. Explain the concept of ideal observer.

20. Explain MLE.

21. Explain the use of the Cramer–Rao inequality. What is the lower bound?

22. Explain Bayes estimator and MMSE criterion.

Answers

Multiple-Choice Questions

1. (d)
2. (a)
3. (b)
4. (d)
5. (a)
6. (b)
7. (c)
8. (b)
9. (c)
10. (b)
11. (b)
12. (a)
13. (a)
14. (a)
15. (b)
16. (a)
17. (d)
18. (b)
19. (a)
20. (c)
21. (d)
22. (b)

5

Fundamentals of Speech Processing

LEARNING OBJECTIVES

- Speech production model
- Nature of speech signal
- Linear time-invariant (LTI) and linear time-varying (LTV) models for speech signal
- Voiced and unvoiced speech
- Voiced/unvoiced (V/UV) decision-making
- Nature of .wav file
- Pitch frequency
- Evaluation of pitch
- Pitch contour
- Evaluation of formants
- Relation between formants and linear predictor coefficients (LPCs)
- Evaluation of mel frequency cepstral coefficients (MFCC)
- Evaluation of LPC

This chapter aims to introduce basic processing concepts and algorithms used for processing of speech signals.

Speech signal is generated by God. It is generated in nature. It is naturally occurring and hence is found to be modeled as a random signal. There are several models put forth by the researchers based on their perception of the speech signal. The speech signal is normally processed using statistical parameters. Even though the speech has a random component, if we consider a short segment of speech, say, about 10 milliseconds speech, we can assume that over the small segment the speech parameters will remain constant. Hence, parametric models are also used. Different parameters extracted from speech signals are pitch frequency, formants, cepstral coefficients, mel frequency cepstral coefficients, linear prediction coefficients (LPC), etc.

We will first go through the generalized human speech production model and then a simple linear time-invariant (LTI) model for speech production. The time-varying nature of the speech signal will then be explained. The format of a simple .wav file is explained. We will focus on voiced speech and unvoiced speech and will indicate different methods to take voiced/unvoiced (V/UV) decision. The speech parameters such as fundamental frequency, formants, pitch contour, MFCC, and linear predictor coefficients (LPC) are introduced. The algorithms for extraction of these parameters will be discussed.

5.1 LTI and LTV Models for Speech Production

We will define and explain LTI and LTV models for speech. Let us start with the LTI model first. The basic assumption of the speech processing system is that the source of excitation and vocal tract system are independent. This assumption of independence will allow us to discuss the transmission function of the vocal tract separately. We can now allow the vocal tract system to get excited from any of the possible source. This assumption is valid for most of the cases except for the production of "p" in the word "pot." On the basis of the assumption, we can put forth the digital model for speech production.

5.1.1 LTI Model for Speech

Speech signal is basically a convolved signal resulting when excitation signal gets generated at the sound box called the larynx. This sound box consists of two thin cords, namely, vocal cords having elasticity to some extent. We can imagine the two vocal cords will form a tuning fork whose length can be varied to some extent. When these vocal cords vibrate with their natural frequency of vibration, a periodic excitation is generated in the larynx. The frequency of vibration of the vocal cords is called fundamental frequency or pitch frequency.

This excitation generated by the vocal cords passes via the vocal tract. This is a circular tube-like structure that is connected to the mouth cavity and nasal cavity. This forms a complex resonant cavity. We can imagine that this circular tube-like structure is a circular wave guide with terminating conditions such that it is closed at one end that is larynx and open at the other end that is mouth cavity. Vocal tract is generally modeled as a filter that may be an all-pole filter or a pole-zero filter. The excitation signal gets convolved with the impulse response of the vocal tract and this again gets convolved with external noise. The resulting convolved signal is the speech signal. The simple LTV model put forth for production of speech signal is as shown in Figure 5.1.

Speech signal consists of voiced segment and unvoiced segment. The impulse train generator shown in Figure 5.1 is used as excitation when a voiced segment is produced; for example, the utterance including any vowel or a composite vowel. When consonants like "s" are uttered, they are generated as a result of turbulence functioning as excitation. The unvoiced segments are generated when the random signal generator is used as the

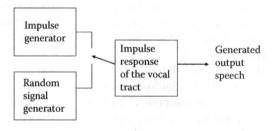

FIGURE 5.1
Generalized LTI model for speech production.

excitation. This model can be used for a short time (10 ms) segment of speech for which we can assume that the parameters of vocal tract remain almost constant.

5.1.2 Nature of Speech Signal

The speech signal is generated by god-gifted components like vocal cords and vocal tract. So far, no researcher has been successful in generating a speech signal. The researchers first record the speech signal, analyze the parameters of the speech signal, and then use some synthesis method to synthesize the speech signal. Hence, this is termed speech synthesis.

The synthesized signal does not exactly match with the naturally occurring signal. Speech is a naturally occurring signal, and hence it has some random component. A random signal is one for which the next signal samples cannot be predicted using any mathematical model. This is interpreted by considering that no mathematical model exists for the generation of the speech signal samples. The deterministic signal can be described by the mathematical equation as the next sample can be predicted knowing the previous samples. The models put forth by different researchers in this field are as per the dominant features extracted by the researcher. There can be as many as 100 parameters describing the speech signal. It is like the story of the elephant and blind people. The researchers are treated as blind people as they do not have a full knowledge of the speech signal. The blind people now describe the elephant. One may say that the elephant is like its tail, others may describe the elephant is like its ear or its leg, and so on. The characteristics of the whole elephant cannot be described by them as they analyze and/or focus only on some part of the elephant. Similarly for the researchers, the speech signal is like the elephant, and their model may be a model of the tail or a leg and so on. Thus, we see that the problem of accurately modeling a naturally occurring signal, such as speech, is very complex.

When we speak, we utter different words. This is possible because we can change the resonant modes for the vocal cavity and we can stretch the vocal cords to some extent for modifying the pitch period for different vowels. It is required that this time-varying nature of the speech signal must be described by the model. The model must therefore be a time-varying model (LTV).

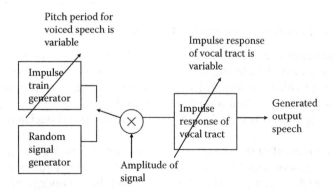

FIGURE 5.2
LTV model for speech.

5.1.3 LTV Model

The basic model represented in Figure 5.1 must be changed so that it will account for the time variations. This is depicted in Figure 5.2. The variation of pitch period and variation of impulse response of the vocal tract are shown. Normally, for a speech signal, these parameters are assumed to be constant for a short segment of speech, say, for a period of 20 ms. The short time window is used to view the signal, and over this small window, the parameters of the speech segment are assumed to remain constant. The parameters over this short segment are described by the LTI model. We will study the short time analysis of speech in Chapter 8.

Concept Check

- What is a source of impulse train generator?
- What is a model used to simulate a vocal tract?
- Why is speech a convolved signal?
- What is the validity of LTI model?
- Why does the speech signal have a time-varying nature?
- How to model a time-varying speech signal?
- What is a random signal?
- Why is a time-varying model used for speech signals?
- What is a deterministic signal?
- How can a person vary the pitch period in the uttered word?
- What is an LTV model?
- What is a segment duration over which pitch period can be assumed to be constant?
- Which parameters of the model vary with time?

5.2 Voiced and Unvoiced Decision-Making

When a speech sentence is recorded and analyzed, there is some silence part in the utterance where there is a grass-like thing, which means that no energy is present or the energy is very low and equivalent to low noise. The unvoiced segment has somewhat higher amplitude than a silence part and a voiced segment has an even higher energy. We will discuss the following methods for taking a voiced and unvoiced decision.

1. Energy measurement of the signal segment: Generally, the log energy is evaluated over a short segment about 20 ms in length. The log energy for voiced segment is found to be much higher than the silence part and the log energy of unvoiced data is usually lower than that for voiced sounds but higher than that for silence. This fact is obvious from the waveforms of voiced and unvoiced segments.

2. Zero crossing rate (ZCR): The zero crossing count is an indicator of the frequency at which the energy is concentrated in the signal spectrum. Voiced speech is produced as a result of excitation of the vocal tract by the periodic flow of air at the glottis, and this segment shows a low zero crossing count. Unvoiced speech is produced due to excitation as turbulence in the vocal tract and shows a high zero crossing count. The zero crossing count of silence is expected to be lower than for unvoiced speech but quite comparable to that for voiced speech. But, for the silence part energy will be quite low compared to the voiced segment.

A MATLAB® program to calculate positive and negative ZCR for a voiced and unvoiced segment is given below.

```
clear all;
fp=fopen('watermark.wav');
fseek(fp,60244,-1);
a=fread(fp,8000);
plot(a);
xlabel('sample number'); ylabel('amplitude');title('plot of speech
segment for 8000 samples');

% file read and plotted. We are reading first 8000 samples after fseek by
60244 samples.

x=0;
figure;
for i=5000:6999,
if (a(i)>128) && (a(i+1)<128)
    x=x+1;
else x=x;

end
end
disp('number of zero crossings for voiced segment=');  disp(x);

% samples between 5000 to 7000 are read and number of zero crossings are
found..

fseek(fp,65044,-1);
b=fread(fp,2000);
plot(b);
xlabel('sample number'); ylabel('amplitude');title('plot of voiced speech
segment');

% We are reading samples between 5000 to 7000 and are plotted. This is
unvoiced segment.

figure;
for i=7000:8999,
if (a(i)>128) && (a(i+1)<128)
    x=x+1;
else x=x;

end
```

```
end
disp('number of zero crossings for unvoiced segment=');  disp(x);

% samples between 7000 to 9000 are read and number of zero crossings are
found..

fseek(fp,67244,-1);
c=fread(fp,1000);
xlabel('sample number');ylabel('amplitude');title('plot of unvoiced
speech segment');
plot(c);

% We are reading samples between 7000 to 8000 and are plotted. This is
unvoiced segment

Output of MATLAB program is -
number of negative zero crossings for voiced segment=
    135

number of negative zero crossings for unvoiced segment=
    193
```

The output plots of the MATLAB program are shown in Figures 5.3 through 5.5. Figure 5.3 depicts the plot of a segment of size 8000 samples. Figure 5.4 depicts the voiced segment and Figure 5.5 shows unvoiced segment of size 1000 samples.

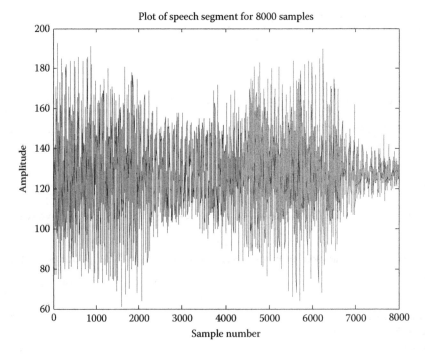

FIGURE 5.3
Plot of speech signal of length 8000 samples.

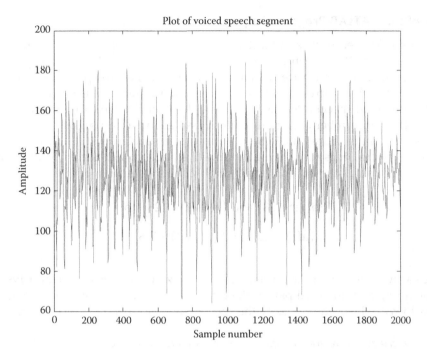

FIGURE 5.4
Plot of voiced speech of length 2000 samples.

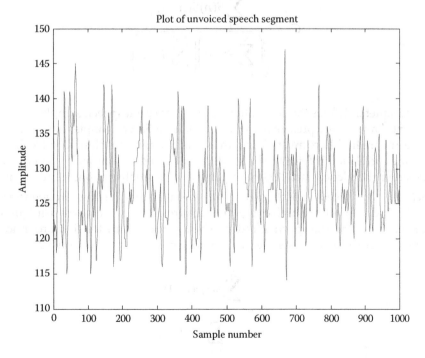

FIGURE 5.5
Plot of unvoiced speech of length 1000 samples.

Analysis of the MATLAB Program Output

We have used the speech file as a "watermark.wav" file. Using fseek command, we enter the speech segment. We concentrate on 8000 samples of the file. We have displayed these 8000 samples as shown in Figure 5.3. We can visually inspect the plot. We find that samples from 5000 to 7000 correspond to voiced speech and samples from 7000 to 8000 correspond to unvoiced speech. We have used samples from 5000 to 7000 for voiced segment and found that the number of negative zero crossings for voiced segment is equal to 135. We have used samples from 7000 to 8000 for unvoiced speech and found that the number of negative zero crossings for unvoiced segment is equal to 192. Similarly, the number of positive zero crossings can be found by changing the statement

$$\text{if } \big(a(i) > 128\big) \text{ and } \big(a(i+1) < 128\big) \text{ to}$$

$$\text{if } \big(a(i) < 128\big) \text{ and } \big(a(i+1) > 128\big)$$

It can be easily verified that the number of positive zero crossings is 146 and 199, respectively, for voiced and unvoiced part of the utterance. We can conclude that the number of zero crossings can be used to classify the speech into voiced and unvoiced segments.

3. *Normalized autocorrelation coefficient:* Normalized autocorrelation coefficient at unit sample delay, C_1, is defined as

$$C_1 = \frac{\sum\limits_{n-1}^{N} S(n)S(n-1)}{\sqrt{\left(\sum\limits_{n-1}^{N} S^2(n)\right)\left(\sum\limits_{n-0}^{N-1} S^2(n)\right)}} \tag{5.1}$$

This parameter defines the correlation between adjacent speech samples. As voiced segments have energy present in low frequencies, adjacent samples of voiced speech waveform are highly correlated, and thus this parameter is close to value 5. In contrast, the correlation is found to be small for unvoiced speech, and it will be further smaller for the silence part. A similar parameter normally used is a spectrum tilt.

4. *Spectrum tilt:* Voiced speech signal has high energy in low frequencies and unvoiced segments have high energy in high frequencies. This will result in opposite spectrum tilts. The spectrum tilt can be represented by the first-order normalized autocorrelation or first reflection coefficient given by Equation 5.2:

$$St = \frac{\sum\limits_{i=1}^{N} s(i)s(i-1)}{\sum\limits_{i=1} s^2(i)} \tag{5.2}$$

This parameter is found to be very reliable as it avoids detection of spikes in low-level signals. It has the ability to detect voiced and unvoiced segments.

A MATLAB program is given that detects the spectrum tilt.

```
%to find spectrum tilt for speech signal
clear all;
fp=fopen('c:\users\anand\desktop\fast_sent1.wav');
fseek(fp,3044,-1);
a=fread(fp,20000,'short');
a=a-128;

subplot(2,1,1);plot(a);title('plot of speech signal');
xlabel('sample no.');ylabel('amplitude');

for j=1:800,
    fseek(fp,44+100*j,-1);
    a=fread(fp,100);
    a=a-128;
    sum(j)=0;
for i=2:100,
    sum(j)=sum(j)+(a(i)*a(i-1));
    end
sum1(j)=0;
for i=2:100,
    if a(i)==0,
        a(i)=0.1;
    end
    sum1(j)=sum1(j)+a(i)*a(i);

end
s(j)=sum(j)/sum1(j);
end

subplot(2,1,2);plot(s);title('plot of spectrum tilt of speech signal');
xlabel('segment number');ylabel('value of spectrum tilt');
```

We have used a speech file containing voiced and unvoiced segments. The plot of spectrum tilt in Figure 5.6 shows that when a signal is voiced, the spectrum tilt has a value very close to 1, whereas for unvoiced segment, the spectrum tilt is less than 1. In the case of the silence part, the spectrum tilt is close to 0. Hence, using spectrum tilt, one can identify the voiced, unvoiced, and silence part of the utterance.

5. *First predictor coefficient:* The process can be listed as follows. We have to take a segment of a speech and evaluate 12 LPCs. It can be shown that the second parameter is higher in voiced segment, quite small for unvoiced speech, and even smaller for silence part of the utterance.

We can use the same voiced and unvoiced part of the speech of size, say, 256 and calculate 12 LPCs. The sample values for voiced segment and unvoiced segment can be obtained using a MATLAB command LPC in the program.

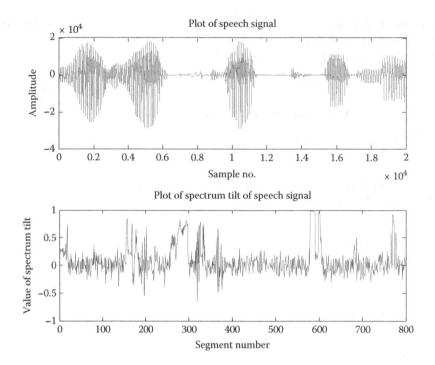

FIGURE 5.6
Plot of speech signal and its spectrum tilt.

LPCs for typical voiced segment are found to be in

     Columns 1 through 7

       5.0000   −5.2491   0.2493   0.1285   −0.0295   −0.0520   −0.0121

     Columns 8 through 13

       −0.0155   −0.0204   0.0175   0.0516   0.0743   −0.1366

LPCs for typical unvoiced segment are found to be in

     Columns 1 through 7

       5.0000   −5.0763   0.0451   0.0488   0.0400   −0.0061   −0.0228

     Columns 8 through 13

       −0.0358   −0.0216   −0.0004   0.0200   0.0231   −0.0096

We can easily verify that the value of second LPC is higher for voiced segment equal to −5.2941 compared to unvoiced segment equal to −5.0763. This result can be interpreted by considering that the successive samples are more correlated in voiced speech.

6. *Presence or absence of pitch:* Different methods of estimation of pitch period can also be used to see if the speech segment has a periodic nature or not. In the case of voiced segment, if one tracks the pitch value continuously, it can be concluded that the pitch value varies but is almost constant. If the pitch value remains almost constant, the pitch information is said to be found. For unvoiced segment, one will find that pitch value obtained will be low and will vary more. Here, we say that pitch information is absent. If pitch information is absent, it must be the unvoiced speech. If repetitive pitch is found, the segment is voiced.

7. *Preemphasized energy ratio:* Voiced and unvoiced segments can be discriminated using normalized preemphasized energy ratio defined by Equation 5.3.

$$\text{Pr} = \frac{\sum_{i=1}^{N} |s(i) - s(i-1)|}{\sum_{i=1} s^2(i)} \tag{5.3}$$

The variance of the difference between adjacent samples for voiced speech is very much smaller and higher for unvoiced speech.

A MATLAB program to plot preemphasized energy is as follows.

```
%to find pre-emphasized energy ratio for speech signal
clear all;
fp=fopen('c:\fast_sent1.wav');
fseek(fp,44,-1);
a=fread(fp);
a=a-128;

subplot(2,1,1);plot(a);title('plot of speech signal');
xlabel('sample no.');ylabel('amplitude');

for j=1:800,
    fseek(fp,44+100*j,-1);
    a=fread(fp,100);
    a=a-128;
    sum(j)=0;
for i=2:100,
    sum(j)=sum(j)+abs(a(i)-a(i-1));

end
sum1(j)=0;
for i=2:100,
    if a(i)==0,
        a(i)=0.1;
    end
    sum1(j)=sum1(j)+a(i)*a(i);
    end
s(j)=sum(j)/sum1(j);
end

subplot(2,1,2);plot(s);title('plot of pre-emphasized energy ratio of
speech signal');
xlabel('segment number');ylabel('value of pre-emphasized energy ratio');
```

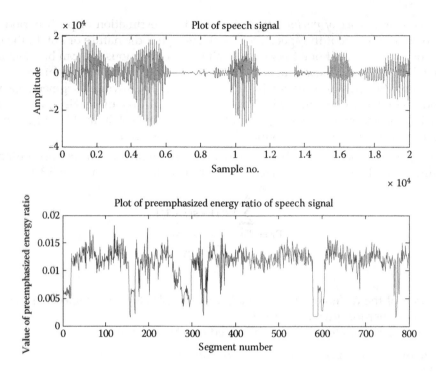

FIGURE 5.7
Plot of speech signal and its preemphasized energy ratio.

We have used a speech file containing voiced and unvoiced segments. The plot of preemphasized energy ratio in Figure 5.7 shows that when a signal is voiced, the preemphasized energy ratio has a value very close to zero, whereas for unvoiced segment, the preemphasized energy ratio is high. In the case of the silence part, the preemphasized energy ratio is zero. Hence, using preemphasized energy ratio, one can identify the voiced, unvoiced, and silence part of the utterance.

8. *Low-band to full-band energy ratio*: Voiced and unvoiced segments can be discriminated using low band to full band energy ratio. In the case of a voiced speech signal, the energy is concentrated in the low band. We can measure a ratio of energy in the first 1 kHz band to full band. This ratio is high for a voiced speech and quite low for an unvoiced speech.

A MATLAB program to calculate low band to full-band energy ratio is as follows.

```
%to find spectrum tilt for speech signal
clear all;
fp=fopen('c:\fast_sent1.wav');
fseek(fp,44,-1);
a=fread(fp);
a=a-128;

subplot(2,1,1); plot(a);title('plot of speech signal');
xlabel('sample no.');ylabel('amplitude');
```

```
for j=1:800,
    fseek(fp,44+100*j,-1);
    a=fread(fp,100);
    a=a-128;
    for i=1:50,
        b=abs(fft(a));
    end
    sum(j)=0;
for i=1:12,
    sum(j)=sum(j)+(b(i));
    end
sum1(j)=0;
for i=1:50,

    sum1(j)=sum1(j)+b(i);

  if sum1(j)==0;
        sum1(j)=1;
    end
end
s(j)=sum(j)/sum1(j);
end

subplot(2,1,2);plot(s);title('plot of low to full band energy ratio of
speech signal');
xlabel('segment number');ylabel('value of low to full band energy
ratio');
```

We have used a speech file containing voiced and unvoiced segments. The plot of low-band to full-band energy ratio in Figure 5.8 shows that when a signal is voiced, the pre-emphasized energy ratio has maximum value, whereas for unvoiced segment, low-to full-band energy ratio is low. In the program, we have considered the ratio of one-fourth band to full band. In the case of silence part, the low-band to full-band energy ratio is zero. Hence, using low-band to full-band energy ratio, one can identify the voiced, unvoiced, and silence part of the utterance.

Concept Check

- What is the characteristic of a voiced speech?
- How will you identify a silence part of the speech signal?
- Can ZCR separate voiced and unvoiced segments?
- What is the meaning of pitch information?
- How autocorrelation will distinguish voiced and unvoiced segments?
- How will you identify voiced and unvoiced segments using spectrum tilt?
- How will you identify voiced, unvoiced, and silence segments using preemphasized energy ratio?

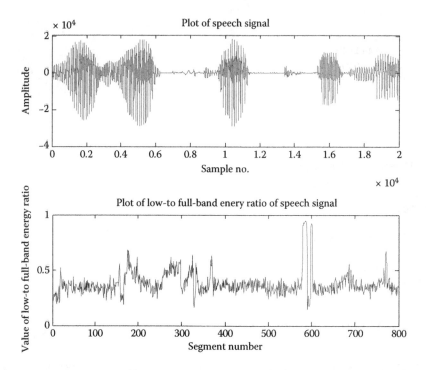

FIGURE 5.8
Plot of signal and its low- to full-band energy ratio.

5.3 Audio File Formats—Nature of .wav File

There are different audio file formats used. They can be classified into three categories.

1. Uncompressed file format
2. Compressed file format (lossless compression)
3. Compressed file format (lossy compression)

The commonly used uncompressed file format is .wav. We will concentrate on the study of the uncompressed file format in this section.

Waveform Audio File Format (*WAVE*, or more commonly known as *WAV* due to its file-name extension) is a Microsoft and IBM audio file format used for storing an audio bit stream in the personal computer (PC). WAV files are probably the simplest of the common formats used for storing audio.

The .WAV file consists of three "chunks," namely, RIFF chunk, FORMAT chunk, and DATA chunk. The RIFF chunk identifies the file as a WAV file. The FORMAT chunk specifies parameters such as sampling rate, bytes per second, and bytes per sample, and the DATA chunk contains the actual data (samples).

RIFF chunk: This has 12 bytes. The first 4 bytes are actually 4 ASCII characters, namely, RIFF. Next 4 bytes indicate the total length of the package to follow in binary. Last 4 bytes are actually 4 ASCII characters, namely, WAVE, as shown in Table 5.1.

FORMAT chunk: This has 24 bytes shown in Table 5.2. The first 4 bytes are "fmt." Next 4 bytes specify the length of FORMAT chunk. It is usually specified as "0 × 10." Specification of 8, 9, 10, and 11th byte is shown in Table 5.3. Bytes 12–15 indicate the sample rate in Hz, and bytes 16–19 and bytes 20–23 represent bytes per second and bits per second, respectively.

DATA chunk: This has a number of bytes depending on the length of the data. The specification of this chunk is represented in Table 5.4. Bytes 4–7 represent the length of the data.

Thus, the length of the header for any WAV file is 44 bytes (12 bytes of FIFF chunk + 24 bytes of FORMAT chunk + 8 bytes of DATA chunk).

TABLE 5.1

RIFF Chunk (12 Bytes in Length Total)

Byte Number	
0–3	"RIFF" (ASCII characters)
4–7	Total length of package to follow (binary, little endian)
8–11	"WAVE" (ASCII characters)

TABLE 5.2

FORMAT Chunk (24 Bytes in Length Total)

Byte Number	
0–3	"fmt_" (ASCII characters)
4–7	Length of FORMAT chunk (binary, always 0×10)
8–9	Always 0×01
10–11	Channel numbers (always 0×01=mono, 0×02=stereo)
12–15	Sample rate (binary, in Hz)
16–19	Bytes per second
20–21	Bytes per sample: 1=8 bit mono, 2=8 bit stereo or 16 bit mono, 4=16 bit stereo
22–23	Bits per sample

TABLE 5.3

DATA Chunk

Byte Number	
0–3	"Data" (ASCII characters)
4–7	Length of data to follow
8–end	Data (samples)

Concept Check

- How many chunks does a .wav file have?
- What is the information present in RIFF chunk?
- How will you identify the sampling rate of the signal?
- How can one find the length of the data in a .wav file?

5.4 Extraction of Fundamental Frequency

This section introduces the significance, meaning, and extraction of one important speech parameter, namely, pitch frequency or fundamental frequency. The fundamental frequency measurement in time domain, spectrum domain, and cepstrum domain will be discussed in this section. We will also deal with pitch and formant measurements in time domain, frequency domain/spectrum domain, and cepstral domain. The formant measurement using power spectrum density estimation is described.

5.4.1 Fundamental Frequency or Pitch Frequency

We will start with the definition of fundamental frequency. The speech signal contains different frequencies that have harmonic relation. The lowest frequency of this harmonic series is called the fundamental frequency. The definition of pitch indicates that it is related to the number of vibrations of the vocal cords. This frequency generated by vocal cords in the form of periodic excitation passes via the vocal tract filter and is convolved with the impulse response of the filter representing a vocal tract to produce a speech signal. Speech is basically a convolved signal convolution of excitation and the impulse response of the vocal tract.

Fundamental frequency is related to a voiced speech segment. Let us have a look at the waveform of voiced speech segment for a file "watermark.wav." The wave file is again plotted here for ready reference. The plot of the vowel part is shown in Figure 5.9. If we have a closer look at the waveform, we can see that a pattern is repeating after sample number 650 with approximately the same number of samples.

Let us concentrate on this repeating waveform and find the number of samples after which it repeats. This number will give the pitch period in terms of the number of samples. If we can find the time corresponding to these samples because the sampling frequency for the speech is known, then we can easily calculate the inverse of time period. This gives the fundamental frequency in Hz. The main principle used in time domain pitch detection algorithms is finding the similarity between original speech signal and its shifted version.

Let us do it practically using different methods such as the method of autocorrelation and the average magnitude difference function (AMDF). Let us first concentrate on the autocorrelation method for finding a pitch period.

5.4.1.1 Autocorrelation Method for Finding Pitch Period of a Voiced Speech Segment

We will start with the definition of autocorrelation. *Autocorrelation* is the correlation of a signal with itself. This will help us to find the similarity between samples as a function of

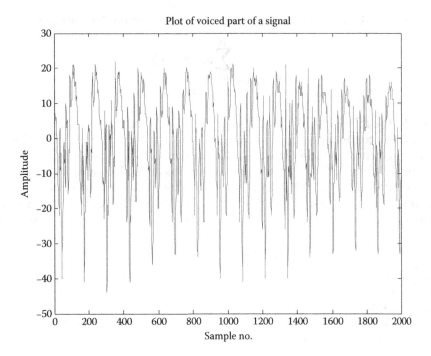

FIGURE 5.9
Plot of voiced part of the utterance.

time separation between them. It can be considered as a mathematical tool for finding if the pattern is repeating. The period can then be found out. Autocorrelation methods need at least two pitch periods to detect pitch.

The algorithm can be described as follows.

1. Take a speech segment equal to at least two pitch periods. We use a speech file with a sampling frequency of 22 kHz. So sampling interval is 1/22 kHz equal to 0.045 ms. In the 20 ms interval, there will be 444.4 samples. Let us consider a speech segment of size 400 samples. We will calculate the autocorrelation for, say, 45 overlapping samples, i.e., two speech segments, one starting from sample number 1 to 45 is correlated to segment from sample number 2 to 46, then 3 to 46, i.e., with the shift of 1, 2, 3 samples, and so on. We use a shift up to 400 samples. We have to find the shift value for which the correlation value has the highest value. The distance between two successive maxima in correlation will give a pitch period in terms of number of samples.

2. Calculate the autocorrelation using the formula given by Equation 5.4:

$$R_{xx}(k) = \lim_{N \to \omega} \frac{\sum_{i=0}^{2N} a(i)a(i+k)}{2N+1} \tag{5.4}$$

The equation is simplified to Equation 2.2 for practical implementation.

$$R_{xx}(k) = \frac{\sum_{i=1}^{45} a(i)a(i+k)}{45} \quad (5.5)$$

3. Refer to the following MATLAB program. The voiced part of the speech signal using fseek command is tracked. When a program is executed on the voiced part of the segment shown in Figure 5.10 (the upper plot), we find that the correlation output shown in Figure 5.10 (the lower plot) has about 135 samples between successive maxima. Thus, 1135 is a pitch period in terms of the number of samples. We can track the maxima and find the actual pitch value. The reader is encouraged to write a program to find a pitch. The correlation is maximum for the offset of zero.

```
%to find pitch period of the signal
clear all;
fp=fopen('watermark.wav');
fseek(fp,224000,-1);
a=fread(fp,2000);
subplot(2,1,1);plot(a);title('plot of voiced part of a signal');
xlabel('sample no.');ylabel('amplitude');

a=a-128;
for k=1:400,
```

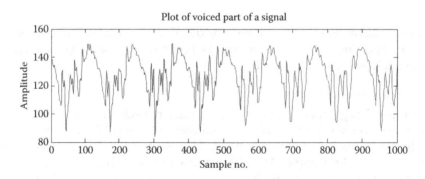

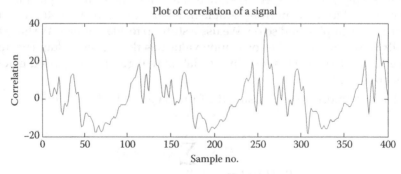

FIGURE 5.10
Plot of voiced speech segment and its autocorrelation function.

```
sum(k)=0;
end
for k=1:400,
for i=1:45,
sum(k)=sum(k)+(a(i)*a(i+k));
sum(k)=sum(k)/45;
end
end
subplot(2,1,2);plot(sum);title('plot of correlation of a signal');
xlabel('sample no.');ylabel('correlation');
```

5.4.1.1.1 Nonlinear Processing

The low amplitude portions of speech contain most of the formant information, and the high amplitude portions contain most of the pitch information. Hence, nonlinear processing that deemphasizes the low amplitude portions can improve the performance of the autocorrelator. Nonlinear processing can be done directly in the time domain. We will discuss two methods of nonlinear processing, namely, center clipping and cubing.

5.4.1.1.2 Center Clipping

Low amplitude portions are removed using a center clipper whose clipping point is decided by the peak amplitude of the speech signal. The clipper threshold is adjusted to remove all portions of the waveform below some threshold *T*. The center-clipped speech is passed through the correlator. The autocorrelation will be zero for most lags and will have sharp peaks at the pitch period with very small secondary peaks. Typically, the clipper will have the input output function given by Equation 2.2a.

$$C(x) = \begin{cases} x-T, & x>T \\ 0, & |x| \leq T \\ x+T & x<-T \end{cases} \tag{5.6}$$

The researchers have reported the value of *T* as 30% of the peak value of the signal. Clippers input output plot is shown in Figure 5.11.

5.4.1.1.3 Cubing

The speech waveform is passed through the nonlinear circuit whose transfer function is $y(t) = x^3(t)$. Cubing tends to suppress the low amplitude portions of the speech. Here, it is not required to keep an adjustable threshold.

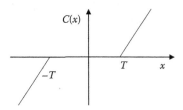

FIGURE 5.11
Transfer characteristic of clipper.

5.4.1.2 AMDF Method for Finding Pitch Period of a Voiced Speech Segment

Let us start with the definition of AMDF. The difference signal *Dm* is formed by delaying the input speech by various amounts, subtracting the delayed waveform from the original, and summing the magnitude of the differences between sample values. We have to take the average of the difference function over the number of samples. The difference signal is always zero at delay=0 and is particularly small at delays corresponding to the pitch period of a voiced sound having a quasi-periodic structure. The main advantage of AMDF method is that it requires only subtractions.

The algorithm can be described as follows.

1. Take a speech segment equal to at least two pitch periods. We use a speech file with sampling frequency of 22 KHz. So sampling interval is 1/22 kHz equal to 0.045 ms. In 20 ms interval, there will be 444.4 samples. Let us consider a speech segment of size 400 samples. We will calculate the AMDF for, say, 45 overlapping samples, i.e., two speech segments: one starting from sample number 1 to 45 is correlated to segment from sample number 2 to 46, then 3 to 46, i.e., with the shift of 1, 2, 3 samples, and so on. We use a shift up to 400 samples. We have to find the shift value for which AMDF is having the smallest value. The distance between two successive minima in AMDF will give a pitch period in terms of the number of samples.

2. Calculate the AMDF using the formula given by Equation 5.7.

$$\text{AMDF}(k) = \frac{\sum_{i=1}^{N} |a(i) - a(i+k)|}{N} \tag{5.7}$$

3. Refer to the following MATLAB program. The voiced part of speech signal is tracked using fseek command. When a program is executed on the voiced part of the segment shown in Figure 5.12 (the upper plot), we find that the AMDF output shown in Figure 5.12 (the lower plot) has about 135 samples between successive minima. Thus, pitch period is 135 in terms of number of samples. We can track the maxima and find the actual pitch value. The reader is encouraged to write a program to find a pitch.

```
%to find pitch period of the signal using AMDF
clear all;
fp=fopen('watermark.wav');
fseek(fp,224000,-1);
a=fread(fp,1000);
subplot(2,1,1);plot(a);title('plot of voiced part of a signal');
xlabel('sample no.');ylabel('amplitude');
for k=1:400,
sum(k)=0;
end
for k=1:400,
for i=1:45,
sum(k)=sum(k)+abs(a(i)-a(i+k));
sum(k)=sum(k)/200;
end
```

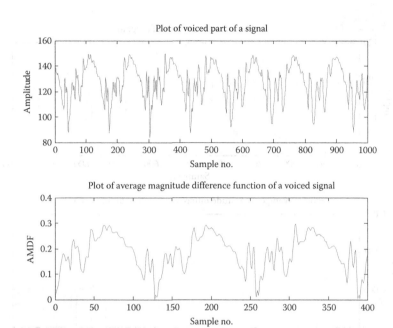

FIGURE 5.12
Plot of voiced part of speech and its AMDF.

```
end
subplot(2,1,2);plot(sum);title('plot of Average magnitude difference
function of a voiced signal');
xlabel('sample no.');ylabel('AMDF');
```

Let us now consider the unvoiced segment in the same speech file. We will take a wav file "watermark.wav" and track the unvoiced segment using fseek command. Let us apply the AMDF calculation for this unvoiced segment using a MATLAB program shown in the following paragraph. The result of MATLAB program is a plot as shown in Figure 5.13. The plot for AMDF indicates that there is no periodicity. If we find and track the points where AMDF becomes the smallest, we will see that it has quite frequent and random occurrence. This indicates that there is no pitch information present in the signal and the signal is unvoiced. This method of pitch calculation can be used for taking V/UV decision.

```
clear all;
fp=fopen('watermark.wav');
fseek(fp,32500,-1);
a=fread(fp,1000);
subplot(2,1,1);plot(a);title('plot of unvoiced part of a signal');
xlabel('sample no.');ylabel('amplitude');
for k=1:400,
sum(k)=0;
end
for k=1:400,
for i=1:45,
sum(k)=sum(k)+abs(a(i)-a(i+k));
sum(k)=sum(k)/200;
```

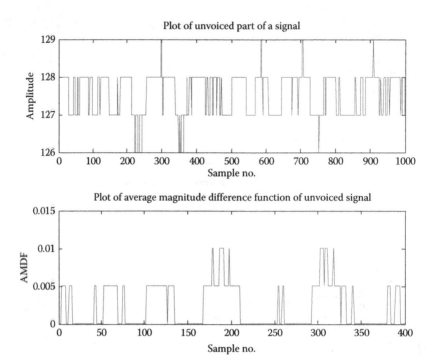

FIGURE 5.13
Plot of unvoiced segment and its AMDF.

```
end
end
subplot(2,1,2);plot(sum);title('plot of Average magnitude difference
function of unvoiced signal');
xlabel('sample no.');ylabel('AMDF');
```

5.4.2 Pitch Contour

Let us understand the meaning of pitch contour. Speech signal has some random component. The voiced segment of speech exhibits periodicity or in fact pseudoperiodicity. Pseudoperiodicity indicates that the waveform has a periodic nature, but the period does not remain constant and varies to a small extent.

The periodic pulse train generated from the vibrations of the vocal cords shown in LTV model acts as an excitation for the voiced speech signal. If we track the pitch period over the complete duration of voiced segment, we find that there is a small variation in the pitch period. The contour of variations of the pitch period is called pitch contour. The pitch contour contains some information related to spoken word and the speaker. Hence, pitch contour is also considered as a feature in speech recognition and speaker verification tasks.

5.4.3 Pitch Period Measurement in Spectral Domain

The time domain approaches have been studied for pitch period measurement in the previous section. This section deals with a frequency domain approach for pitch period measurement. The frequency domain pitch detection algorithms operate in the spectrum domain. The periodic signal has a harmonic structure. The algorithm tracks the distance between successive harmonics. The main drawback of frequency domain methods is that computational complexity increases. Let us study the use of fast Fourier transform (FFT) for pitch measurement. FFT is a fast algorithm for the computation of discrete Fourier transform (DFT). DFT coefficient $X(k)$ of any sampled signal (sequence) $x(n)$ is given by Equation 5.8.

$$X(k) = \sum_{n=0}^{N-1} x(n)W^{nk} \tag{5.8}$$

where W is given by Equation 5.9:

$$W = e^{-j2\pi/N} \tag{5.9}$$

Computation of DFT coefficient requires complex multiplications and additions. FFT algorithms simplify the computations. There are two FFT algorithms that are commonly used. They are decimation in frequency algorithm and decimation in time algorithm. If one computes 512-point FFT for a sequence containing 512 samples, then we get a plot that is found to be symmetric about the center value, i.e., 256 in this case because of complex conjugate property of DFT coefficients. This plot is a graph of magnitude of FFT plotted against frequency bin number. In the case of 512-point FFT, if the sampling frequency of the signal is 22,100 Hz, then 0–22,100 Hz range gets divided in 512 frequency components and each frequency point now corresponds to a frequency of 22,100 Hz/512 equal to 43.16 Hz. We can easily calibrate the DFT coefficient number axis to actual frequency value in Hz. The spectrum domain analyzes the magnitude of each spectral component of a signal. It has a spectral magnitude plotted with respect to frequency.

Let us use the voiced segment of speech containing 512 samples. Let us go through a MATLAB program to take FFT of the signal and plot it. We will then track the first peak in the FFT output to find the fundamental frequency. The resolution for fundamental frequency measurement is decided by the number of points in FFT. If we take 2048-point FFT, the frequency resolution will improve to 22,100/2048 equal to 10.79 Hz. The measurement of fundamental frequency cannot be accurate using FFT analysis.

A MATLAB program for spectrum calculation is as follows.

```
%to find pitch period of the signal using spectrum of speech
clear all;
fp=fopen('watermark.wav');
fseek(fp,40500,-1);
a=fread(fp,1024);
subplot(2,1,1);plot(a);title('plot of voiced part of a signal');
xlabel('sample no.');ylabel('amplitude');
a=a-128;
b=abs(fft(a));
f=22100/1024:22100/1024:22100;
subplot(2,1,2);plot(f,b);title('plot of fft of a signal');
xlabel('frequency');ylabel('amplitude');
disp(b);
```

The signal plot and the output plot of FFT are depicted in Figure 5.14. It indicates that the first peak has a value equal to 12,560 and it occurs at the 16th position, so fundamental frequency is $10.79 \times 16 = 172.26$ Hz. This pitch frequency value is valid, and hence, the speech segment is a voiced speech segment. Sometimes, the first peak is missing. A more practical method is as follows. It detects all the harmonic peaks and measures the fundamental frequency as the common divisor of these harmonics. From Figure 5.14, we find that we get peaks at regular intervals for voiced segment. This method for pitch measurement is also called *harmonic peak detection* method. There is one more method called *spectrum similarity* method. This method assumes that the spectrum is fully voiced and has peaks located at multiples of fundamental frequency. This method has a stored template of the spectrum for different values of pitch frequency. To find the pitch frequency, one will compare its spectrum with stored templates to find a match. When a match is found, the corresponding value of pitch frequency is read.

5.4.3.1 Spectral Autocorrelation Method for Pitch Measurement

Pitch frequency can also be measured by calculating the spectral autocorrelation. The spectrum autocorrelation is the highest for frequency interval equal to pitch frequency in the case of a voiced segment. This method is called *spectral autocorrelation* method. A plot of spectral autocorrelation in Figure 5.15 shows spectral peaks at regular intervals. A MATLAB program to find spectral autocorrelation is as follows.

A MATLAB program to find pitch frequency using spectrum method for unvoiced segment is as follows.

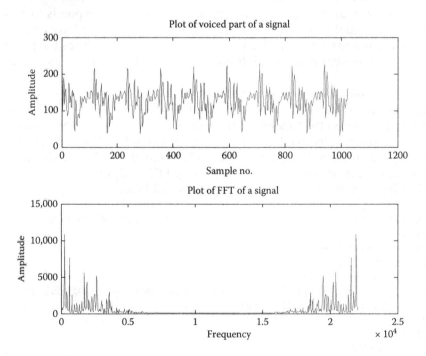

FIGURE 5.14
Plot of voiced signal and its FFT output.

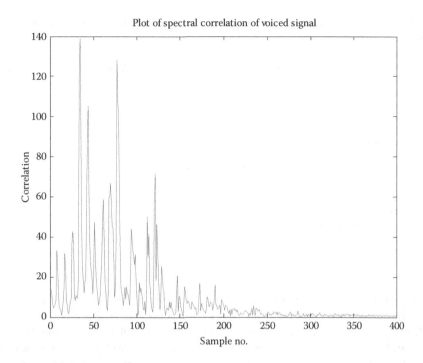

FIGURE 5.15
Plot of spectral correlation for voiced speech signal.

```
%to find pitch period of the signal using spectral autocorrelation of
voiced speech
clear all;
fp=fopen('watermark.wav');
fseek(fp,40500,-1);
a=fread(fp,1024);
subplot(2,1,1);plot(a);title('plot of voiced part of a signal');
xlabel('sample no.');ylabel('amplitude');
a=a-128;
b=abs(fft(a));
f=22100/1024:22100/1024:22100;
subplot(2,1,2);plot(f,b);title('plot of fft of a signal');
xlabel('frequency');ylabel('amplitude');
for k=1:400,
sum(k)=0;
end
for k=1:400,
for i=1:45,
sum(k)=sum(k)+(b(i)*b(i+k));
sum(k)=sum(k)/52000;
end
end
figure;
plot(sum);title('plot of spectral correlation of voiced signal');
xlabel('sample no.');ylabel('correlation');
```

A MATLAB program to find pitch frequency using spectrum method for unvoiced segment is as follows.

```
%to find pitch period of the signal using spectral autocorrelation of
voiced speech
clear all;
fp=fopen('watermark.wav');
fseek(fp,12000,-1);
a=fread(fp,1024);
subplot(2,1,1);plot(a);title('plot of unvoiced part of a signal');
xlabel('sample no.');ylabel('amplitude');
a=a-128;
b=abs(fft(a));
c=log(b.*b);
f=22100/1024:22100/1024:22100;
subplot(2,1,2);plot(f,c);title('plot of fft of a signal');
xlabel('frequency');ylabel('amplitude');
disp(b);
```

The signal plot for unvoiced speech and the output plot of FFT are shown in Figure 5.16. The figure indicates that the first peak has a value of 801 and it occurs at the first position, so the fundamental frequency is 10.79 Hz. This cannot be the value of valid pitch frequency, and hence the signal is unvoiced. We can note from Figure 5.16 that in the case of unvoiced segments, we do not get peaks at regular intervals. For spectrum similarity experiment for unvoiced segment, it can be found that we do not get a match with any synthesized spectrum.

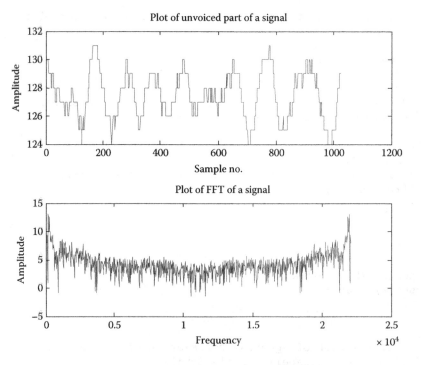

FIGURE 5.16
Plot of unvoiced speech signal and its FFT.

A MATLAB program for the calculation of spectral autocorrelation can be written on the same lines. Append the following lines to the previous program. We can even take the V/UV decision based on spectral similarity. The plot of spectral autocorrelation for unvoiced segment is shown in Figure 5.17. It indicates no regular peaks.

```
for k=1:400,
sum(k)=0;
end
for k=1:400,
for i=1:45,
sum(k)=sum(k)+(b(i)*b(i+k));
sum(k)=sum(k)/52000;
end
end
figure;
plot(sum);title('plot of spectral correlation of unvoiced signal');
xlabel('sample no.');ylabel('correlation');
```

5.4.4 Cepstral Domain

This section describes cepstral domain approach for pitch period measurement. We will first define a cepstrum. A cepstrum is obtained when we take Fourier transform (FT) of the log spectrum. The term "cepstrum" is derived by reversing the first four letters of the "spectrum." We can have a power cepstrum or a complex cepstrum.

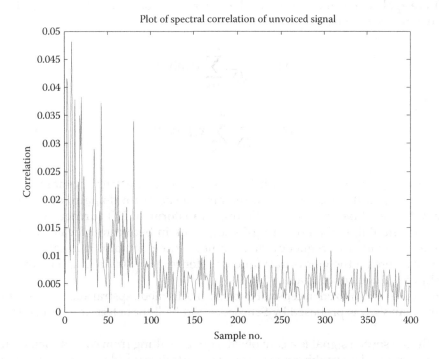

FIGURE 5.17
Plot of spectral autocorrelation of unvoiced speech signal.

We will now define a power cepstrum. It is a squared magnitude of the FT of the log of squared magnitude of FT of a signal. We can algorithmically define it as

signal → FT → abs() → square → log → FT → abs() → square → power cepstrum

Let us also define a complex cepstrum. The complex cepstrum was defined by Oppenheim in his development of homomorphic system theory. It may be defined as follows. The complex cepstrum of any signal is the FT of the logarithm obtained with the unwrapped phase of the FT of a signal. Sometimes, it is called the spectrum of a spectrum. We will define it algorithmically.

signal → FT → abs() → log → phase unwrapping → FT → cepstrum   (5.10)

The real cepstrum will use the log function defined for real values. It is related to the power cepstrum, and the relation can be specified as Equation 5.11.

$$4(\text{Real spectrum})^2 = \text{power spectrum} \tag{5.11}$$

In different texts, the process of taking cepstrum is defined algorithmically as

$$\text{FT} \rightarrow \text{abs}() \rightarrow \log \rightarrow \text{IFT} \tag{5.12}$$

This equation involves IFT instead of FT. However, this is not relevant because when FT and IFT are applied to complex signals, the difference is only of a phase sign. FT and IFT are defined by the following equations.

$$X(k) = \frac{1}{\sqrt{N}} \sum_{n=0}^{N-1} x(n) W^{nk} \tag{5.13}$$

$$x(n) = \frac{1}{\sqrt{N}} \sum_{k=0}^{N-1} X(k) W^{-nk} \tag{5.14}$$

When the FT of a signal is taken, it is converted to a spectrum domain. Taking the absolute value and the log, the signal remains in spectrum domain, i.e., x axis is frequency. When we take inverse fast Fourier transform (IFFT) further, we come back to time domain. But, is the presence of a log block in between indicates that the output domain is neither a frequency domain nor a time domain. It is called quefrency domain by reversing the words in frequency. Hence, a cepstrum graph of a signal will be amplitude vs quefrency graph.

When a signal is analyzed in cepstral domain, it is termed cepstral analysis. A short-time cepstrum analysis was proposed by Schroeder and Noll for determining pitch in human speech.

Basically, the speech signal is a convolved signal resulting from convolution of impulse response of the vocal tract with either periodic pulse train or random signal/the turbulence for voiced and unvoiced speech. When one tries to find pitch frequency as a first peak in the spectrum, the first formant and fundamental frequency may overlap. In such a situation, it is difficult to separate the first formant from the pitch frequency. The isolation of these two components is possible in the cepstral domain. Hence, the cepstral domain is preferred for pitch measurement. We will see how this is made possible in the next section.

5.4.5 Pitch Period Measurement Using Cepstral Domain

Consider a block schematic for the evaluation of a pitch period using cepstrum as shown in Figure 5.18.

Let us find the function of each block. The speech signal given as input to a system consists of the periodic excitation convolved with the impulse response of a vocal tract, which is a slowly varying function. FFT block takes DFT of a signal, and spectrum of the signal is obtained. When we take the log magnitude, the amplitude is calculated in dB, and the periodic excitation is a rapidly varying function, whereas the vocal tract response is an envelope and a slowly varying function. When we take IFFT, we find that a slowly varying function of the vocal tract clusters near the origin and a rapidly varying function appears as regular pulses away from the origin. We can now use a cepstral window, allowing the pitch information (rapidly varying function) to pass through. The FFT output of this windowed cepstrum is a spectrum, with only a rapidly varying function. If we track the peak of this spectrum, we can find the pitch frequency. The slowly varying function of the vocal tract is isolated, and hence, the possibility of the first formant overlapping with the pitch frequency is eliminated.

Let us do it practically by considering a voiced speech signal and unvoiced speech signal. Let us write a MATLAB program to execute the block schematic on 2048 samples of speech and observe the results. The speech signal given at the input is shown in Figure 5.19. We have used Hamming window. The plot of the window function is also shown in Figure 5.19. The windowed signal is shown in Figure 5.20. The output of FFT block is a spectrum as shown in Figure 5.20. The log magnitude block operates on spectrum and will generate the amplitude in dB. Figure 5.21 depicts the log spectrum. Now we take IFFT again. We have entered the cepstrum domain and the output obtained is shown in Figure 5.21. After passing the cepstrum via a window, the cluster near the origin becomes zero, as shown in Figure 5.22. We have made first 40 and last 40 symmetrical points zero. The FFT of the windowed cepstrum is depicted in Figure 5.22. The peak occurs at the 17th position in the FFT of windowed cepstrum indicating a fundamental frequency of $17 \times 22{,}400/1028 = 366.88\,\text{Hz}$. The following MATLAB program listing executes all the functions described.

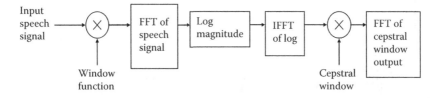

FIGURE 5.18
Block schematic for pitch calculation of voiced segment.

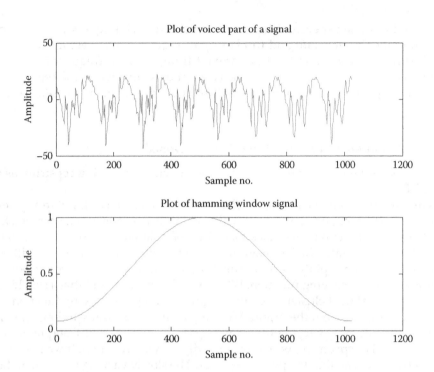

FIGURE 5.19
Plot of voiced speech of 1024 samples and Hamming window function.

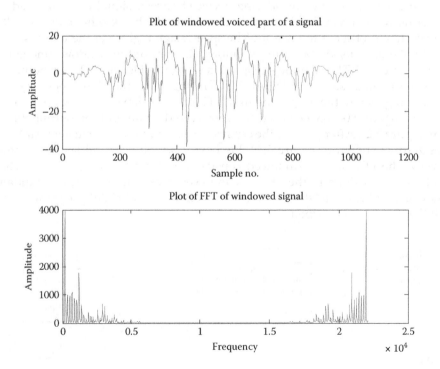

FIGURE 5.20
Plot of 1024 samples of Hamming windowed voiced speech and its FFT.

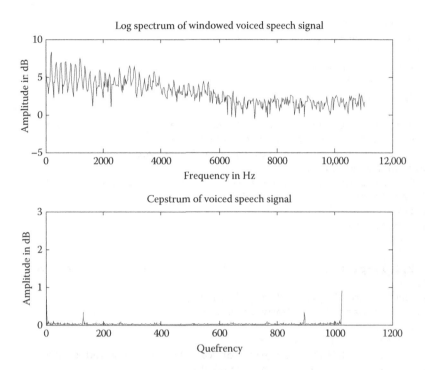

FIGURE 5.21
Plot of log spectrum of 1024 samples of voiced speech and its cepstrum.

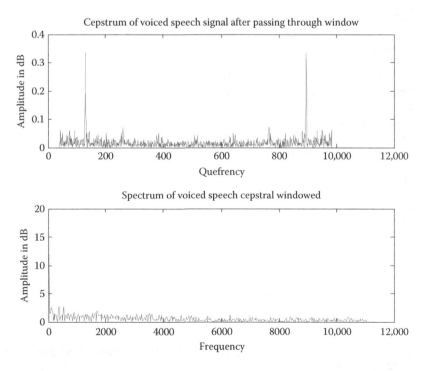

FIGURE 5.22
Plot of windowed cepstrum of 1024 samples of voiced speech and its FFT.

```
%to find pitch period
of the voiced signal using cepstrum of speech signal
clear all;
fp=fopen('watermark.wav');
fseek(fp,224000,-1);
a=fread(fp,1024);
a=a-128;
subplot(2,1,1);plot(a);title('plot of voiced part of a signal');
xlabel('sample no.');ylabel('amplitude');
%2048 samples of the voiced signal are displayed
a1=window(@Hamming,2048);
subplot(2,1,2);plot(a1);title('plot of Hamming window signal');
xlabel('sample no.');ylabel('amplitude');
for i=1:1024,
    a11(i)=a(i)*a1(i);
end
figure;
subplot(2,1,1);plot(a11);title('plot of windowed voiced part of a
signal');
xlabel('sample no.');ylabel('amplitude');
%the signal is passed via Hamming window and displayed.
b=abs(fft(a11));
f=22100/1024:22100/1024:22100;
subplot(2,1,2);plot(f,b);title('plot of fft of windowed signal');
xlabel('frequency');ylabel('amplitude');
c=log(b);
figure;
for i=1:512,
    d(i)=c(i);
end
f=22100/1024:22100/1024:11050;
subplot(2,1,1);plot(f,d);
title('log spectrum of windowed voiced speech signal');
xlabel('Frequency in Hz');ylabel('amplitude in dB');
e=abs(ifft(c));
subplot(2,1,2);plot(e);
title('cepstrum of voiced speech signal');
xlabel('Quefrency');ylabel('amplitude in dB');
for i=1:40,
    h(i)=0;
end
for i=41:983,
    h(i)=e(i);
end
for i=984:1024,
    h(i)=0;
end
% for i=1:1024,
%     h(i)=g(i);
% end
figure;
subplot(2,1,1);plot(e); title('cepstrum of voiced speech signal after
passing through window');
```

```
xlabel('Quefrency');ylabel('amplitude in dB');
g=abs(fft(h));
for i=1:512,
    k(i)=g(i);
end
f=22100/1024:22100/1024:11050;
subplot(2,1,2);
plot(f,k);title('spectrum of voiced speech cepstral windowed');
xlabel('frequency');ylabel('amplitude in dB');
disp(k);
```

5.4.5.1 Interpretation of the Result

The input signal to a pitch calculation module is a voiced speech segment. A total of 1024 samples of the signal are taken. The signal plot indicates that the signal has a repetitive nature. This signal is generated as a result of convolution of excitation generated from vocal cords and impulse response of vocal tract. The signal is passed via a Hamming window in order to avoid the Gibbs phenomenon. If there is abrupt truncation in signal magnitude, it will introduce oscillations, i.e., high frequency components in the signal. Windowing a signal will avoid this end effect.

The Hamming window is selected because the spectrum of it has the highest side lobe attenuation. When we multiply a signal in time domain by a rectangular window function, the frequency domain responses convolve. This convolution results in a ripple in passband and large oscillations in the stopband. The stopband oscillations are termed spectral leakage and the oscillations near the band edge of the filter are called the Gibbs phenomenon. The oscillations near the band edge get reduced by using a smooth window function. There are different window functions such as Hamming, Hanning, Blackman, and Bartlett. The spectrum of these window functions has a broader main lobe and smaller side lobes. Smooth windows reduce Gibbs phenomenon but lead to larger transition widths. If we go through the comparison of these window functions as shown in Table 5.4, we see that we have use compromise between the side lobe amplitude and transition width. M in Table 5.4 stands for the filter order. For the same transition width of $8\pi/M$, the stopband attenuation is the highest for the Hamming window, and hence we selected it. The equation of the Hamming window is given by Equation 5.15.

$$w(n) = 0.54 - 0.46\cos\left(\frac{2\pi n}{M-1}\right)$$

(5.15)

TABLE 5.4

Performance Comparison of Different Window Functions

Name of Window	Transition Width	Minimum Stopband Gain (dB)
Rectangular	$4\pi/M$	−13
Bartlett	$8\pi/M$	−27
Hanning	$8\pi/M$	−32
Hamming	$8\pi/M$	−43
Blackman	$12\pi/M$	−58

The windowed signal is passed via FFT block. It has the first peak corresponding to the fundamental frequency. We are taking the log of the spectrum in order to represent the amplitude of the spectral lines in dB. The log of FFT block shows rapid variations superimposed on the slowly varying envelope. The rapid variations correspond to the fundamental frequency, and slowly varying envelope corresponds to the vocal tract parameters. The IFFT of the log spectrum is a cepstrum. The plot of the cepstrum shows peaks at regular intervals indicating the voiced nature of the signal. When we take IFFT of the log spectrum considering it as a signal, its slowly varying components will appear as a cluster near the origin. The vocal tract information closer to the origin is removed by a window of length 40 samples from both sides as the cepstrum is also symmetric. FFT of this cepstral windowed signal returns the signal back in the log spectrum domain. The final output plot indicates rapid variations corresponding to the fundamental frequency. We can track the first peak. It will not overlap with the first formant now, as the cluster near the origin corresponding to the vocal tract information is removed.

We will now use the same program to execute this process on unvoiced segment. We will track the unvoiced part of the utterance. Figure 5.23 shows a plot of the unvoiced segment and window function. The plot of windowed unvoiced utterance and its FFT are depicted in Figure 5.24. The plot of the log spectrum of unvoiced segment is calculated. Taking IFFT of the log spectrum takes us in the cepstral domain. Figure 5.25 shows plots of the log spectrum and its cepstrum. Figure 5.26 shows the plot of windowed cepstrum and the corresponding FFT.

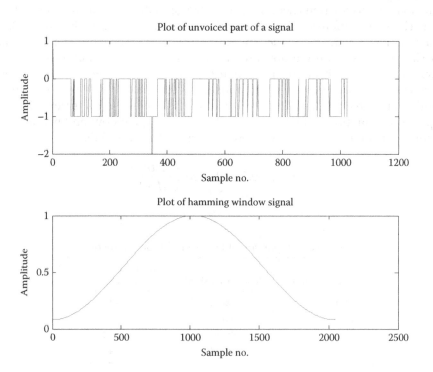

FIGURE 5.23
Plot of unvoiced segment and a plot of window function.

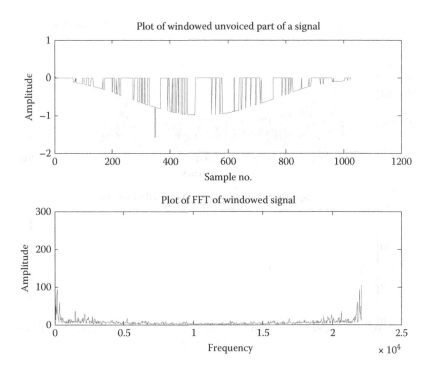

FIGURE 5.24
Plot of windowed unvoiced utterance and its FFT

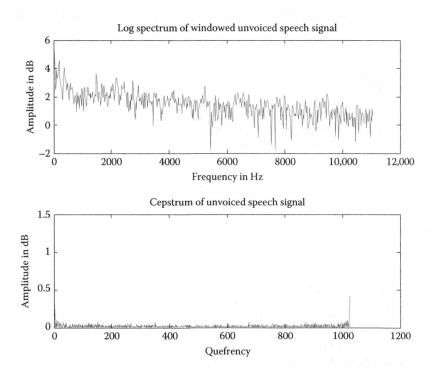

FIGURE 5.25
Plot of log spectrum of unvoiced segment and its cepstrum.

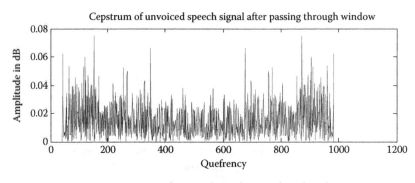

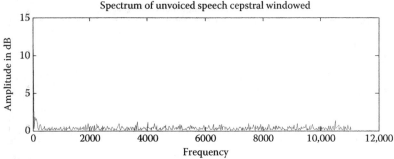

FIGURE 5.26
Plot of windowed cepstrum and its FFT.

```
%to find pitch period of the unvoiced signal using cepstrum of speech
signal
clear all;
fp=fopen('watermark.wav');
fseek(fp,23044,-1);
a=fread(fp,1024);
a=a-128;
subplot(2,1,1);plot(a);title('plot of unvoiced part of a signal');
xlabel('sample no.');ylabel('amplitude');
%2048 samples of the unvoiced signal are displayed

a1=window(@Hamming,1024);
subplot(2,1,2);plot(a1);title('plot of Hamming window signal');
xlabel('sample no.');ylabel('amplitude');
for i=1:1024,
    a11(i)=a(i)*a1(i);
end
figure;
subplot(2,1,1);
plot(a11);title('plot of windowed unvoiced part of a signal');
xlabel('sample no.');ylabel('amplitude');
%the signal is passed via Hamming window and displayed.
b=abs(fft(a11));
f = 22100/1024:22100/1024:22100;
subplot(2,1,2);plot(f,b);title('plot of fft of windowed signal');
xlabel('frequency');ylabel('amplitude');
```

```
c=log(b);
figure;
for i=1:512,
    d(i)=c(i);
end
f = 22100/1024:22100/1024:11050;
subplot(2,1,1);plot(f,d);
title('log spectrum of windowed unvoiced speech signal');
xlabel('Frequency in Hz');ylabel('amplitude in dB');
e=abs(ifft(c));

subplot(2,1,2);plot(e);
title('cepstrum of unvoiced speech signal');
xlabel('Quefrency');ylabel('amplitude in dB');
for i=1:40,
    h(i)=0;
end
for i=41:983,
    h(i)=e(i);
end
for i=984:1024,
    h(i)=0;
end
figure;
subplot(2,1,1);plot(h); title('cepstrum of unvoiced speech signal after
passing through window');
xlabel('Quefrency');ylabel('amplitude in dB');

g=abs(fft(h));

for i=1:512,
    k(i)=g(i);
end
f = 22100/1024:22100/1024:11050;
subplot(2,1,2);
plot(f,k);title('spectrum of unvoiced speech cepstral windowed');
xlabel('frequency');ylabel('amplitude in dB');
disp(k);
```

5.4.5.2 Interpretation of the Result

The input signal is an unvoiced speech segment. We have taken 1024 samples of the signal. The signal plot shows that the signal has no repetitive nature. This signal is generated as a result of the convolution of random signal generated due to turbulence and impulse response of vocal tract. The signal is passed via a Hamming window to avoid the end effects.

The windowed signal is passed via the FFT block. It has a first peak very close to the origin indicating a fundamental frequency of the order of tens of hertz. This indicates that the segment is unvoiced. The log of the spectrum is taken to represent the magnitude of the spectral lines in dB. The log of FFT block shows random variations superimposed on the slowly varying envelope. The random variations correspond to the absence of fundamental frequency, and slowly varying envelope corresponds to the vocal tract parameters. The IFFT of the log spectrum is a cepstrum. The plot of cepstrum indicates no peaks as the voicing information is not present. When we take the IFFT of the log spectrum considering it as a signal, its slowly varying components appear as a cluster near the origin. The vocal tract information closer to the origin is erased by a window of length 40 from both sides as the cepstrum is also symmetric. FFT of this cepstral windowed signal will return the signal in the log spectrum domain. The final output plot is to depict random variations.

Concept Check

- What is a pitch frequency?
- How is a pitch frequency related to vocal tract?
- What is autocorrelation?
- If autocorrelation is plotted with respect to offset, then at what offset will the auto-correlation be maximum?
- What is AMDF?
- How will you calculate pitch period using AMDF?
- What will be the nature of AMDF graph for unvoiced speech signal?
- What must the value of AMDF at zero offset?
- What is a parallel processing approach?
- What is the function of low-pass filter (LPF)?
- What is pitch contour?
- What is a spectrum domain?
- Define DFT of a signal?
- How will you calibrate DFT coefficient number in terms of frequency in Hz?
- How will you find pitch period using first peak of spectrum?
- What is a cepstrum?
- What is the x-axis in cepstral domain?
- Why is cepstral domain preferred for pitch measurement?
- What are the operations required to enter cepstral domain?
- What is the function of a window function block in cepstral measurement?
- Why is a Hamming window used?
- How will you interpret the graph of the log spectrum?
- We see a slowly varying envelope in the log spectrum. What does it represent?
- If we see periodic peaks in the cepstrum, what is your interpretation?
- If there are no peaks seen in the cepstrum, what is the interpretation?
- Why do we eliminate the cluster near the origin in cepstral domain?

5.5 Formants and Relation of Formants with LPC

Pitch frequency or pitch period is related to vocal cords. Let us now discuss the parameters related to the vocal tract. The vocal tract can be imagined as a circular waveguide with proper terminating conditions, i.e., closed at the larynx end and open at mouth end. In the case of voiced and unvoiced sound, the excitation differs. Excitation is a periodic pulse train for voiced speech generation. For unvoiced signal, the excitation comes from the random noise generator or turbulence. When the excitation passes via the vocal tract, it resonates at certain resonating frequencies decided by the mode of excitation in the cavity, i.e., TE10, TE20, TM11 mode, etc. These resonating frequencies are termed as formants.

Using these resonating frequencies, one can write the transfer function of the vocal tract in Z domain as shown in Equation 5.16.

$$H(Z) = \frac{Z^M}{(Z - P_1)(Z - P_2)\ldots\ldots(Z - P_M)} \tag{5.16}$$

We represent, for example, M number of poles of the transfer function. This model of the vocal tract is called the all-pole model or autoregressive (AR) model. Here, we assume that there are zeros only at $Z=0$, i.e., origin. If we multiply the factors of the denominator, we can write a denominator polynomial as given in Equation 2.15.

$$H(Z) = \frac{Z^M}{Z^M + a_1 Z^{M-1} + a_2 Z^{M-2} + \ldots + a_{M-1} Z + a_M} \tag{5.17}$$

where a_1, a_2, etc., are the coefficients of the polynomial called LPC (to be discussed in the next section). If we factorize the denominator polynomial again, we will get the poles called formant frequency locations. This defines the relation between the formants and LPCs.

5.5.1 Practical Considerations

Stability of the vocal tract transfer function is decided by the locations of poles. If all poles are within the unit circle in Z domain, the transfer function is stable. When we evaluate the formants, we must take care to see that the formants are within unit circle in Z domain. If any of the poles is outside the unit circle, then one must invert the pole about the unit circle by replacing P1 by 1/P5. This gives us a modified pole location. Thus, the poles are required to be shifted in order to see that the transfer function is stable. This point must be taken care of when one is using a formant synthesizer or LPC synthesizer, which will be discussed in Chapter 7. When LPCs are calculated, one must find the roots of the polynomial and calculate the location of formants. If the formants are outside the unit circle in Z domain, then the poles must be inverted with respect to the unit circle. This will ensure the stability of the vocal tract transfer function.

Concept Check

- What are LPCs?
- What are formants?
- How will you relate LPCs with formants?
- How will you find out whether a transfer function of the vocal tract is stable?
- What corrective measure is required to make the transfer function stable?

5.6 Evaluation of Formants

We will discuss the evaluation of formant frequencies in this section using the log spectrum method and the cepstrum method.

5.6.1 Evaluation of Formants Using Cepstrum

Consider a block schematic for evaluation of formants using the cepstrum, as shown in Figure 5.27. The block schematic is the same as for evaluation of pitch except that the cepstral window is different. Amplitude of the formant frequency, position of the formant, namely, the formant frequency, and band width of the formant are the parameters extracted from formant information. There parameters are useful for speech recognition, speaker verification, etc.

5.6.1.1 Evaluation of Formants for Voiced Speech Segment

Let us analyze function of each block. The speech signal given as input to a system consists of the periodic excitation convolved with the impulse response of a vocal tract, which is a slowly varying function. FFT block takes DFT of a signal, and spectrum of the signal will be obtained. When we take the log| |, we see that the periodic excitation is a rapidly varying and vocal tract response, which is a slowly varying envelope function. When we take IFFT, we find that a slowly varying function of vocal tract clusters near the origin and a rapidly varying function is seen as regular pulses away from the origin. We can now design a cepstral window, allowing the formant information

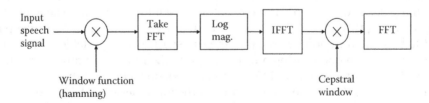

FIGURE 5.27
Block schematic for formant calculation.

(slowly varying function) to pass through. The FFT output of this windowed cepstrum is a spectrum with only a slowly varying function. If we track the peaks of this spectrum, we can find the formant frequencies. The rapidly varying function of vocal cords is now isolated, and hence, the possibility of the first formant overlapping with the pitch frequency is removed.

Let us do it practically by considering a voiced speech signal and unvoiced speech signal. We will write a MATLAB program to execute the block schematic on 2048 samples of speech and observe the results. The speech signal given at the input is shown in Figure 5.28. We have used the Hamming window. The plot of window function is shown in Figure 5.28. The windowed signal is shown in Figure 5.29. The output of FFT block is a spectrum as shown in Figure 5.29. The log magnitude| block operates on the spectrum and generates the amplitude in dB. Figure 5.30 depicts the log spectrum. Now we take the IFFT again. We have entered the cepstrum domain and the output obtained is shown in Figure 5.30. After passing the cepstrum via a window that retains only 40 samples of the cluster near the origin, as shown in Figure 5.31, we retained the first 40 and last 40 symmetrical points. The FFT of the windowed cepstrum is depicted in Figure 5.31. The MATLAB program is as follows.

```
%to find formants from the voiced signal using cepstrum of speech signal
clear all;
fp=fopen('watermark.wav');
fseek(fp,224000,-1);
a=fread(tp,1024);
a=a-128;
subplot(2,1,1);plot(a);title('plot of voiced part of a signal');
```

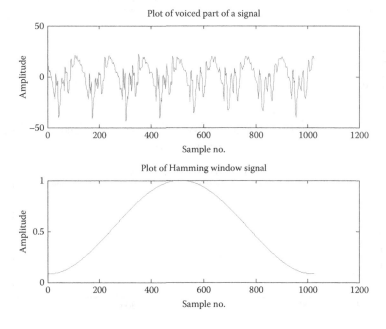

FIGURE 5.28
Plot of voiced speech segment and Hamming window.

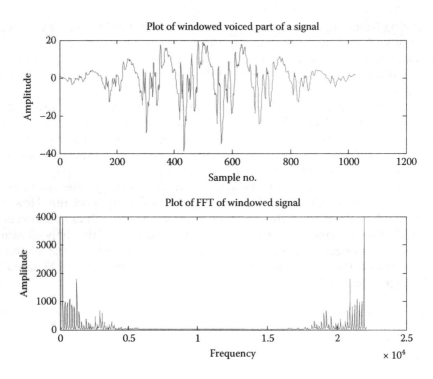

FIGURE 5.29
Plot of windowed voiced speech segment and FFT.

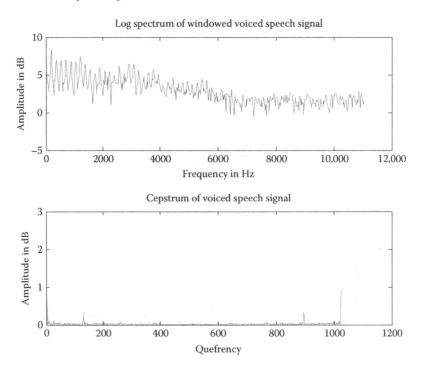

FIGURE 5.30
Plot of log spectrum of voiced speech segment and cepstrum.

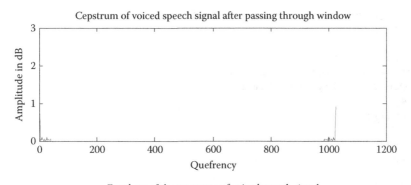

Cepstrum of voiced speech signal after passing through window

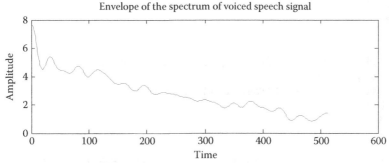

Envelope of the spectrum of voiced speech signal

FIGURE 5.31
Plot of windowed cepstrum and FFT of cepstrum.

```
xlabel('sample no.');ylabel('amplitude');
%2048 samples of the voiced signal are displayed

a1=window(@Hamming,1024);
subplot(2,1,2);plot(a1);title('plot of Hamming window signal');
xlabel('sample no.');ylabel('amplitude');
for i=1:1024,
    a11(i)=a(i)*a1(i);
end
figure;
subplot(2,1,1);
plot(a11);title('plot of windowed voiced part of a signal');
xlabel('sample no.');ylabel('amplitude');
%the signal is passed via Hamming window and displayed.
b=abs(fft(a11));
f = 22100/1024:22100/1024:22100;
subplot(2,1,2);plot(f,b);title('plot of fft of windowed signal');
xlabel('frequency');ylabel('amplitude');
c=log(b);
figure;
for i=1:512,
    d(i)=c(i);
end
f = 22100/1024:22100/1024:11050;
subplot(2,1,1);plot(f,d);
title('log spectrum of windowed voiced speech signal');
```

```
xlabel('Frequency in Hz');ylabel('amplitude in dB');
e=abs(ifft(c));

subplot(2,1,2);plot(e);
title('cepstrum of voiced speech signal');
xlabel('Quefrency');ylabel('amplitude in dB');
for i=1:40,
    h(i)=e(i);
end
for i=41:983,
    h(i)=0;
end
for i=984:1024,
    h(i)=e(i);
end
figure;
subplot(2,1,1);plot(h); title('cepstrum of voiced speech signal after
passing through window');
xlabel('Quefrency');ylabel('amplitude in dB');
w=abs(fft(h));
for i=1:512,
w1(i)=w(i);
end
subplot(2,1,2);plot(w1); title('envelope of the spectrum of voiced speech
signal');
xlabel('time');ylabel('amplitude');
```

5.6.1.1.1 Interpretation of the Result

The input signal to a formant calculation module is a voiced speech segment. We have taken 1024 samples of the signal. The signal plot clearly shows that the signal has a repetitive nature. This signal is generated as a result of convolution of excitation generated from vocal cords and impulse response of vocal tract. The signal is passed via a Hamming window to avoid the end effects. If there is an abrupt truncation, it will introduce high frequency components in the signal. Windowing a signal avoids this end effects.

The Hamming window is selected for the reason stated in the previous section. The windowed signal is passed via FFT block. It has first peak corresponding to the fundamental frequency. We are taking the log of the spectrum in order to represent the amplitude of the spectral lines in dB. The log of the FFT block shows rapid variations superimposed on the slowly varying envelope. The rapid variations correspond to the fundamental frequency and slowly varying envelope corresponds to the vocal tract parameters. The IFFT of the log spectrum is a cepstrum. The plot of cepstrum shows peaks at regular intervals, indicating the voiced nature of the signal. When we take IFFT of a log spectrum considering it as a signal, its slowly varying components will appear as a cluster near the origin. The vocal tract information closer to the origin is retained by a window of length 40 from both sides as the cepstrum is also symmetric. The FFT of this cepstral windowed signal will return the signal back in the log spectrum domain. The final output plot shows a slowly varying envelope corresponding to the formants. We can now track the peaks. The first four peaks corresponding to the first four formants can be tracked by observation, and we note that they are near 800, 1900, 2400, and 4000 Hz.

5.6.1.2 Evaluation of Formants for Unvoiced Segment

We will execute the process mentioned in the previous section on an unvoiced segment now. The listing of the program is identical except that we have tracked the unvoiced part of the utterance. Figure 5.32 shows a plot of unvoiced segment of speech and a plot of Hamming window function. The plot of windowed unvoiced utterance and its FFT are depicted in Figure 5.33. Plot of the log spectrum of unvoiced segment is calculated. Taking IFFT of the log spectrum transforms it in the cepstral domain. Figure 5.34 shows plots of the log spectrum and its cepstrum. Figure 5.35 shows the plot of the windowed cepstrum and the corresponding FFT.

```
%to find formants from the unvoiced signal using cepstrum of speech
signal
clear all;
fp=fopen('watermark.wav');
fseek(fp,23044,-1);
a=fread(fp,1024);
a=a-128;
subplot(2,1,1);plot(a);title('plot of unvoiced part of a signal');
xlabel('sample no.');ylabel('amplitude');
%2048 samples of the voiced signal are displayed

a1=window(@Hamming,1024);
subplot(2,1,2);plot(a1);title('plot of Hamming window signal');
```

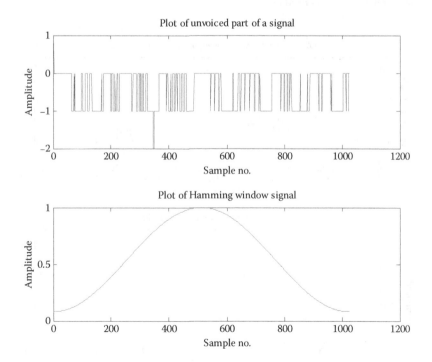

FIGURE 5.32
Plot of unvoiced segment and Hamming window.

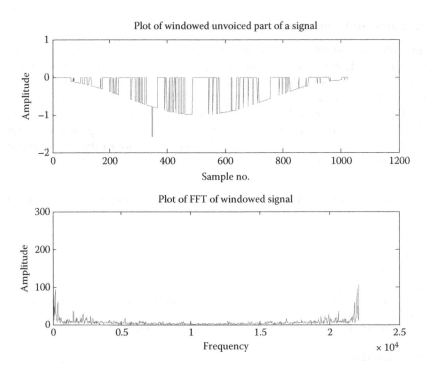

FIGURE 5.33
Plot of windowed unvoiced segment and its FFT.

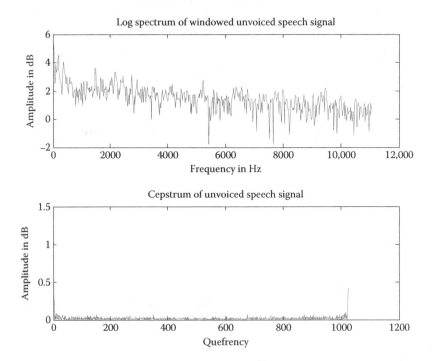

FIGURE 5.34
Plot of log spectrum of windowed unvoiced segment and cepstrum.

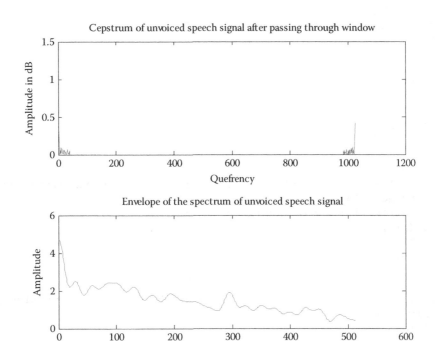

FIGURE 5.35
Plot of windowed cepstrum and its FFT.

```
xlabel('sample no.');ylabel('amplitude');
for i=1:1024,
    a11(i)=a(i)*a1(i);
end
figure;
subplot(2,1,1);
plot(a11);title('plot of windowed unvoiced part of a signal');
xlabel('sample no.');ylabel('amplitude');
%the signal is passed via Hamming window and displayed.
b=abs(fft(a11));
f = 22100/1024:22100/1024:22100;
subplot(2,1,2);plot(f,b);title('plot of fft of windowed signal');
xlabel('frequency');ylabel('amplitude');
c=log(b);
figure;
for i=1:512,
    d(i)=c(i);
end
f = 22100/1024:22100/1024:11050;
subplot(2,1,1);plot(f,d);
title('log spectrum of windowed unvoiced speech signal');
xlabel('Frequency in Hz');ylabel('amplitude in dB');
e=abs(ifft(c));

subplot(2,1,2);plot(e);
title('cepstrum of unvoiced speech signal');
```

```
xlabel('Quefrency');ylabel('amplitude in dB');
for i=1:40,
    h(i)=e(i);
end
for i=41:983,
    h(i)=0;
end
for i=984:1024,
    h(i)=e(i);
end
figure;
subplot(2,1,1);plot(h); title('cepstrum of unvoiced speech signal after
passing through window');
xlabel('Quefrency');ylabel('amplitude in dB');
w=abs(fft(h));
for i=1:512,
    w1(i)=w(i);
end

subplot(2,1,2);plot(w1); title('envelope of the spectrum of unvoiced
speech signal');
xlabel('time');ylabel('amplitude');
```

5.6.1.2.1 Interpretation of the Result

The input signal to a formant calculation module is an unvoiced speech segment. We have taken 1024 samples of the signal. The signal plot clearly shows that the signal is random. This signal is generated as a result of convolution of excitation generated from turbulence and impulse response of vocal tract. The signal is passed via a Hamming window in order to avoid the end effects.

The Hamming window is selected for the reason stated in the previous section. The windowed signal is passed via the FFT block. The FFT output shows no repetitions. We are taking the log of the spectrum in order to represent the amplitude of the spectral lines in dB. The log of the FFT block shows rapid, but random, variations superimposed on the slowly varying envelope. The rapid random variations correspond to the turbulence as excitation and slowly varying envelope correspond to the vocal tract parameters. The IFFT of the log spectrum is a cepstrum. The plot of cepstrum shows no peaks, indicating the absence of voicing information. When we take IFFT of the log spectrum considering it as a signal, its slowly varying components will appear as a cluster near the origin. The vocal tract information closer to the origin is retained by a window of length 40 from both sides as the cepstrum is also symmetric. FFT of this cepstral windowed signal will return the signal back in the log spectrum domain. The final output plot shows a slowly varying envelope corresponding to the formants. We can now track the peaks. The first four peaks corresponding to the first four formants can be tracked by observation, and we see that they are near 800, 1200, 1900, and 2400 Hz.

5.6.2 Evaluation of Formants Using the Log Spectrum

Formants can also be evaluated from the log spectrum. The block schematic for evaluation of the log spectrum is as shown in Figure 5.36.

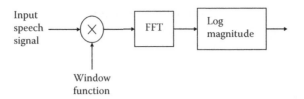

FIGURE 5.36
Block schematic for formant calculation using log spectrum.

5.6.2.1 Evaluation of Formants for Voiced Speech Segment

Let us see the function of each block. The speech signal given as input to a system consists of the periodic excitation convolved with the impulse response of a vocal tract that is a slowly varying function of time. When we take the log magnitude|, the periodic excitation is rapidly varying and the vocal tract response is the envelope and is a slowly varying function. The formant information (seen as a slowly varying function) lies in the envelope of the log spectrum.

Let us do it practically by considering a voiced speech signal and an unvoiced speech signal. We will write a MATLAB program to execute the block schematic on 1024 samples of speech and observe the results. The speech signal is given at the input in Figure 5.37. A Hamming window is used. The windowed signal is shown in Figure 5.37. The output

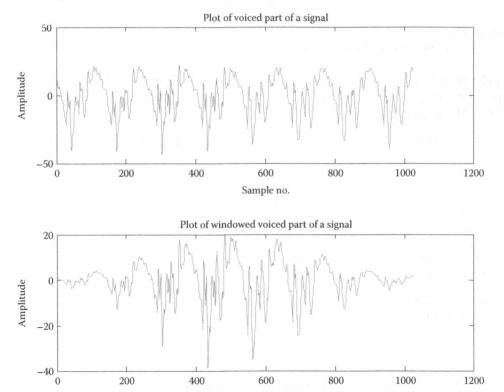

FIGURE 5.37
Plot of voiced speech and windowed signal.

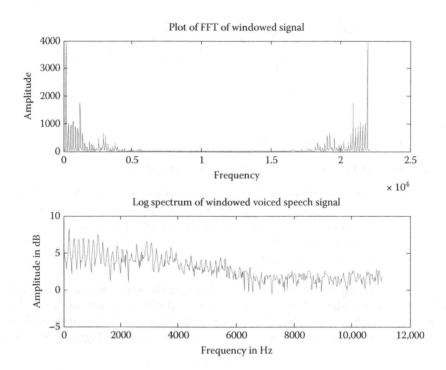

FIGURE 5.38
Plot of FFT of windowed signal and log spectrum.

of FFT block is a spectrum as shown in Figure 5.38. The log magnitude block operates on spectrum and generates the amplitude in dB. Figure 5.38 depicts the log spectrum. The MATLAB program is as follows.

```
%to find formants from the voiced signal using log spectrum of speech
signal
clear all;
fp=fopen('watermark.wav');
fseek(fp,224000,-1);
a=fread(fp,1024);
a=a-128;
subplot(2,1,1);plot(a);title('plot of voiced part of a signal');
xlabel('sample no.');ylabel('amplitude');
%2048 samples of the voiced signal are displayed

a1=window(@Hamming,1024);
for i=1:1024,
    a11(i)=a(i)*a1(i);
end

subplot(2,1,2);
plot(a11);title('plot of windowed voiced part of a signal');
xlabel('sample no.');ylabel('amplitude');
figure;
%the signal is passed via Hamming window and displayed.
b=abs(fft(a11));
```

```
f = 22100/1024:22100/1024:22100;
subplot(2,1,1);plot(f,b);title('plot of fft of windowed signal');
xlabel('frequency');ylabel('amplitude');
c=log(b);

for i=1:512,
    d(i)=c(i);
end
f = 22100/1024:22100/1024:11050;
subplot(2,1,2);plot(f,d);
title('log spectrum of windowed voiced speech signal');
xlabel('Frequency in Hz');ylabel('amplitude in dB');
e=abs(ifft(c));
```

5.6.2.1.1 Interpretation of the Result

The input signal is a voiced speech segment. We have taken 1024 samples of the signal. The signal plot clearly shows that the signal has a repetitive nature. This signal is generated as a result of convolution of excitation generated from vocal cords and impulse response of the vocal tract. The signal is passed via a Hamming window in order to avoid the end effects.

The windowed signal is passed via the FFT block. It has a first peak corresponding to the fundamental frequency. After taking the log of the spectrum, the amplitude of the spectral lines is represented in dB. The log of the FFT block shows rapid variations and these are superimposed on the slowly varying envelope. The rapid variations correspond to the fundamental frequency and slowly varying envelope corresponds to the vocal tract parameters. The final output plot of the log spectrum indicates a slowly varying envelope corresponding to the formants. We can now track the peaks. The first four peaks corresponding to the first four formants can be tracked by observation and we see that they are near 800, 1200, 1900, 3000, and 3800 Hz. The very first peak at about 200 Hz is corresponding to the fundamental frequency.

5.6.2.2 Evaluation of Formants for Unvoiced Segment

We will execute the aforementioned process on an unvoiced segment too. The listing of the program is identical except that we have tracked the unvoiced part of the utterance. Figure 5.39 shows a plot of unvoiced segment and a plot of windowed signal. The plots of the FFT of unvoiced segment and the log spectrum are depicted in Figure 5.40. The program listing for a MATLAB program is given below.

```
%to find formants from the unvoiced signal using clog spectrum of speech
signal
clear all;
fp=fopen('watermark.wav');
fseek(fp,23044,-1);
a=fread(fp,2048);
a=a-128;
subplot(2,1,1);plot(a);title('plot of unvoiced part of a signal');
xlabel('sample no.');ylabel('amplitude');
%2048 samples of the voiced signal are displayed

a1=window(@Hamming,1024);
for i=1:1024,
    a11(i)=a(i)*a1(i);
```

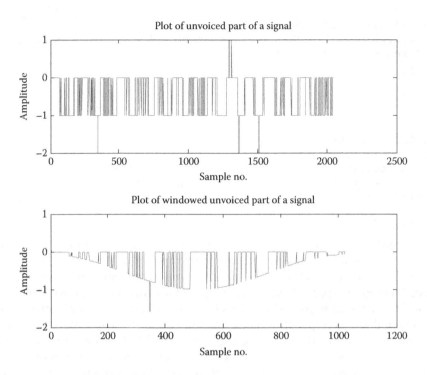

FIGURE 5.39
Plot of unvoiced segment and windowed segment.

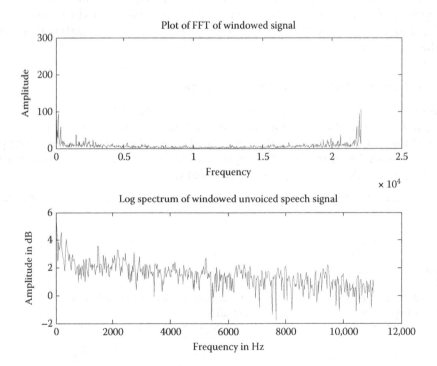

FIGURE 5.40
Plot of FFT of windowed speech and log spectrum.

```
end

subplot(2,1,2);
plot(all);title('plot of windowed unvoiced part of a signal');
xlabel('sample no.');ylabel('amplitude');
figure;
%the signal is passed via Hamming window and displayed.
b=abs(fft(all));
f = 22100/1024:22100/1024:22100;
subplot(2,1,1);plot(f,b);title('plot of fft of windowed signal');
xlabel('frequency');ylabel('amplitude');
c=log(b);

for i=1:512,
    d(i)=c(i);
end
f = 22100/1024:22100/1024:11050;
subplot(2,1,2);plot(f,d);
title('log spectrum of windowed unvoiced speech signal');
xlabel('Frequency in Hz');ylabel('amplitude in dB');
e=abs(ifft(c));
```

5.6.2.2.1 *Interpretation of the Result*

The input signal to a formant calculation module is the unvoiced speech segment. We have taken 1024 samples of the signal. The signal plot clearly shows that the signal is random. This signal is generated as a result of convolution of excitation generated from turbulence and impulse response of the vocal tract. The signal is passed via a Hamming window in order to avoid the end effects.

The Hamming window is selected. The windowed signal is passed via the FFT block. We take the log of the spectrum to represent the amplitude of the spectral lines in dB. The log of FFT block shows rapid, but random, variations that are superimposed on the slowly varying envelope. The rapid random variations correspond to the turbulence used as excitation and slowly varying envelope corresponds to the vocal tract response. We can now track the peaks. The first four peaks corresponding to the first four formants can be tracked by observation and we see that they are near 800, 1200, 2200, and 2400 Hz.

Concept Check

- What is the method to enter a cepstral domain?
- If we take the FFT of the windowed cepstral, which retains the cluster near the origin, how will the peaks in the FFT output be interpreted?
- What is your observation about the cepstrum of voiced and unvoiced speech?
- What is the method to evaluate the log spectrum?
- How will the log spectrum graph be interpreted?
- Can you write a program to find the peaks in the log spectrum?

- Can you identify the difference between the log spectrum for voiced speech and log spectrum for the unvoiced speech?

5.7 Evaluation of MFCC

This section will describe the spectral analysis of speech signals. For the analysis of a speech signal, which is a convolved signal, linear filters cannot be used. We will see how homomorphic processing can be used to convert convolved components into additive components. We will then discuss the mel scale and bark scale for frequency and discuss the evaluation of MFCC.

5.7.1 Homomorphic Processing

Let us first consider the definition of homomorphic processing. The term homomorphic processing is applied to a class of systems that obey a generalized principle of superposition, which can be stated as follows. If $x_1(n)$ and $x_2(n)$ are inputs to a homomorphic system, $y_1(n)$ and $y_2(n)$ are the respective outputs, and c is a scaling factor, then if

$$y_1(n) = T(x_1(n))$$

$$y_2(n) = T(x_2(n))$$

then

$$T[x_1(n) + x_2(n)] = T[x_1(n)] + T[x_2(n)] \tag{5.18}$$

and

$$T[c \times x_1(n)] = cT[x_1(n)]$$

The importance of this type of system lies in the fact that the operator T of homomorphic processing can be decomposed into a cascade of operations, as shown in Figure 5.41.

The systems A0 and A0^{-1} shown in Figure 5.41 are the inverse systems, also called canonical systems. The system L is the LTI system.

Let us develop a canonical system for speech signal. Speech is a convolved signal. We need to develop a canonical system for speech. To convert convolution operation into addition, convert convolution operator into multiplication first using any transform-like FT and then convert the multiplication operator into addition operator using the log operator. Hence, the canonical system A0 for speech signal will consist of FT followed by the log. The

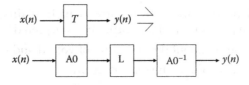

FIGURE 5.41
General diagram for homomorphic processing system.

inverse canonical system will then be exp operation followed by IFFT. The image signal is a multiplicative signal resulting from multiplication of illumination signal and reflectance signal. The homomorphic system for multiplication consists of just the log block that will convert the multiplicative signals into additive components and we can use LTI filter in the log domain. The inverse system will be the exp block.

We have seen that voiced or unvoiced speech is produced when either pulse train generator output or random generator output convolve with the impulse response of vocal tract. If one wants to separate the response of vocal cords from the response of vocal tract, one cannot use any linear filter as both these responses are in the convolved state. The theory of linear filter is well developed and studied. We can make use of these linear filters if we can convert the convolved components in the additive form. This type of system that will convert convolved components into additive ones is called the homomorphic processing system for convolution. The block schematic for the homomorphic system for convolution can be drawn as shown in Figure 5.42.

Consider the function of each block. Let us consider the input to the system is a convolved speech signal. Let it be a voiced speech signal. The voiced speech signal is generated as a result of convolution of pulse train generator and impulse response of vocal tract. Pulse train generator represents the natural frequency of vibrations of vocal cords. The aim of the experiment is to separate the impulse response of the vocal tract from the frequency of vocal cords. The convolved signal passes via the FFT block. The output of the FFT block is the multiplication of the FFT of pulse train generator and the FFT of the impulse response of the vocal tract. This is because the convolution in the time domain is the multiplication in transform domain according to the convolution property of DFT.

The next block is the log block. The output of the log block will be the addition of the log of FFT of pulse train generator and the log of the FFT of the impulse response of the vocal tract. This is because the log of two multiplicative components is the addition of their logs as per the log rule.

Now we have the two components in the additive form. This will allow us to use the LTI filters. If we use a LPF allowing slowly varying envelope of vocal tract response to pass through, the vocal cord response, i.e., pulse train generator, will be eliminated. The processed output is then passed via exp and the IFFT block to undo operations and return the signal in the time domain. The resulting output in the time domain will be the impulse response of the vocal tract. If we use a high-pass filter in place of a low-pass one, we will eliminate the vocal tract response and get the pulse train generated by vocal cords.

We will do it practically by taking a voiced speech segment of size 1024 samples. A MATLAB program for the execution of homomorphic processing is given below. Figure 5.43 shows the plot of a signal and its log spectrum. This log spectrum contains the components in the additive form. A LPF is designed to pass the envelope of the log spectrum considering it as a signal. The filtered log spectrum shows only the envelope as depicted in the output plot in Figure 5.44. The exp block and the IFFT block will undo

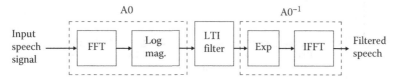

FIGURE 5.42
Homomorphic system for convolution: Generalized block diagram.

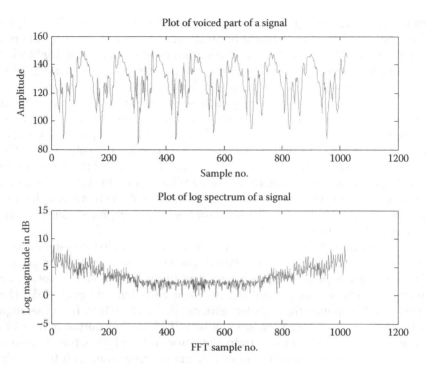

FIGURE 5.43
Plot of voiced speech segment and its log magnitude.

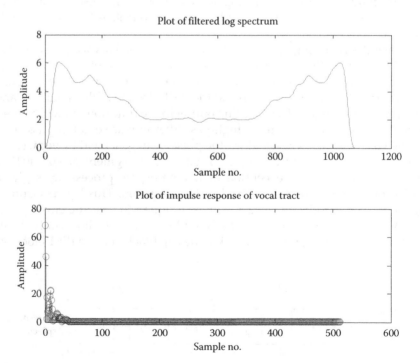

FIGURE 5.44
Plot of the filtered log spectrum and impulse response of vocal tract.

the operation and the final output is the impulse response of the vocal tract as shown in Figure 5.44.

```
%Homomorphic processing for voiced speech to find impulse response of
vocal
%tract
clear all;
fp=fopen('watermark.wav');
fseek(fp,224000,-1);
a=fread(fp,1024);
subplot(2,1,1);plot(a);title('plot of voiced part of a signal');
xlabel('sample no.');ylabel('amplitude');
b=fft(a);
b1=abs(b);
c=log(b1);
subplot(2,1,2);plot(c);title('plot of log spectrum of a signal');
xlabel('fft sample no.');ylabel('log magnitude in dB');
w=fir1(50,0.01,'low');
x=conv(c,w);
%a=[1 0.2 0.1 0.01];b=[1]
%d=filter(a,b,c);
figure;
subplot(2,1,1);plot(x);title('plot of filtered log spectrum');
xlabel('sample no.');ylabel('amplitude');
y=exp(x);
z=ifft(y);
z1=abs(z);
for i=1:512,
    z11(i)=z1(i);
end
 subplot(2,1,2);stem(z11);title('plot of impulse response of vocal
tract');
xlabel('sample no.');ylabel('amplitude');
```

5.7.2 The Auditory System as a Filter Bank

Majority of the systems have converged to a feature vector as cepstral vector derived from a filter bank designed per some model of the auditory system. Let us first study different auditory models. Harvey Fletcher experimented extensively on human hearing. He did some work on the masking phenomenon.

He performed the following experiment. The listener was presented with a tone with wide band noise. Initially, the intensity of the tone was kept low and then it was slowly increased until the tone was perceived by the listener. The intensity after which the tone was perceived was noted as the threshold intensity. Then the bandwidth of noise was slowly deceased and it was observed that there was no change in the threshold until bandwidth equal to a critical band was reached. The threshold intensity was decreased when bandwidth of noise was deceased below the critical band. These experiments indicated that there exists an auditory filter in the vicinity of the tone that effectively blocks the noise and detects the tone. The experiments also indicated that the width of the critical band increases if the frequency of the tone is increased. This gives clear indication that the critical band filters must be designed; however, the shape of the filer is not clear from the experimentation.

There are two commonly used scales, namely, mel scale and bark scale.

The mel scale is based on the pitch perception and uses triangular filters as shown in Figure 5.45. The scale is approximately linear below 1000 Hz and nonlinear and logarithmic after 1000 Hz. The filter bank will mimic the critical band, which will represent different perceptual effect at different frequency bands. Additionally, the edges are placed so that they coincide with the center frequencies in adjacent filters.

A common model for the relation between frequencies in mel and linear scales given in Equation 5.19. Figure 5.46 shows frequency warping using mel scale.

$$\text{Mel Frequency} = 2595 \times \log(1 + \text{linear frequency}/700) \tag{5.19}$$

A MATLAB program to convert frequency to mel scale is as follows.

```
%program to convert frequency in Hz to Mel Scale
clear all;
for i=1:10000,
    m(i)=2595*log(1+i/700);
end
```

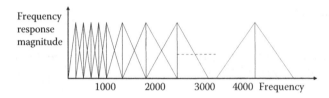

FIGURE 5.45
Mel scale filter bank.

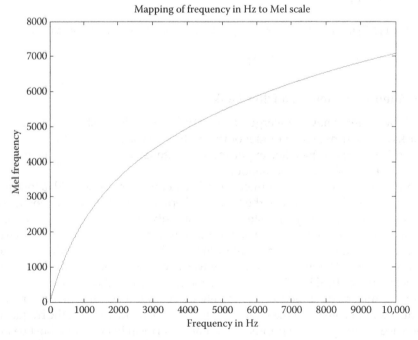

FIGURE 5.46
Frequency warping using mel scale.

```
plot(m);title('mapping of frequency in Hz to Mel scale');
xlabel('frequency in Hz');ylabel('Mel frequency');
```

The *bark scale* is a psychological scale proposed by Eberhard Zwicker in 1961. It is named after Heinrich Barkhausen for proposing the first subjective measurements of loudness. The scale ranges from 1 to 24 and corresponds to the first 24 critical bands of hearing. The subsequent band edges are (in Hz) 20, 100, 200, 300, 400, 510, 630, 770, 920, 1080, 1270, 1480, 1720, 2000, 2320, 2700, 3150, 3700, 4400, 5300, 6400, 7700, 9500, 12,000, 15,500, as shown in Table 3.1. The bark scale is related to, but somewhat less popular than, the mel scale. Figure 5.47 shows a plot of the bark scale. It uses trapezoid-shaped filters.

The bark scale is defined as a table. The cursive values are the center frequencies of the critical bands. Table 5.5 specifies the bark scale.

To convert a frequency f (Hz) into bark, we can use the formula stated in Equation 5.20.

$$\Omega(\omega) = 6\ln\left(\frac{\omega}{1200\pi} + \left[\left(\frac{\omega}{1200\pi}\right)^2 + 1\right]^{0.5}\right) \tag{5.20}$$

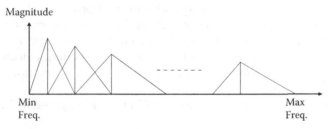

FIGURE 5.47
Bark scale plot for the filter bank.

TABLE 5.5

Bark Scale—Bark Against Frequency in Hz

Bark	Frequency	Bark	Frequency	Bark	Frequency
	50		1000		3400
1	100	9	1080	17	3700
	150		1170		4000
2	200	10	1270	18	4400
	250		1370		4800
3	300	11	1480	19	5300
	350		1600		5800
4	400	12	1720	20	6400
	450		1850		7000
5	510	13	2000	21	7700
	570		2150		8500
6	630	14	2320	22	9500
	700		2500		10,500
7	770	15	2700	23	12,000
	840		2900		13,500
8	920	16	3150	24	15,500

As can be observed from Figure 5.47, the distance of center to the left edge is different from the distance of center to the right edge.

A MATLAB program to convert frequency in bark scale is as follows.

```
Clear all;
for i=1:10000,
    f(i)=frq2bark(i);
end
plot(f);title('plot of bark frequency Vs frequency');
xlabel('frequency in Hz');ylabel('frequency in bark scale');
```

A plot of frequency in the bark scale to frequency in Hz is depicted in Figure 5.48.

The three constants define the filter bank, namely, the number of filters, the minimum frequency, and the maximum frequency. The minimum and maximum frequencies determine the frequency range that is spanned by the filter bank. These frequencies depend on the sampling frequency. In the case of the telephone channel, the band is between 300 and 3700 Hz. These will define minimum and maximum frequencies for the filter bank to be used. For clean speech, the minimum frequency should be higher than about 100 Hz because there is no speech information below 100 Hz. It can be noted that by setting the minimum frequency above 50/60 Hz, we can get rid of the hum resulting from the alternating current (ac) power. The maximum frequency has to be lower than the Nyquist frequency. To eliminate high frequency noise, maximum frequency of 6000 Hz is considered as appropriate, as no information is present above 6000 Hz in speech signal.

Typical values for the constants defining the filter bank are given in Table 5.6.

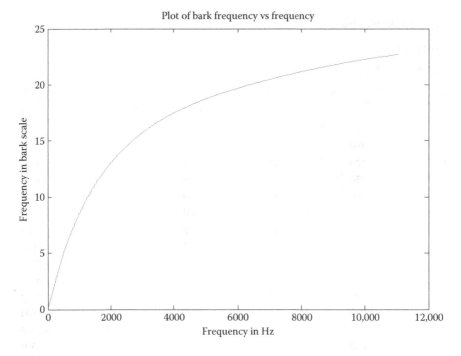

FIGURE 5.48
Conversion of frequency to bark scale frequency.

TABLE 5.6

Three Constants Defining a Filter Bank for Different Sampling Frequencies

Sample rate (Hz)	16,000	11,025	8000
No. of filters	40	36	31
Minimum frequency in Hz	130	130	200
Maximum frequency in Hz	6800	5400	3500

5.7.3 Mel Frequency Cepstral Coefficients

The mel-frequency cepstrum (MFC) can be defined as the short time power spectrum of speech signal, which is calculated as a linear cosine transform of a log power spectrum on a nonlinear mel scale of frequency. MFCC are the coefficients evaluated in the MFC representation. Let us see the difference between the cepstrum and the mel cepstrum. In the case of MFC, the frequency bands are essentially equally spaced on the mel scale. This mel scale approximates the human auditory system's response more closely than the linearly spaced frequency bands. The mel scale warps the frequency and allows better representation that is similar to human auditory system.

MFCCs can be calculated as follows:

- Take the FFT of a windowed signal. Compute its squared magnitude. This gives a power spectrum.

- Preemphasize the spectrum to approximate the unequal sensitivity of human hearing at different frequencies.

- Next, integrate the power spectrum within overlapping critical band filter response. The integration is to be done with a triangular overlapping windows (mel filters). This reduces the frequency sensitivity over the original spectral estimate especially at higher frequencies. The higher frequencies are emphasized because of the wider band.

- Compress the spectral amplitudes by taking logarithm. Optionally, the integration of log power spectrum can be done.

- Take the inverse discrete Fourier transform (IDFT). This gives cepstral coefficients.

- Perform spectral smoothing by truncating the cepstrum. Typically, lower 12 or 14 coefficients are used from 20 or more coefficients.

The block schematic for the computation of MFCC as per these steps can be drawn, as shown in Figure 5.49.

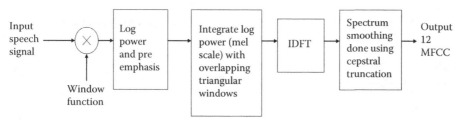

FIGURE 5.49
Block schematic for MFCC calculation.

The length of the output array is equal to the number of filters created. The triangular mel filters in the filter bank are placed in the frequency axis so that each filter's center frequency is as given by the mel scale.

Let us do it practically.

1. We will take a speech file and use 256 samples from the voiced part of the utterance for calculation of MFCC. Take the FFT and its square magnitude to find the power spectrum.

2. Implement the integration of the FFT point over triangular filters with 50% overlap as shown in Figure 5.45. We prefer to design the filters in mel scale as the unequally spaced filters on frequency axis in Hz will be equally spaced on mel scale. First, convert the frequency axis in Hz to frequency axis in mel scale using Equation 5.19. Plot the log spectrum for each fft bin in mel scale. Use a triangular filter of width equal to 300 mel. So the first filter will be from 300 to 600 mel. Next will be from 450 to 750 mel to have 50% overlap between the filters. Design 28 such filters. This spans the band up to 4500 mel. To integrate the fft bin values over these filters, find the number of fft bin falling in the filter width and multiply the magnitude of each with the weight corresponding to the height of the triangle at that position and add the result for all bins falling in the triangle width. One gets one integration value per filter, i.e., 28 values are obtained.

3. Take IDFT to get cepstral coefficients.

4. Perform the spectrum smoothing by truncating 14 values and retain 14 cepstral coefficients. These are the required MFCC values.

A MATLAB program to calculate MFCC for a speech segment of size 512 is given. The reader is encouraged to track the steps by going through the comments given in the program.

```
%program to convert frequency in Hz to Mel Scale
clear all;
fp=fopen('watermark.wav');
fseek(fp,224000,-1);
a=fread(fp,512);
%plot 512 points of voiced speech
subplot(2,1,1);plot(a);title('plot of voiced part of a signal');
xlabel('sample no.');ylabel('amplitude');
%find 512 point FFT
b=fft(a);
b1=(abs(b));
for i=1:512,
    b1(i)=b1(i)*b1(i); %calculation of squared power spectrum
end

c=log10(b1);
%calculate frequency in Hz for every FFT bin
for i=1:256,
    f(i)=22100/512*i;
end
for i=1:256,
```

```
    c1(i)=c(i);
end
 %calculate mel scale frequency for each frequency
 %in HZ corresponding to FFT bin
 for i=1:256,
    m(i)=2595*log(1+f(i)/700);

 end
 %plot log spectrum in mel scale foe each FFT bin
subplot(2,1,2);stem(m,c1);title('plot of log spectrum in Mel scale for
voiced speech');
xlabel('Frequency in Mel scale');ylabel('Amplitude in dB');

%divide mel scale in equally spaced triangular filter each of width equal
to 300 mel with 50% overlap. first filter is from 300 mel to 600 mel.
Next is from 450 mel to 750 mel and so on. integratpe log power spectrum
bins over the filter width
for j=1:28,
    sum(j)=0;
    for i=1:58,

        if ((m(i)>300+(j-1)*150) && (m(i)<600+(j-1)*150))
            if (m(i)<450+(j-1)*150)
                g(i)=((m(i)-(300+150*(j-1)))*1/150);
             else
                g(i)=((600+150*(j-1)-m(i))*1/150);
            end

        sum(j)=sum(j)+c1(i)*g(i);

        end

    end
end
%find ifft of the integrated output. This is a cepstral domain.
d=ifft(sum);
d=abs(d);

figure;

for i=1:14,
    x(i)=d(i);
end
 %plot first 14 MFCC coefficients by cepstral truncation.
stem(x);title('plot of MFCC for voiced speech');
xlabel('Frequency in Mel scale');ylabel('Amplitude in dB');
```

Figure 5.50 shows the plot of the signal and its log power spectrum for each fft bin on the mel scale. Figure 5.51 shows a plot of corresponding MFCC points.

We can calculate MFCC using melcepst command available in the voice box. The program to calculate MFCC is given below. Plot of voiced speech and its log spectrum in mel scale and MFCC values are shown in Figures 5.52 and 5.53, respectively.

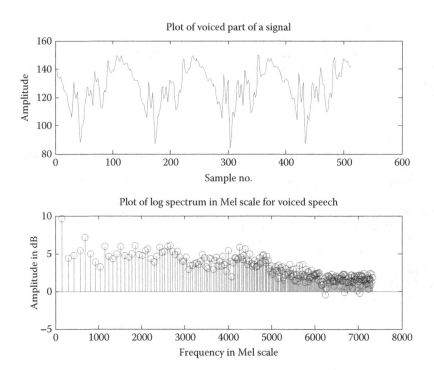

FIGURE 5.50
Plot of speech and its log power spectrum fft point on mel scale.

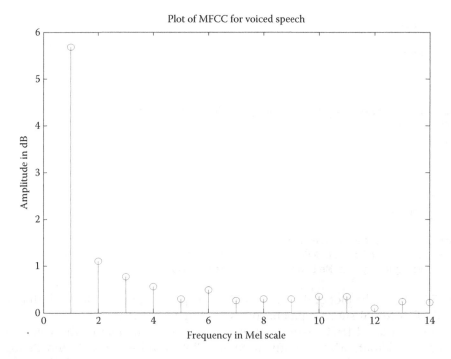

FIGURE 5.51
Plot of 14 MFCC after truncation.

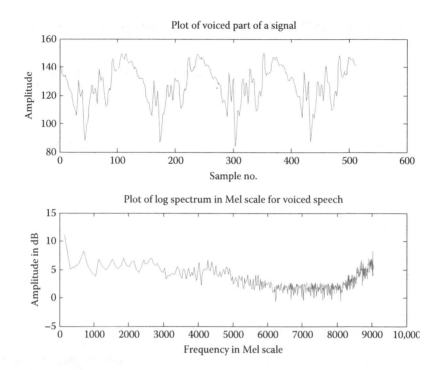

FIGURE 5.52
Plot of voiced speech and its log spectrum in mel scale.

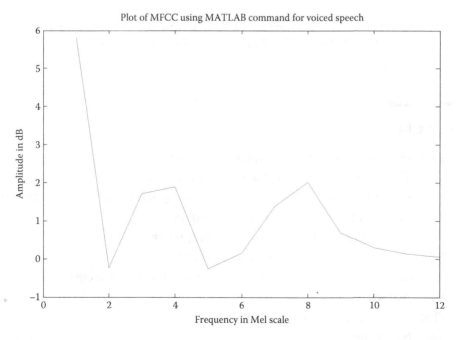

FIGURE 5.53
Plot of 12 MFCC obtained using melcepst.

Melcepst command uses DCT in place of IFFT to convert spectrum coefficients to cepstral coefficients.

```
%program to calculate MFCC for voiced speech signal using MFCC command
 clear all;
fp=fopen('watermark.wav');
fseek(fp,224000,-1);
a=fread(fp,256);
subplot(2,1,1);plot(a);title('plot of voiced part of a signal');
xlabel('sample no.');ylabel('amplitude');
b=fft(a);
b1=abs(b);
c=log(b1);
for i=1:256,
    f(i)=22100/256*i;
end
for i=1:256,
    c1(i)=c(i);
end

 for i=1:256,
     m(i)=2595*log(1+f(i)/700);
 end
subplot(2,1,2);plot(m,c1);title('plot of log spectrum in Mel scale for
voiced speech');
xlabel('Frequency in Mel scale');ylabel('Amplitude in dB');

x=melcepst(a);

figure;
plot(x);title('plot of MFCC using MATLAB command for voiced speech');
xlabel('Frequency in Mel scale');ylabel('Amplitude in dB');
```

Concept Check

- What is the meaning of homomorphic processing?
- How will you convert convolved signals in the additive form?
- Will you be in a position to convert multiplicative signals into additive components?
- What are the properties of homomorphic systems?
- Can you tell the application of homomorphic processing for speech?
- Define a masking threshold using experiments?
- How is the existence of the auditory filter around the tone detected?
- Define mel scale.
- What is the property of the mel scale filters?
- Define bark scale.

- What is the use of auditory filter banks?
- Define MFCC.
- What is the meaning of preemphasis filter?
- How will you reduce the frequency sensitivity of the spectrum?
- How will you convert spectrum coefficients to cepstral coefficients?
- Why are the coefficients called MFCC?
- How is a spectrum smoothing carried out?
- How does a melcepst command convert spectrum coefficients to cepstral coefficients?
- Can you convert unequal spaced filters to equally spaced filters?

5.8 Evaluation of LPC

This section describes LPCs and their evaluation. We will first understand linear prediction and will deal with the autocorrelation method for finding the LPCs using Levinson–Durbin algorithm.

5.8.1 Forward Linear Prediction

Let us consider forward and backward linear prediction for a random process in this section. Let us first concentrate on forward linear prediction. Basically, it is a problem of predicting a future value of the random process using the past values. We will focus on one-step prediction, which involves the prediction of the next sample from a weighted linear combination of past values $x(n-1)$, $x(n-2)$, etc. The linearly predicted value of $x(n)$ is given by

$$x(n) = -\sum_{k=1}^{P} a_p(k)x(n-k) \text{ for } k = 1 \text{ to } P \tag{5.21}$$

Here, the weights $a_p(k)$ are called the prediction coefficients for a one-step predictor of order P. The negative sign is for mathematical convenience. The forward linear predictor can be represented as shown in Figure 5.54. Here, we have shown only the first-order predictor.

Consider the Pth order predictor. The difference between $x(n)$ and $x(\hat{n})$ is called forward prediction error denoted as $f_p(n)$ and given by

$$f_p(n) = x(n) - x(\hat{n}) = x(n) + \sum_{k=1}^{P} a_p(k)x(n-k)$$

$$= \sum_{k=0}^{P} a_p(k)x(n-k) \tag{5.22}$$

Linear prediction can be viewed as linear filtering, where the prediction is embedded in the feedback loop as shown in Figure 5.54. This is termed as prediction error filter with input sequence $x(n)$ and output sequence as $f_p(n)$. Putting a prediction filter in the feedback loop will convert all pole prediction to finite impulse response (FIR) prediction error filter. FIR filter is a linear phase filter and is stable.

The system represented using Equation 5.22 can be seen as an FIR filter with the system transfer function given by

$$A_p(Z) = \sum_{k=0}^{P} a_p(k)Z^{-k} \tag{5.23}$$

By definition $a_p(0) = 1$.

Let us see how we can use this model for speech signals. Refer to an LTV model discussed in Chapter 1. This model is drawn with little modification as shown in Figure 5.55.

In the case of voiced speech, the signal is generated as a result of convolution of impulse response of the vocal tract with the impulse train. Impulse response coefficients or filter coefficients of vocal tract filter are replaced by the LPCs. Let us justify it. We will assume all-pole autoregressive (AR) model for a filter. This assumption is valid for all natural sounds except nasals and fricatives. If the order of the predictor is sufficiently large, then all-pole AR model is a good approximation to almost all sounds. The attractive part of the model is that the parameters of the model, i.e., gain and filter coefficients, can be evaluated with less computational cost using linear predictive analysis.

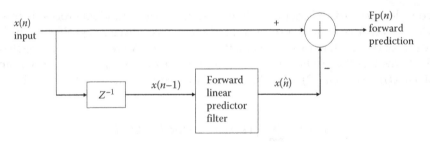

FIGURE 5.54
Forward linear prediction filter.

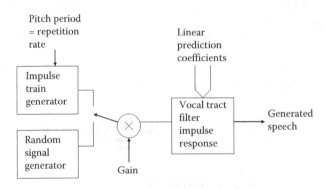

FIGURE 5.55
LTV Model for speech production in terms of LPC.

Speech signal has some random component. It has high autocorrelation. Hence, it is possible to predict the next sample from previous samples, even though exact prediction is not possible. The error obtained in the prediction is called prediction error. Consider a voiced speech segment. The speech signal sample can be expressed as

$$x(n) = -\sum_{k=1}^{P} \alpha_p(k)x(n-k) + GU(n) \tag{5.24}$$

where $U(n)$ is a unit sample sequence or impulse representing excitation for voiced signal. G is a gain represented in the gain contour for impulse excitation. Thus, $GU(n)$ is the input to a system and α_p is the prediction coefficient. Transferring Equation 5.23 in Z domain, we get

$$X(Z) = \frac{G}{1 + \sum_{k=1}^{P} \alpha_p(k)Z^{-k}} \tag{5.25}$$

We define a linear predictor to estimate a sample. The Z domain transfer function of the predictor is given by

$$P(Z) = -\sum_{k=1}^{P} \alpha_p(k)Z^{-k} \tag{5.26}$$

We will use this predictor in the feedback loop of Figure 4.4. We can define the prediction error as

$$e(n) = x(n) - x(\hat{n}) = x(n) + \sum_{k=1}^{P} \alpha_p(k)x(n-k) = \sum_{k=0}^{P} \alpha_p(k)x(n-k) \tag{5.27}$$

$$A(Z) = 1 + \sum_{k=1}^{P} \alpha_k Z^{-k} = \sum_{k=0}^{P} \alpha_k Z^{-k}$$

Let $A(Z)$ represent a transfer function of the forward error prediction system. $A(Z)$ represents the transfer function of the FIR filter. So, the approach is as follows. Calculate the set of linear prediction parameters that will minimize the mean squared error of the prediction over a short duration segment of speech. The resulting parameters calculated will then be assumed to be the filter coefficients.

There are three different approaches for finding LPC.

1. Autocorrelation method
2. Covariance method
3. Lattice structure method

Let us calculate the mean squared value of forward prediction error $f_p(n)$. Let us denote it as error. Error can be written as

$$\text{Error} = E\left[f_p(n)\right]^2 = E\left[x(n) + \sum_{k=1}^{P}\alpha_p(k)x(n-k)\right]^2$$

$$= E\left[x(n)x(n)\right] + 2E\left[\sum_{k=1}^{P}\alpha_p(n)x(n)x(n-k)\right] + E\left[\sum_{k=1}^{P}\alpha_p(k)x(n-k)\sum_{l=1}^{P}\alpha_p(l)x(n-l)\right]$$

$$= \gamma_{xx}(0) + 2Re\left[\sum_{k-1}^{P}\alpha_p(k)\gamma_{xx}(k)\right] + \sum_{k=1}^{P}\sum_{l=1}^{P}\alpha_p(k)\alpha_p^*(l)\gamma_{xx}(l-k) \qquad (5.28)$$

Here, γ_{xx} in Equation 5.27 stands for the autocorrelation function and E stands for the expected value. Let us find how to minimize error. The error function is differentiated with respect to each of the predictor coefficients and equated to zero. We will get P equations in P unknowns, and these are to be solved simultaneously. This will lead to a second set of updated predictor coefficients. The procedure is to be repeated for the number of iterations until the mean squared error is the least. This procedure is called adaptive procedure. Let us differentiate Equation 4.25 and equate it to zero.

$$\frac{d}{d\alpha_p(k)}\text{error} = \frac{d}{d\alpha_p(k)}\left[\gamma_{xx}(0) + 2Re\left[\sum_{k-1}^{P}\alpha_p(k)\gamma_{xx}(k)\right] + \sum_{k=1}^{P}\sum_{l=1}^{P}\alpha_p(k)\alpha_p^*(l)\gamma_{xx}(l-k)\right] = 0$$

$$0 + 2\gamma_{xx}(k) + 2\sum_{l=1}^{P}\alpha_p(l)\gamma_{xx}(l-k) = 0 \qquad (5.29)$$

$$\gamma_{xx}(k) = -\sum_{l=1}^{P}\alpha_p(l)\gamma_{xx}(l-k) \quad \text{for} \quad k = 1 \text{ to } P$$

The set of equations in Equation 5.29 represents *normal equations*. For the derivation of normal equations, a main assumption is that the signal of the model is stationary. In the case of speech signal, the model is valid only for a short segment of speech. Hence, the calculation of expected value is not feasible. We have to limit the summation over a finite number of samples for the calculation of expected value or γ_{xx}.

5.8.2 Autocorrelation Method

The autocorrelation approach of solving the normal equations assumes that the signal is zero outside interval $0 \le n \le N$. Hence, the prediction error will not be zero for predicted samples; similarly, the beginning of the frame will be affected by the same inaccuracy in case of a previous frame. The limits for calculation of autocorrelation can be specified as between 0 and $N - 1 - |l - k|$.

We can write the normal equations for autocorrelation method as

$$\gamma_{xx}(k) = -\sum_{l=1}^{P} \alpha_p(l)\gamma_{xx}(l-k) \quad \text{for} \quad k = 1 \text{ to } P$$

$$\sum_{l=1}^{P} \alpha_p(l)\gamma_{xx}(l-k) = E_P^f \text{ for } k = 0 \tag{5.30}$$

$$\sum_{n=0}^{N-1-|l-k|} x(n)x(n-k) = -\sum_{n=0}^{N-1-|l-k|}\sum_{l=1}^{P}\sum_{k=1}^{P} \alpha_p(k)x(n)x(n-|l-k|)$$

We can write Equation 4.28 in the matrix form as

$$\begin{bmatrix} \gamma_{xx}(1) \\ \gamma_{xx}(2) \\ -- \\ \gamma_{xx}(P) \end{bmatrix} = -\begin{bmatrix} \alpha_P(1) \\ -- \\ -- \\ \alpha_P(P) \end{bmatrix}\begin{bmatrix} \gamma_{xx}(0) & \gamma_{xx}(1) & -- & \gamma_{xx}(P-1) \\ \gamma_{xx}(1) & \gamma_{xx}(0) & -- & \gamma_{xx}(P-2) \\ -- & -- & -- & -- \\ \gamma_{xx}(P-1) & -- & -- & \gamma_{xx}(0) \end{bmatrix} \tag{5.31}$$

We can observe that the diagonal elements are $\gamma_{xx}(0)$. When the displacement is 0, naturally the correlation is 1. The off-diagonal elements represent the correlation with offset as 1, 2, ..., etc., ... P. We can verify that the matrix element in the second row and first column is the same as the element in the first row and second column; the matrix element in Pth row and first column is the same as the element in the first row and the Pth column, etc. The matrix is symmetric. If the matrix is symmetric and diagonal elements are equal, then it is called the Toeplitz matrix. By exploiting the property of the Toeplitz matrix, a recursive procedure can be formulated for the calculation of predictor coefficients. This recursive algorithm is a well-known Levinson–Durbin algorithm. *Levinson recursion* or *Levinson–Durbin recursion* are procedures in the linear algebra to recursively calculate the solution to any equation involving a Toeplitz matrix.

Let us find how the recursive algorithm is developed. Let us first consider $P=1$. The matrix equation for $P=1$ reduces to

$$\gamma_{xx}(0)\alpha_1(1) = -\gamma_{xx}(1)$$

$$\alpha_1(1) = -\frac{\gamma_{xx}(1)}{\gamma_{xx}(0)}$$

$$E_1^f = \gamma_{xx}(0) + \alpha_1(1)\gamma_{xx}(-1) \tag{5.32}$$

$$= \gamma_{xx}(0)[1-|\alpha_1(1)|^2]$$

$$= \gamma_{xx}(0)[1-K_1^2]$$

$$E_0^f = \gamma_{xx}(0) = 1$$

Referring to Section 4.1 on the first-order lattice structure for FIR filter, we can verify that for the first-order predictor

$$\alpha_1(1) = K_1 \tag{5.33}$$

The predictor coefficient for the first-order predictor is equal to the reflection coefficient of the first-order lattice.

Let $P = 2$. The matrix equation can be written as

$$\begin{bmatrix} \gamma_{xx}(0) & \gamma_{xx}(1) \\ \gamma_{xx}(1) & \gamma_{xx}(0) \end{bmatrix} \begin{bmatrix} \alpha_2(1) \\ \alpha_2(2) \end{bmatrix} = -\begin{bmatrix} \gamma_{xx}(1) \\ \gamma_{xx}(2) \end{bmatrix} \tag{5.34}$$

If we expand the matrix equations, we can write

$$\gamma_{xx}(0)\alpha_2(1) + \gamma_{xx}(1)\alpha_2(2) = -\gamma_{xx}(1)$$
$$\gamma_{xx}(1)\alpha_2(1) + \gamma_{xx}(0)\alpha_2(2) = -\gamma_{xx}(2) \tag{5.35}$$

Eliminating the value of $\gamma_{xx}(1)$ from Equation 5.33, we get

$$\gamma_{xx}(1) = \alpha_1(1)\gamma_{xx}(0)$$
$$\gamma_{xx}(0)\alpha_2(1) + \gamma_{xx}(0)\alpha_1(1)\alpha_2(2) = -\gamma_{xx}(0)\alpha_1(1)$$
$$\gamma_{xx}(0)\alpha_1(1)\alpha_2(1) + \gamma_{xx}(0)\alpha_2(2) = -\gamma_{xx}(2) \tag{5.36}$$

We can solve these equations to find the values of predictor coefficients of the second-order lattice. We know $\alpha_2(2) = K_2$, the reflection coefficient of the second-order lattice. The reader can verify that

$$\alpha_2(1) = \alpha_1(1) + \alpha_2(2)\alpha_1^*(1)$$
$$\alpha_2(1) = K_1 + K_2 K_1^* \tag{5.37}$$

In general for the mth order predictor, we can write

$$\alpha_m(k) = \alpha_{m-1}(k) + K_m \alpha_{m-1}^*(m-k)$$
$$\text{for } k = 1, 2, \ldots, m-1 \text{ and } m = 1, 2, \ldots, P \tag{5.38}$$

Let us eliminate $\alpha_2(1)$ from Equation 5.36,

$$\gamma_{xx}(1) = -\alpha_1(1)\gamma_{xx}(0)$$
$$\gamma_{xx}(1)\alpha_2(1) + \gamma_{xx}(0)\alpha_2(2) = -\gamma_{xx}(2)$$
$$\gamma_{xx}(0)[\alpha_1(1) + \alpha_2(2)\alpha_1(1)] + \gamma_{xx}(0)\alpha_2(2) = -\gamma_{xx}(2)$$
$$\alpha_2(2) = -\frac{\gamma_{xx}(2) + \alpha_1(1)\gamma_{xx}(0)}{\gamma_{xx}(0)[1 - |\alpha_1(1)|^2]} \tag{5.39}$$

$$k_2 = -\frac{\gamma_{xx}(2) + \alpha_1(1)\gamma_{xx}(0)}{E_1^f} = -\frac{\displaystyle\sum_{i=0}^{2-1}\alpha_i(i)\gamma_{xx}(2-i)}{E_1^f} \text{ with } \alpha_0(0) = 1$$

In general, for the mth order predictor, we can write

$$K_m(m) - \frac{\sum_{i=0}^{m-1} \alpha_i(i)\gamma_{xx}(m-i)}{E_{m-1}^f} \quad \text{with} \tag{5.40}$$

$$E_m^f = E_{m-1}^f[1 - |K_m|^2]$$

We can see that the error reduces when we execute iterations of the algorithm for the predictor of order 2, 3, …, P as the value of $|K_m| < 1$.

To evaluate the predictor coefficients of the Pth order lattice, the algorithm starts with the first-order predictor and finds the predictor coefficient as the reflection coefficient and uses Equation 5.39 to calculate the parameters of second, third, … Pth order predictor. We can say that the Levinson–Durbin algorithm produces reflection coefficients for the optimum lattice prediction filter as well as the coefficients of the optimum direct form FIR predictor.

The Levinson–Durbin algorithm can be stated as follows.

1. Take 128 samples of the speech signal.
2. Find values for autocorrelation coefficients for segment length equal to 128-$|l–k|$ where $(l–k)$ denotes the offset. Use the formula to find normalized autocorrelation given by

$$\gamma_{xx}(i) = \frac{\sum_{i=1}^{N-i} x(k)x(k+i)}{\sqrt{\sum_{i=1}^{N-i} x(k)x(k) + \sum_{i=1}^{N-i} x(k-i)x(k-i)}} \tag{5.41}$$

3. Use Equation 5.40 to find the reflection coefficients. Find error after each iteration using Equation 5.39.
4. Use a recursive relation 5.39 to find the filter coefficients for further higher values of prediction order.

Convert the problem of finding the filter coefficients of infinite impulse response (IIR) filter. It is basically all-pole filter (namely the vocal tract filter). If we want to find the filter coefficients of predictor error filter which is the inverse filter, the stability of the filter is not a problem. This is because it is FIR filter.

Let us do it practically. We have taken 128 samples of speech signal. The autocorrelation coefficients can be calculated and Levinson–Durbin recursion on the array of autocorrelation elements can be applied. The first predictor coefficient is initialized to 1. The variable rccj1 in the program below finds the reflection coefficient for each filter section using Equation 4.38. Finally, the vector g represents the reflection coefficients. Implement Equation 5.39 to find the LPCs and use Equation 5.40 to find the squared error e. Here, "rc" represents the reflection coefficients. We can verify that e reduces as the iterations proceed. Figure 5.56 shows a plot of a signal and predictor coefficients. Figure 5.57 shows a plot of reflection coefficient and the error. A MATLAB program to find lpc is given below.

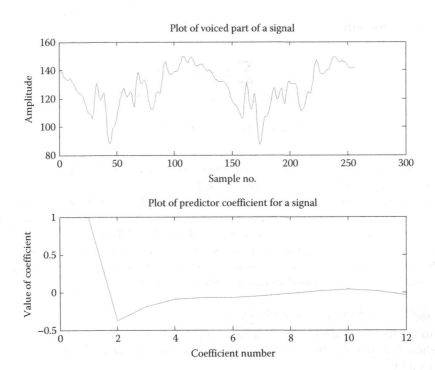

FIGURE 5.56
Plot of signal and the predictor coefficients.

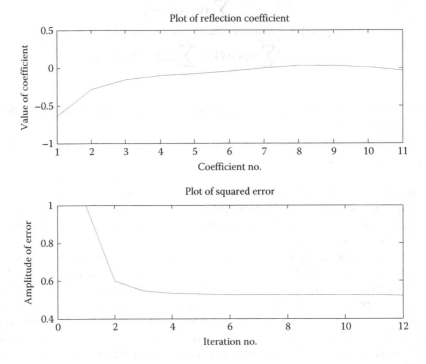

FIGURE 5.57
Plot of reflection coefficients and prediction error.

```
%program to calculate lpc coefficients, reflection coefficients using
%levinson Durbin algorithm.

clear all;
fp=fopen('watermark.wav');
fseek(fp,224000,-1);
x=fread(fp,128);
subplot(2,1,1);plot(x);title('plot of voiced part of a signal');
xlabel('sample no.');ylabel('amplitude');
x=x-128;

sum=0;
i=1;
for k=1:128,
    sum=sum+x(k)*x(k);
end
r(1)=1;

for i=2:12,% a loop to find time domain autocorrelation, r is the array
of autocorrelation coefficients
sum1=0;
    sum2=0;
    for k=1:128-i,
        sum1=sum1+x(k)*x(k+i);
        sum2=sum2+x(k+i)*x(k+i);

    end
   r(i)=sum1/sqrt(sum*sum+sum2*sum2);


end
 %Lvinson Durbin recursion
ap = 0;
    g = [];
    aj(1) = 1;
    ej = r(1);
    e = [ej];
    for j=1:11,
        aj1 = zeros(j+1, 1);
        aj1(1) = 1;
        rcj = r(j+1);
        for i=2:j,
            rcj = rcj + aj(i)*r(j-i+2);
        end
        rccj1 = -rcj/ej;
        g = [g ; rccj1];
        for i=2:j,
            aj1(i) = aj(i)+rccj1*(aj(j-i+2)');
        end
        aj1(j+1) = rccj1;
        ej1 = ej*(1-abs(rccj1)^2);
        e = [e ; ej1];
        aj = aj1;
        ap = aj1;
```

```
        ej = ej1;
    end
    b0 = sqrt(ej1);
 subplot(2,1,2);plot(ap);title('plot of predictor coefficient for a
signal');
xlabel('coefficient number');ylabel('value of coefficient');
figure;
subplot(2,1,1);plot(g);title('plot of reflection coefficient');
xlabel('coefficient no.');ylabel('value of coefficient');
subplot(2,1,2);plot(e);title('plot of squared error');
xlabel('iteration no.');ylabel('amplitude of error');
```

Concept Check

- What is forward linear prediction?
- What is prediction error filter?
- Why is a prediction filter put in a feedback loop to generate prediction error filter?
- What is the assumption in the derivation of normal equations?
- What is the assumption used in the autocorrelation method?
- Why can the prediction error be nonzero in this method?
- When will you call a matrix as the Toeplitz matrix?
- Why is the stability of the filter not considered as a problem?

Summary

We have discussed the definitions of pitch frequency and formants. We have dealt with different methods for pitch period measurement and formant measurement including time domain, spectrum domain, and cepstrum domain methods. We have also dealt with the speech parameters such as MFCC and LPC.

1. We have first discussed the LTI model for speech. The need for time varying model is emphasized and the LTV model is presented. Different methods such as ZCR, passband to full-band energy ratio, and spectral tilt for taking a V/UV decision are described. The format for a .wav format is explained. The parameters for speech, namely, the pitch period and its evaluation and measurement for voiced as well as unvoiced segment, are discussed. Pitch period is defined and its measurement using autocorrelation method and AMDF method is explained. The MATLAB programs for both the methods are added, and the results are interpreted for voiced as well as unvoiced segment. The pitch measurement in frequency domain is also described. Pitch period measurement in the spectrum domain is explained. The pitch period measurement using cepstrum is illustrated using a block schematic. The method of entering a cepstrum domain is first discussed, and the correct way

to use a cepstrum window to eliminate the vocal tract information is pointed out. The peaks can be seen in the cepstrum in the case of voiced segment. This fact can be used for taking voiced and unvoiced decisions. The meaning of pitch contour is explained.

2. The formants are defined and their relation to LPC parameters is explained. The practical aspect of stability of the vocal tract model is discussed. The chapter then proceeds with the measurement of formants using again the cepstrum domain; just by properly selecting a cepstrum window, one can extract the formants. The extraction of formants is illustrated for voiced and unvoiced segments. The log spectrum method for formant estimation is discussed. Here, there is a possibility of overlapping of first formant with the fundamental frequency. The formant tracking is explained for voiced and unvoiced segments. The calculation of power spectrum and estimation of power spectrum density are then clearly explained. First, the periodogram method is discussed. The estimation of power spectrum density using method of periodograms leads to large variance in the estimation. The variance can be reduced by reducing the window size. But, this decreases the frequency resolution. This fact is clearly explained using one example for each method and then comparing the results.

3. The next section first deals with homomorphic processing. The speech signal is a convolved signal and the LTI system is not useful for processing speech directly. The use of homomorphic system for evaluation of impulse response of the vocal tract is explained in detail. Then, it discusses the evaluation of MFCC. We have first explained the mel scale and the bark scale. The complete block schematic for evaluation of MFCC is drawn and each block is explained in detail. The algorithm for evaluation of MFCC is discussed.

4. The further section deals with the linear prediction. The block schematic for the linear prediction is explained in detail. The normal equations are derived. We then concentrated on autocorrelation method for calculation of filter coefficients. We derived the normal equations and showed that the matrix equation representing normal equations leads to the autocorrelation matrix, which is Toeplitz. We used the properties of Toeplitz matrix to find the solution of the matrix equation. Levinson–Durbin recursion equations were derived and the algorithm is described in detail. A MATLAB program for Levinson algorithm is explained and implementation results are shown.

Key Terms

Fundamental frequency	Cepstrum window
Autocorrelation	Quefrency
AMDF	Power spectrum density
Voiced/unvoiced	PSD estimate
Parallel processing elements	Periodogram
Harmonics	Welch method
Log spectrum	Variance
Cepstrum	Frequency resolution

Linear prediction coefficients	Normal equations
Formants	Auditory filters
Pitch period	Canonical system
Formants	Convolutional system
MFCC	Multiplicative system
LPC	LTI filter
Forward linear prediction	Impulse response of vocal tract
Mel scale	Train of impulse
Bark scale	Cepstral features
Homomorphic processing	V/UV decision
Levinson–Durbin algorithm	

Multiple-Choice Questions

1. Autocorrelation will be maximum for the offset equal to
   a. Pitch frequency
   b. Pitch period in ms
   c. Half the pitch period
   d. Pitch period in number of samples.

2. AMDF is evaluated for measurement of
   a. Formants
   b. Pitch frequency
   c. Similarity
   d. Pitch period

3. The parallel processing is used to
   a. Process two speech segments simultaneously
   b. Process the impulses at maxima and minima positions
   c. Process pitch calculation for impulses produced at maxima and minima positions
   d. Process the impulses in the waveform

4. The cepstrum is
   a. IFFT of log spectrum
   b. FFT of spectrum
   c. Windowing of spectrum
   d. Windowed FFT of the log spectrum.

5. If the speech segment is voiced
   a. We observe repetition in the cepstrum
   b. We observe peaks at regular intervals in the cepstrum
   c. We observe random nature in the cepstrum
   d. We observe no peaks in the cepstrum

6. Formants are evaluated as
   a. Peaks in the windowed cepstrum
   b. Position of peaks in the windowed cepstrum
   c. Position of peaks in the FFT of the windowed cepstrum
   d. Peaks in FFT of cepstrum
7. The formant frequencies can be found from
   a. FFT of the signal
   b. IFFT of the signal
   c. Spectrum of the signal
   d. Position of peaks in the log spectrum of the signal
8. Welch method uses
   a. Segments with small size
   b. Segments with 50% overlap
   c. Segments of large size
   d. Segments with any overlap
9. Periodogram method is
   a. Having a drawback of increased variance in the estimate
   b. Having the advantage of increased variance in the estimate
   c. Having lass frequency resolution
   d. Having less accuracy
10. The autocorrelation approach of solving the normal equations assumes that
    a. The signal is zero outside interval $0 \leq n \leq N$
    b. The signal is zero between $P+1$ and N
    c. The signal is zero below n equal to zero
    d. The signal is zero in the interval P to $N-1$
11. Toeplitz matrix is symmetric about the diagonal and
    a. The diagonal elements are zero
    b. The diagonal elements are equal
    c. The diagonal elements are unequal
    d. The diagonal elements are less than one
12. A Homomorphic system for convolution has a canonical system consisting of
    a. Log
    b. FFT followed by LOG
    c. LOG followed by DFT
    d. FFT
13. Impulse response of vocal tract can be obtained from cepstral domain by using
    a. A Hamming window
    b. A window that passes a cluster near the origin
    c. A window that passes peaks in cepstral domain
    d. A window that truncates the cepstrum

14. Cepstral domain features are extracted by taking
    a. Log of signal
    b. FFT of signal
    c. LOG of squared or simple magnitude of FFT of a signal
    d. IFFT of log of power spectrum of a signal.
15. Mel scale is
    a. Linear up to 1000 Hz and logarithmic ahead
    b. Nonlinear throughout
    c. Nonlinear up to 100 Hz
    d. Logarithmic
16. Auditory filters of human ear are simulated using
    a. Mel and bark scale filters
    b. Triangular filter bank
    c. Trapezoidal filter bank
    d. None of such filters

17. MFCC uses mel scale filters, which are
    a. Trapezoidal filters
    b. Rectangular filters
    c. Triangular filters
    d. Equally spaced filters
18. In case of MFCC spectral, smoothing is done by
    a. Cepstral truncation
    b. A low-pass window
    c. A high-pass window
    d. Cepstral windowing

Review Questions

1. Explain the autocorrelation method for calculation of pitch frequency.
2. What is AMDF? How will you use AMDF for pitch period measurement?
3. Explain the use of parallel processing approach for accurate pitch period estimation
4. Draw a block schematic for parallel processing approach for pitch detection. Explain the function of each block.
5. What is the role of rundown circuit in pitch period computation?
6. What is a cepstrum? How will you evaluate the cepstrum of a speech segment?
7. How will you take a V/UV decision form the cepstrum plot?
8. How will you interpret the cepstrum for vocal tract and vocal cord information?

9. Explain the measurement of formant frequencies using a cepstrum.

10. Define formant frequencies. How will you relate it to vocal tract?

11. Define the relation between formants and LPCs.

12. Explain the procedure for finding formant frequencies using a log spectrum.

13. What is a periodogram? Explain the use of periodogram for power spectrum density estimation.

14. What is a drawback of the periodogram method? Explain how Welch method eliminates this drawback. What is the drawback of Welch method?

15. Explain Welch method for estimation of power spectrum density.

16. Explain the method of formant estimation using filter bank method.

17. Draw a block schematic for LTV model for speech and interpret the vocal tract filter coefficients in terms of predictor coefficients.

18. Draw a block schematic for a forward prediction error filter and show that error filter is FIR filter.

19. Derive the normal equations.

20. State the properties of autocorrelation matrix.

21. Explain the Levinson–Durbin recursion.

22. Explain Levinson–Durbin algorithm for calculation of predictor coefficients.

23. Explain the homomorphic processing system for convolution with a block schematic. Justify the use of LTI filter for a convolved signal.

24. Explain the method to separate the impulse response of the vocal tract from the speech signal. Draw the block schematic and explain the function of each block.

25. Define a cepstral domain. How will you find the cepstral coefficients for a speech frame of size 256 samples?

26. What is the auditory filter bank? How will you convert the frequency in Hz to frequency on mel scale. Describe the mel scale filters.

27. How is mel scale different from bark scale? How are the bark scale filers specified?

28. How will you convert power spectrum to mel scale? Explain the procedure for calculation of MFCC with a block schematic. Clearly explain how the integration of power is done on mel scale filters. How will you compress the amplitude of the power spectrum? How is spectral smoothing done?

Problems

1. Record a speech waveform for different words using .wav file format. Write a MATLAB program to find pitch period using autocorrelation method and take V/UV decision.

2. Record a speech waveform for a sentence and take V/UV decision for a speech segment using AMDF.

3. Write a MATLAB program for tracking pitch contour using any pitch calculation method.

4. Record any word and find formants using log spectrum by executing a MATLAB program.

5. Try to write a MATLAB program for estimation of power spectrum density using Welch method.

6. Write a program to find the impulse response coefficients for a vocal tract. Execute it and interpret your results.

7. Write a program to find the cepstral coefficients for a speech segment of size 512 samples. Execute it and interpret your results. Record a speech file with sampling frequency of 11,050 Hz.

8. Write a program to covert the frequency in Hz to frequency in the mel scale and the bark scale. Execute it and interpret your results.

9. Write a program to find the MFCC for a speech segment of size 512. Execute it and interpret your results. Record a speech file with sampling frequency of 11,050 Hz.

10. Write a program to track unvoiced part of utterance and use the program for Levinson–Durbin recursion to find LPC for unvoiced signal.

Answers

Multiple-Choice Questions

1. (d)
2. (d)
3. (c)
4. (a)
5. (b)
6. (c)
7. (d)
8. (b)
9. (a)
10. (a)
11. (b)
12. (b)
13. (b)
14. (c)
15. (a)
16. (a)
17. (c)
18. (a)

Problems

1. See sample programs in the text.
2. See the program for AMDF.
3. Use any pitch calculation program from the sample programs and execute it by updating the file pointer by pitch period.
4. Refer to a sample program.
5. Use the algorithm given in the text for writing a program.
6. Refer to program in the text.
7. Record a speech file with sampling frequency of 11,050 Hz. Here, the information is available up to 5525 Hz. Refer to a program in text to find cepstral coefficients.
8. Refer to program in the text.
9. Record a speech file with sampling frequency of 11,050 Hz. Here, the information is available up to 5525 Hz. Design 28 filters with equal spacing on mel scale over the entire range, as we are interested in frequencies up to 5000 Hz.
10. Refer to the program in text. Only modification to be done is as follows. You have to plot the entire wav file and see by observation where the unvoiced part is. Use fseek command to track the unvoiced part and use the program given in the text on this unvoiced part.

6

Spectral Estimation of Random Signals

LEARNING OBJECTIVES

In this chapter, we will learn about the following concepts.

- Autoregressive (AR) model estimation
- Energy density spectrum
- Power density spectrum
- Spectrum estimation using discrete Fourier transform (DFT)
- Bartlett method
- Welch method
- Blackman–Tukey method
- Parametric methods
- Spectral estimation and detection of harmonic signals
- Music algorithm
- Maximum entropy method (MEM)
- Auto regressive moving average (ARMA) model estimation
- Moving average (MA) model estimation
- Minimum variance estimation
- Eigen analysis algorithm for spectrum estimation
- Applications of power spectrum estimation: formant estimation
- Cepstrum evaluation
- Pitch period measurement using cepstrum
- Cumulant estimation
- Bispectrum estimation

In this chapter, we will discuss the classification of signals into two classes, namely, energy signals and power signals. The exact evaluation of the spectrum is not possible. Hence, the spectrum of random signals like speech needs to be estimated. Different methods for spectrum estimation such as the periodogram method, Welch method, and the Blackman–Tukey method will be described in this chapter. We will introduce the signal models AR, MA, and ARMA. Power spectrum estimation using parametric methods such as MA model parameters, AR model parameters, and ARMA model parameters will be explained. We will also deal with other spectrum evaluation methods like minimum variance method and eigenvalue algorithm method. The cepstrum domain is described, and its use for pitch period measurement of speech signal is illustrated. Finally, higher order spectrum (HOS) estimation called cumulant spectra is described. The evaluation of cumulant spectra and HOS estimation is dealt with in detail.

6.1 Estimation of Density Spectrum

In this section, we will discuss the estimation of density spectrum for two classes of signals—energy and power. Spectrum estimation is not without problems. First, we only have a finite length of data record from which we have to estimate the spectrum; second, we do not know the probability density functions of naturally occurring signals like speech, and electrocardiogram (ECG), electroencephalogram (EEG) signals.

6.1.1 Classification of Signals

Signals can be classified in the following ways.

1. Periodic and aperiodic signals: A periodic signal $x(t)$ is a function for which there exists some value of T that satisfies $x(t+T) = x(t)$ for all t. Here, T denotes the period. For example, consider a sinusoidal signal given by $x(t) = A\sin(t)$. Here, the signal is periodic with a period of 2π. For some signals, no such value of T exists for which $x(t+T) = x(t)$ holds good. Such signals are termed aperiodic signals.

2. Energy and power signals: For a signal $x(t)$, x represents the amplitude or magnitude of the signal. The instantaneous power is given by $|x(t)|^2$. The instantaneous power when integrated over its duration gives the total energy. The generalized format for representing the total energy of a signal $x(t)$ is

$$E = \int_{-\infty}^{\infty} |x(t)|^2 \, dt \qquad (6.1)$$

or

$$E = \lim_{T\to\infty} \int_{-T}^{T} |x(t)|^2 \, dt. \qquad (6.2)$$

Power is nothing but energy per unit time. To calculate power over a period, say, $2T$, we divide the energy over the period $2T$ by the value $2T$. The average power is calculated as the limit of power over the period by allowing the period to tend to infinity. Mathematically, power can be written as

$$P = \lim_{T\to\infty} \frac{1}{2T} \int_{-T}^{T} |x(t)|^2 \, dt. \qquad (6.3)$$

Energy signals have finite energy and zero average power. Power signals have infinite energy but the average power is finite. Signals for which both energy and average power are infinite are neither energy signals nor power signals. An example of such a signal is white noise (idealized form of noise; its power density is independent of operating frequency). The adjective "white" is used as white light contains equal amounts of all frequencies within the visible band of electromagnetic radiation. Periodic signals are power signals. They exist over infinite time and the total energy is infinite. The average power over a period is finite. For example, a sinusoidal signal is a power signal and a short-lived aperiodic signal is an energy signal.

6.1.2 Power Spectral Density and Energy Spectral Density

In this section, energy spectrum and power spectrum are explained. Naturally occurring signals have some random component. We cannot predict the next sample from the previous samples using a mathematical model. Moreover, naturally occurring signals are always of infinite duration and we have only finite duration signals available with us. Using this finite duration signal, we try to predict the characteristics of the signal such as mean and variance.

6.1.2.1 Computation of Energy Density Spectrum of Deterministic Signals

Consider a deterministic signal having finite energy and zero average power. This signal is called an energy signal. We will estimate the energy density spectrum of this energy signal. Let us consider $x(t)$ as a finite duration signal. The problem is to estimate the true spectrum from a finite duration record. If $x(t)$ is a finite energy signal, the energy E of the signal is finite and is given by

$$E = \int_{-\infty}^{\infty} |x(t)|^2 \, dt < \infty. \tag{6.4}$$

The Fourier transform (FT) exists and is given by

$$X(F) = \int_{-\infty}^{\infty} x(t) e^{-j2\pi Ft} \, dt. \tag{6.5}$$

We know that energy is neither created nor destroyed. Hence, the energy of the time domain signal is the same as the energy of the Fourier domain signal. We can write

$$E = \int_{-\infty}^{\infty} |x(t)|^2 \, dt = \int_{-\infty}^{\infty} |X(F)|^2 \, dF. \tag{6.6}$$

The symbol $|X(F)|^2$ represents the distribution of signal energy with respect to frequency and is called the energy density spectrum. It is denoted as $S_{XX}(F)$. The total energy is calculated as the integration of $S_{XX}(F)$ over all frequencies F. We will show that $S_{XX}(F)$ is the FT of an autocorrelation function.

The autocorrelation function of the finite energy signal is given by

$$R_{XX}(\tau) = \int_{-\infty}^{\infty} x^*(t) \, x(t+\tau) dt, \tag{6.7}$$

where $x^*(t)$ is the complex conjugate of $x(t)$. Let us calculate the FT of this autocorrelation.

$$\int_{-\infty}^{\infty} R_{XX}(\tau) e^{-j2\pi F\tau} \, d\tau = \int_{-\infty}^{\infty} \int_{-\infty}^{\infty} x^*(t) x(t+\tau) e^{-j2\pi F\tau} \, dt \, d\tau. \tag{6.8}$$

Put $t' = t+\tau$, $dt' = d\tau$.

$$\int_{-\infty}^{\infty} x^*(t) e^{j2\pi Ft} \, dt \int_{-\infty}^{\infty} x(t') e^{-j2\pi Ft'} \, dt' = X^{*(F)} X(F) = |X(F)|^2 = S_{XX}(F). \tag{6.9}$$

Hence, $R_{XX}(\tau)$ and $S_{XX}(F)$ are FT pairs.

We can prove the result for discrete time (DT) signals as well. Let $r_{XX}(k)$ and $S_{XX}(f)$ denote the autocorrelation and energy density for a DT signal.

$$r_{XX}(k) = \sum_{n=-\infty}^{\infty} x^*(n) x(n+k). \tag{6.10}$$

Put $n+k = n'$. The FT of the autocorrelation is given by

$$S_{XX}(f) = \sum_{k=-\infty}^{\infty} r_{XX}(k) e^{-j2\pi kf} = \sum_{n=-\infty}^{\infty} x^*(n) e^{j2\pi nf} \sum_{k=-\infty}^{\infty} x(n') e^{-j2\pi n'f} = |X(f)|^2. \tag{6.11}$$

We will list two different methods for computing energy density spectrum, namely, the direct method and the indirect method.

Direct method: In the direct method, one computes the FT of the sequence $x(n)$ and then calculates the energy density spectrum as

$$S_{xx}(f) = |X(F)|^2 = \left| \sum_{n=-\infty}^{\infty} x(n) e^{-j2\pi fn} \right|^2. \tag{6.12}$$

Indirect method: In the indirect method, the autocorrelation is computed first. Then, the FT is found to get the energy density spectrum.

$$S_{xx}(f) = \sum_{k=-\infty}^{\infty} \gamma_{xx}(k) e^{-j2\pi kf}. \tag{6.13}$$

There is a practical problem involved in the computation of the energy density spectrum using Equations 6.12 and 6.13. The sequence is available only for a finite duration. Availability of only a finite sequence is equivalent to multiplication of the sequence by a rectangular window. This results in spectral leakage due to Gibbs phenomenon as discussed in Chapter 5.

6.1.2.2 Estimation of Power Density Spectrum of Random Signals

Stationary random signals are a class of signals of infinite duration and infinite energy. The average power is finite. Such signals are classified as power signals. We will estimate the power density spectrum for such signals. For a random process, the autocorrelation is given by

$$\gamma_{xx}(\tau) = E[x^*(t)x(t+\tau)], \tag{6.14}$$

where E denotes expected value. This cannot be evaluated if the probability density function (pdf) is not known. For a random process, pdf is not known. We can only estimate the value of autocorrelation as

1. We have records of only finite duration.
2. pdf is not known.

We assume that the process is wide sense stationary and also ergodic. We calculate the time average autocorrelation and equate it to ensemble autocorrelation in case of an ergodic process. Hence, we calculate

$$\gamma_{xx}(\tau) = \lim_{\tau \to \infty}[R_{xx}(\tau)]. \tag{6.15}$$

The estimate $P_{xx}(F)$ of the power density spectrum is calculated as the FT of autocorrelation, and the actual value of the power density spectrum is then the expected value of $P_{xx}(F)$ in the limit as $T_0 \to \infty$.

$$
\begin{aligned}
P_{xx}(F) &= \int_{-T_0}^{T_0} R_{xx}(\tau)e^{-j2\pi F\tau}\,d\tau \\
&= \frac{1}{2T_0} \int_{-T_0}^{T_0}\left[\int_{-T_0}^{T_0} x^*(t)x(t+\tau)dt\right]e^{-j2\pi F\tau}\,d\tau \\
&= \frac{1}{2T_0}\left|\int_{-T_0}^{T_0} x(t)e^{-j2\pi Ft}\,dt\right|^2.
\end{aligned}
\tag{6.16}
$$

The actual power density spectrum is obtained as

$$\Gamma_{xx}(F) = \lim_{T_0 \to \infty} E(P_{xx}(F)). \tag{6.17}$$

There are two approaches for computation of power density spectrum, namely, the direct method as given by Equation 6.14 or the indirect method by first calculating the autocorrelation and then finding the FT of it to get the power density spectrum.

In practice, we have only finite samples of the signals. The time average autocorrelation is computed using the equation

$$\gamma_{xx}(m) = \frac{1}{n - |m|} \sum_{n=|m|}^{N-1} x^*(n)x(n+m) \quad m = -1, -2, \ldots, 1 - N. \tag{6.18}$$

The normalization factor results in a mean value of autocorrelation, which is computed as autocorrelation. We find the FT of the mean value of autocorrelation given by Equation 6.16 to get an estimate of the power spectrum, which can also be written as

$$P_{xx}(f) = \frac{1}{N} \left| \sum_{n=0}^{N-1} x(n)\, e^{-j2\pi f n} \right|^2 = \frac{1}{N} |X(f)|^2. \tag{6.19}$$

Equation 6.17 represents the well-known form of the power density spectrum estimate and is termed as periodogram. To find the pdf, one draws the histogram for the signal with finite duration. Then, curve fitting tests such as chi-square test or Kolmogorov–Smirnov (K–S test) are used to fit a standard probability distribution for the given signal. The mathematical equation for that standard pdf will be used in calculations.

Concept Check

- What is energy density spectrum? Why is it estimated?
- What are power signals?
- How do we estimate power density spectrum?
- What is the relation between autocorrelation and power density spectrum?
- How will you fit a standard pdf to a random signal?

6.2 Nonparametric Methods

Nonparametric methods are also used to estimate the power spectrum. These methods do not make any assumption about how the data are generated. To be practicable, the methods use finite duration data record, and its use is equivalent to assuming that the data pass via a window of length, say, N. The spectral resolution is limited to $1/N$. However, the limitation on spectral resolution is compensated by the smaller variance for the estimate achieved by these methods. Later in this chapter, we will realize that the product of spectral resolution and variance of the estimate is constant.

We will now discuss power spectral estimation using discrete Fourier transform (DFT). We will also describe different periodogram methods such as the Bartlett method, the Welch method, and the Blackman–Tukey method.

6.2.1 Use of DFT for Power Spectrum Estimation

In this section, we will discuss the basic method of estimating the power spectrum, namely, using DFT. Consider N data samples. We list the algorithm to find the samples of the periodogram as follows.

1. Read N data values.
2. Find N point DFT.
3. The computation of DFT gives N points of the periodogram.

$$P_{XX}\left(\frac{k}{N}\right) = \frac{1}{N}\left|\sum_{n=0}^{N-1} x(n)e^{-j2\pi nk/N}\right|^2 \text{ for } k = 0,1,...,N-1. \tag{6.20}$$

4. Increase the length of the sequence by padding extra zeros to interpolate the spectrum.

The block schematic for using DFT to find the periodogram is shown in Figure 6.1. Speech samples (N) are taken, DFT is calculated, and N points of the periodogram are found. We will then increase the length of the sequence by padding zeros and observe the effect on the plot of the periodogram. Figure 6.2 shows the plot of speech signal samples and the plot of its periodogram. If we pad zeros to the signal, the effect of padding zeros on the periodogram is shown in Figure 6.3. We can observe that the new periodogram is only an interpolation of the earlier periodogram. The spectral resolution does not change. The frequency resolution will depend only on the length of the data samples.

```
%a program to find periodogram using DFT
clear all;
fp=fopen('watermark.wav');
fseek(fp,224000,-1);
a=fread(fp,1024);
a=a-128;
subplot(2,1,1);plot(a);title('plot of voiced part of a signal');
xlabel('sample no.');ylabel('amplitude');
%1024 samples of the voiced signal are displayed
b=fft(a);
b1=abs(b.*b);
subplot(2,1,2);plot(b1);
title('plot of periodogram using DFT');
xlabel('sample no.');ylabel('squared amplitude');
for i=1:2048,
    if i<1025,
        a1(i)=a(i);
```

FIGURE 6.1
Block schematic for finding a signal's periodogram.

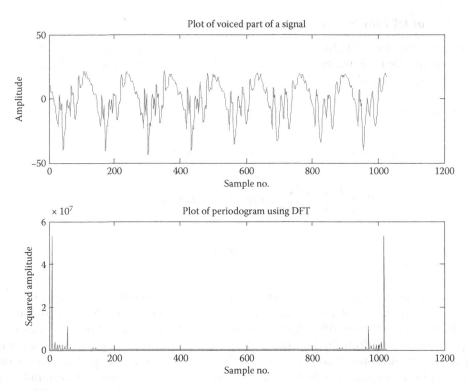

FIGURE 6.2
Plot of signal and its periodogram.

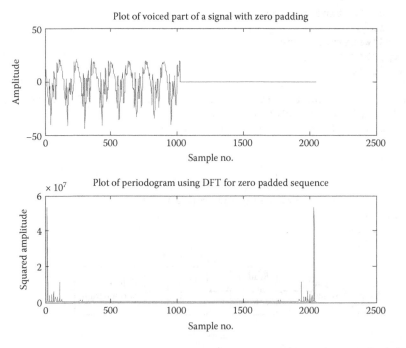

FIGURE 6.3
Plot of signal and its periodogram after zero padding.

```
        else
            a1(i)=0;
        end
end
    figure;
  subplot(2,1,1);plot(a1);title('plot of voiced part of a signal with
zero padding');
xlabel('sample no.');ylabel('amplitude');
b=fft(a1);
b1=abs(b.*b);
subplot(2,1,2);plot(b1);
title('plot of periodogram using DFT of zero padded sequence');
xlabel('sample no.');ylabel('squared amplitude');
```

6.2.2 Bartlett Method

The Bartlett method or the method of averaging periodograms reduces the variance of the power density spectrum estimate. It involves the following three steps.

1. Read N points of data. Divide it in K nonoverlapping segments each of length M.
2. Compute the periodogram for each segment.

$$P_{XX}{}^i\left(\frac{k}{M}\right) = \frac{1}{M}\left|\sum_{n=0}^{N-1} x(n)e^{-j2\pi nk/M}\right|^2 \text{ for } k = 0,1,...,N-1. \quad (6.21)$$

3. Lastly, average the periodograms for the k segments to get the Bartlett power spectrum estimate.

$$P_{XX}^B\left(\frac{k}{M}\right) = \frac{1}{K}\sum_{i=0}^{K-1} P_{XX}^i\left(\frac{k}{M}\right). \quad (6.22)$$

The block schematic for the Bartlett method is shown in Figure 6.4. It is observed that the true spectrum convolves with the frequency characteristic of Bartlett window; hence, the name Bartlett method. The data length is effectively reduced by a factor K. The spectral width increases by the same factor K, and it reduces the frequency resolution. The reduction in frequency resolution results in reduced variance of the power density spectrum estimate by the same factor K. The result of execution of the following MATLAB program is shown in Figure 6.5. The figure is a plot of the signal and its periodogram using the Bartlett method.

```
%a program to find periodogram using Bartlett method
clear all;
fp=fopen('watermark.wav');
fseek(fp,224000,-1);
a=fread(fp,1024);
a=a-128;
subplot(2,1,1);plot(a);title('plot of voiced part of a signal');
xlabel('sample no.');ylabel('amplitude');
```

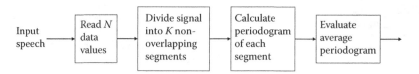

FIGURE 6.4
Block schematic for finding the periodogram using the Bartlett method.

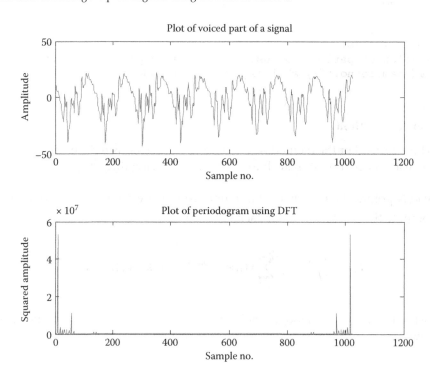

FIGURE 6.5
Plot of signal and its periodogram using the Bartlett method.

```
%1024 samples of the voiced signal are displayed
for i=1:256,
    a1(i)=a(i);
end
for i=257:512,
    a2(i-256)=a(i);
end
for i=513:1024,
    a3(i-512)=a(i);
end
for i=1025:1536,
    a4(i-1024)=a(i);
end
b1=abs(fft(a1));
b2=abs(fft(a2));
b3=abs(fft(a3));
b4=abs(fft(a4));
b11=abs(b1.*b1);
```

```
b22=abs(b2.*b2);
b33=abs(b3.*b3);
b44=abs(b4.*b4);
for i=1:256,
    b(i) = (b11(i)+b22(i)+b33(i)+b44(i))/4;
end
subplot(2,1,2);plot(b);
title('plot of periodogram using Bartlett method');
xlabel('sample no.');ylabel('squared amplitude');
```

6.2.3 Welch Method

This method is also called the method of modified averaging periodogram. The Welch method has two basic modifications on the Bartlett method. It allows the segments to overlap. The overlap used is usually 50%. The use of overlap preserves the statistical properties of the signal. The second modification suggested by Welch is to window the segments prior to computing the periodograms. When a segment passes via a window, the time resolution and frequency resolution are fixed according to the Heisenberg uncertainty principle. If the window length reduces, time resolution improves at the cost of the frequency resolution. It is because of these modifications that the method is also termed the method of modified periodograms. The power spectrum is normalized by a factor equal to the window energy. The Welch estimate is calculated as the average of the periodograms for all segments. As mentioned earlier, the method is at the cost of the frequency resolution but reduces the variance of the power spectrum estimate. The algorithm can be stated as follows.

1. Read the data values. Divide the data segment into L overlapping segments (say, 50% overlap)
2. Pass each of the data segment via a window.
3. Find the periodogram for each segment and divide it by the window energy.
4. Find the average periodogram.

A block schematic for finding the periodogram using the Welch method is shown in Figure 6.6.
A MATLAB program to find the periodogram using the Welch method is given here. We have used a Bartlett window for windowing each segment. Figure 6.7 shows a plot of a signal and its periodogram using the Welch method.

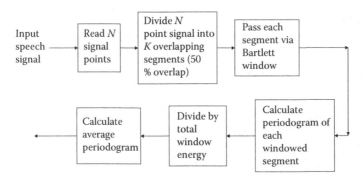

FIGURE 6.6
Block schematic for finding periodogram using the Welch method.

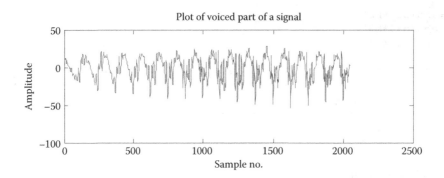

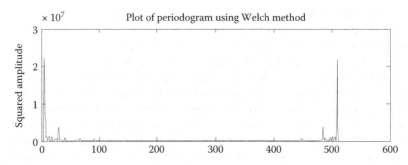

FIGURE 6.7
Plot of signal and periodogram using the Welch method.

```
%a program to find periodogram using Welch method
clear all;
fp=fopen('watermark.wav');
fseek(fp,228000,-1);
a=fread(fp,2048);
a=a-128;
subplot(2,1,1);plot(a);title('plot of voiced part of a signal');
xlabel('sample no.');ylabel('amplitude');
%Divide the data in 50 % overlapping segments and pass it via a window
w=window(@bartlett,512);% Bartlett window function
for i=1:512,
    a1(i)=a(i)*w(i);
end
for i=257:768,
    a2(i-256)=a(i)*w(i-256);
end
for i=513:1024,
    a3(i-512)=a(i)*w(i-512);
end
for i=769:1280,
    a4(i-768)=a(i)*w(i-768);
end
for i=1025:1536,
    a5(i-1024)=a(i)*w(i-1024);
end
for i=1281:1792,
    a6(i-1080)=a(i)*w(i-1280);
```

```
end
for i=1537:2048,
    a7(i-1536)=a(i)*w(i-1536);
end
b1=abs(fft(a1));
b2=abs(fft(a2));
b3=abs(fft(a3));
b4=abs(fft(a4));
b5=abs(fft(a5));
b6=abs(fft(a6));
b7=abs(fft(a7));
b11=abs(b1.*b1);
b22=abs(b2.*b2);
b33=abs(b3.*b3);
b44=abs(b4.*b4);
b55=abs(b*b5);
b66=abs(b6.*b6);
b77=abs(b7.*b7);
W=0;
for i=1:512,
    W=W+w(i);%calculation of window energy for normalization
end
for i=1:512,
    b(i)=(b11(i)+b22(i)+b33(i)+b44(i))+b55(i)+b66(i)+b77(i)/W*7;
end
subplot(2,1,2);plot(b);
title('plot of periodogram using Welch method');
xlabel('sample no.');ylabel('squared amplitude');
```

6.2.4 Blackman–Tukey Method

The Blackman–Tukey method uses the method of autocorrelation to find the power spectrum. It passes the autocorrelation via a smooth window to compensate error while calculating autocorrelation. When a lag is large, the estimation of autocorrelation is less reliable. This happens because a smaller number of data points enters into the estimate when the lag is large. The algorithm can be stated as follows.

1. Read the data sequence.
2. Find the autocorrelation of the data sequence.
3. Pass the autocorrelation via a window, say a Bartlett window.
4. Find the FT of the windowed autocorrelation.

A block schematic for finding the periodogram using the Welch method is shown in Figure 6.8.

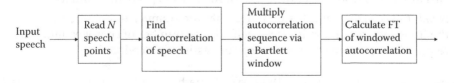

FIGURE 6.8
Block schematic for finding periodogram using the Blackman–Tukey method.

A MATLAB program to find the power spectrum estimate using the Blackman–Tukey method is given here.

```
%to find power spectrum using Blackman-Tukey method
%to find autocorrelation of the signal
clear all;
fp=fopen('watermark.wav');
fseek(fp,228500,-1);
a=fread(fp,2100);
subplot(2,1,1);plot(a);title('plot of voiced part of a signal');
xlabel('sample no.');ylabel('amplitude');

a=a-128;
for k=1:2048,
sum(k)=0;
end
for k=1:2048,
for i=1:45,
sum(k)=sum(k)+(a(i)*a(i+k));
sum(k)=sum(k)/45;
end
end
subplot(2,1,2);plot(sum);title('plot of autocorrelation of a signal');
xlabel('sample no.');ylabel('correlation');
w=window(@bartlett,2048);% Bartlett window function
for i=1:2048,
    sum1(i)=sum(i)*w(i);
end
b=abs(fft(sum1));
figure;
plot(b);title('plot of power density spectrum of a signal');
xlabel('sample no.');ylabel('power density');
```

Figure 6.9 shows a plot of a signal and its autocorrelation and Figure 6.10 shows a plot of FT of the windowed autocorrelation. This is the periodogram using the Blackman–Tukey method.

6.2.5 Performance Comparison of Nonparametric Methods

To compare the performance of the nonparametric methods for estimating power spectrum, the following parameters are used.

1. *Quality*: It is defined as the ratio of the square of mean of the power spectrum estimate and its variance.
2. *Variability*: The reciprocal of quality is defined as the variability. It is the ratio of the variance and the square of the mean of the power spectrum estimate.
3. *Number of computations*: It is assumed that the number of data points N is fixed; the frequency resolution is also specified. The number of complex multiplications required to compute the power spectrum estimate is counted.

The comparison of the Bartlett method, Welch method, and Blackman–Tukey method in terms of quality and number of computations is given in Table 6.1.

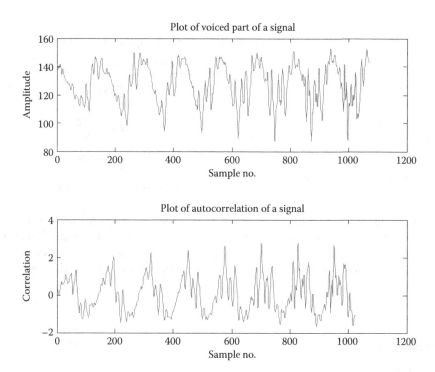

FIGURE 6.9
Plot of signal and its autocorrelation.

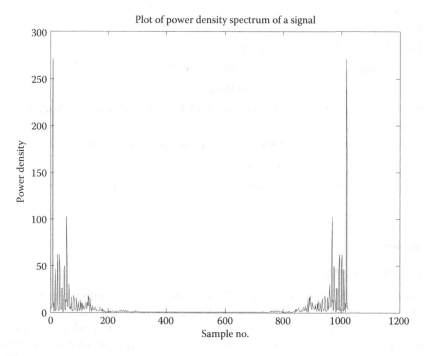

FIGURE 6.10
Plot of FT of windowed autocorrelation (periodogram).

TABLE 6.1
Quality and Number of Computations for Three Nonparametric Methods

Method	Quality	Number of Complex Multiplications
Bartlett method	$1.11N\Delta f$	$\dfrac{N}{2}\log_2\dfrac{0.9}{\Delta f}$
Welch method	$1.39N\Delta f$	$N\log_2\dfrac{5.12}{\Delta f}$
Blackman–Tukey method	$3.34N\Delta f$	$N\log_2\dfrac{1.28}{\Delta f}$

Referring to Table 6.1, we can see that the Welch method requires little more computational cost than the other methods. Quality is highest for the Blackman–Tukey method. The quality increases as N, the number of points in a sequence, increases. For a given quality, the frequency resolution can be improved by increasing the data length N of a sequence.

Concept Check

- What is a periodogram?
- How will you use DFT for power spectrum calculation?
- What is the method of averaging periodogram?
- Why is 50% overlap used in the Welch method?
- Why is frequency resolution poor in case of the Welch method?
- What is the main advantage of the Welch method?
- What is the difference between the Welch method and the Blackman–Tukey method?
- Name the parameters used for comparing nonparametric methods.
- Which method performs the best?
- Which method uses autocorrelation for computation of periodograms?

6.3 Parametric Methods

Nonparametric models have two basic limitations. First, the autocorrelation estimate is assumed to be zero outside the signal, that is, for $m > N$. The second assumption that limits these methods is that the signal is periodic with a period of N samples. Both these assumptions are nonrealistic. Parametric methods use no such assumptions. They extrapolate the autocorrelation. Extrapolation is possible because the signal has some predictability.

Parametric methods depend on the process of signal generation. Consider a speech signal. We know that there is high correlation between successive speech samples. Further samples can be predicted if previous samples are available. This indicates that the limitation of a finite data record does not impede the parametric approach. We can predict further samples when we have already calculated the predictor coefficients. We have discussed linear prediction in detail in Chapter 5. Lattice method and autocorrelation methods can be used to compute predictor coefficients using the Levinson–Durbin algorithm, Burg algorithm, etc. We have also seen how to calculate the prediction error. The reader may go through these concepts again before going through the parametric methods. Parametric methods include AR model parameters, MA model parameters, and ARMA model parameters. AR model is the most commonly used model.

6.3.1 Power Spectrum Estimation Using AR Model Parameters

The AR model is the most commonly used model for signals like speech. An AR system has only poles, and zeros exist only at zero. It is also called an all-pole model. We will study how the model parameters are used for computation of the power spectrum. Parametric methods are based on modeling signal data sequence as the output of a linear system characterized by a transfer function of the form

$$H(Z) = \frac{\displaystyle\sum_{k=0}^{M} b_k Z^{-k}}{1 + \displaystyle\sum_{k=1}^{N} a_k Z^{-k}}. \tag{6.23}$$

The power density spectrum of the data sequence can be written as

$$\Gamma_{XX}(f) = |H(f)|^2 \, \Gamma_{WW}(f). \tag{6.24}$$

If we assume the input sequence is a zero mean white noise sequence, the autocorrelation is given by

$$\gamma_{WW}(n) = \sigma_W^2 \delta(n). \tag{6.25}$$

$$\Gamma_{XX}(f) = |H(f)|^2 \, \sigma_W^2. \tag{6.26}$$

The Yule–Walker method estimates the autocorrelation from data values and uses the estimate to find the AR model parameters. We have discussed the Levinson–Durbin algorithm for computation of AR model parameters in Chapter 5. The power density spectrum can be estimated as

$$P_{XX}^{YW}(f) = \frac{\sigma_{WP}^2}{\left|1 + \displaystyle\sum_{k=1}^{P} a(k) e^{-j2\pi fk}\right|^2}. \tag{6.27}$$

P is the order of the predictor and $\sigma_{WP}^2 = E_P^f$ is the forward error of the predictor. The algorithm for power spectrum estimation can be written as follows.

1. Divide the input data sequence into segments of size, say, 128 samples each.
2. Estimate the autocorrelation from the data.
3. Use the Yule–Walker method to find the predictor coefficients.
4. Find the forward error for the predictor.
5. Use Equation 6.25 to evaluate the power spectrum estimation.

The selection of the order of the predictor is important. The AR method of estimation of power spectrum is a block processing one. This means that the method calculates the AR process parameters for a block and estimates its power density spectrum. It then uses the next block for processing. The method may also change or update the coefficients as the next sample arrives. Let us estimate the power density spectrum using the Yule–Walker method. We have used a direct command in MATLAB. A MATLAB program is given here. Figure 6.11 shows a plot of a signal and its power spectrum estimation using the Yule–Walker method.

```
%to find power spectrum using Yule-Walker method.
 clear all;
fp=fopen('watermark.wav');
fseek(fp,224000,-1);
a=fread(fp,1024);
a=a-128;
subplot(2,1,1);plot(a);title('plot of voiced part of a signal');
xlabel('sample no.');ylabel('amplitude');
%1024 samples of the voiced signal are displayed
subplot(2,1,2);
Fs = 22100;    t = 0:1/Fs:.296;
h=spectrum.yulear
d = psd(h,a,'Fs',Fs);
%Calculate the power spectrum estimate using Yule-Walker method
plot(d);
```

We can use the Levinson–Durbin algorithm to find the linear predictor coefficients (LPC) and use a random signal as input to the AR system and find its output. We can plot the power spectrum estimate using the command pyulear in MATLAB.

6.3.2 Burg's Method for Power Spectrum Estimation (Maximum Entropy Method—MEM)

The algorithm uses forward and backward prediction errors and minimizes the sum of squared errors. The LPC can be predicted using Burg's algorithm and the power spectrum can be found using the LPC obtained based on the AR model, same as in Section 6.3.1. The input data samples are from the random process of speech signal, and hence, the pdf for the signal is not known. The process of calculating the power spectrum estimate involves evaluating the FT of autocorrelation. The estimation of autocorrelation can be accurately done if the pdf is known. Normally, the histogram of the signal is compared with known standard density functions using the K–S test, as discussed in Chapter 1, to get the pdf for the unknown data samples. If the exact fit for the known density function is obtained,

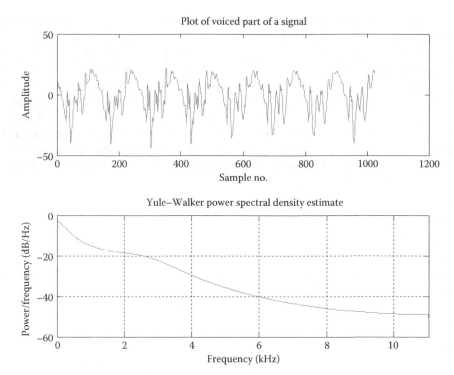

FIGURE 6.11
Plot of signal and its power spectrum using the Yule–Walker method.

which is a very rare situation, then the entropy of the signal is maximized. The power spectrum estimation obtained will then be termed the maximum entropy power spectrum estimation.

Autocorrelation calculation is done for finite data samples. The autocorrelation is then interpolated to find further data samples, which were initially not available. Burg postulated that the interpolation that gives the flattest spectrum will result in the maximum entropy. This is due to the fact that obtaining a flat spectrum is a rare situation and hence has minimum probability of occurrence, which confirms that information is maximized for such a situation. The method of power spectrum estimation based on this postulation is called maximum entropy method (MEM) for power spectrum estimation.

Burg's method uses the lattice filter structure for the system and evaluates reflection coefficients using forward and backward prediction. The power spectrum estimation can be found as follows.

$$P_{XX}^B(f) = \frac{E_P}{\left|1 + \sum_{k=1}^{P} a(k)e^{-j2\pi fk}\right|^2}.$$

(6.28)

$$E_P = \sum_{n=P}^{N-1} |f_P(n)|^2 + |g_P(n)|^2 \text{ forward \& backward error.}$$

Burg's method results in a stable filter, gives high resolution, and is found to be computationally efficient. However, it exhibits line splitting at high signal-to-noise ratio. The actual single spectral line shows two closely spaced peaks. In case of higher order models, it shows spurious peaks. For a sinusoidal signal, the spectrum is sensitive to the initial phase. Modifications, namely, windowing and weighting, are found to reduce the drawbacks.

The MATLAB command pburg finds the power spectral density (PSD) estimate using Burg's method. The frequency is expressed in units of radians/sample. ORDER is the order of the autoregressive (AR) model that is used to produce the PSD. pburg uses a default fast fourier transform (FFT) length of 256, which in turn determines the length of P_{XX}. A MATLAB program to find PSD using Burg's method is given here. Figure 6.12 shows a plot of the signal and its PSD.

```
clear all;
fp=fopen('watermark.wav');
fseek(fp,224000,-1);
a=fread(fp,1024);
a=a-128;
subplot(2,1,1);plot(a);title('plot of voiced part of a signal');
xlabel('sample no.');ylabel('amplitude');
%1024 samples of the voiced signal are displayed
subplot(2,1,2);
Fs = 22100;    t = 0:1/Fs:.296;
[p t]=pburg(a,12);
%Calculate the power spectrum estimate using Burg's method
plot(p); title('plot of PSD using Burg's method');
xlabel('sampling time');ylabel('amplitude');
```

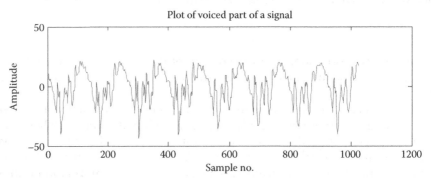

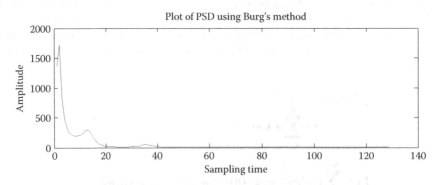

FIGURE 6.12
Plot of signal and its PSD using Burg's method.

6.3.3 Power Spectrum Estimation Using ARMA Model

The ARMA model has poles as well as zeros. It improves the power spectrum estimate using less number of coefficients. This model is suitable when the signal is corrupted by noise. If the signal is generated using AR model and it has additive noise, the overall system will become an ARMA model. The model parameters can be estimated using the least squares technique. The estimated power spectrum can be written as

$$P_{XX}^{ARMA}(f) = \frac{P_{VV}^{MA}(f)}{\left|1 + \sum_{k=1}^{P} a(k)e^{-j2\pi fk}\right|^2}.$$

$$P_{VV}^{MA}(f) = \sum_{m=-q}^{q} r_{VV}(m)e^{-j2\pi fm}.$$

$\gamma_{VV}(m)$: an estimate of autocorrelation for the MA model.   (6.29)

6.3.4 Power Spectrum Estimation Using MA Model

The power spectrum for an MA model can be estimated using MA model parameters, which represent parameters of an all-zero model. The power spectrum can be written as

$$P_{XX}^{MA}(f) = \sum_{m=-q}^{q} r_{XX}(m)e^{-j2\pi fm}.$$

(6.30)

$\gamma_{XX}(m)$: an estimate of autocorrelation for the MA model.

The parameters of the MA model can be determined using least squares inverse design.

Concept Check

- List some methods to find autocorrelation coefficients.
- How will you use AR parameters for estimating the power spectrum?
- If the signal is corrupted with noise, which model will you use for power spectrum estimation

---

6.4 Other Spectral Estimation Methods

This section describes other methods for power spectrum estimation, namely, minimum variance spectral estimation and eigen analysis algorithm for power spectrum estimation.

6.4.1 Minimum Variance Power Spectrum Estimation

Consider an finite impulse response (FIR) filter with coefficients a_k $0 \le k \le P$. We will not constrain a_0 to be unity. Consider an input data sequence $x(n)$. This data sequence is passed via this FIR filter. The output of the filter can be written as

$$y(n) = \sum_{k=0}^{P} a_k x(n-k) = X^T(n)A,$$ (6.31)

where $X^T(n) = [x(n)\, x(n-1)...x(n-P)]$: data vector and $A = [a_0\, a_1, ..., a_P]$: filter coefficient vector.

If we assume the expected value of the signal to be zero, the variance of the output sequence can be written as

$$E(y(n)^2) = E[a^{*T}X^*(n)X^t(n)a]$$

$$= A^{*T}\Gamma_{xx}A,$$ (6.32)

where Γ_{xx} is the autocorrelation matrix of $x(n)$. The filter coefficients are selected so that the response of the filter at every frequency is normalized to unity. If the variance of the output is now minimized [1], the filter passes that frequency, say, f_1, undistorted and attenuates other frequencies away from f_1. Changing f_1 over the entire range, we can find the power spectrum estimate with minimum variance.

6.4.2 Eigen Analysis Algorithm for Spectrum Estimation

Let us first discuss some concepts that are required to understand this method. Consider a square matrix of dimension $N \times N$ represented as A. The matrix A is said to have an eigenvector x and corresponding eigenvalue λ if

$$Ax = \lambda x.$$ (6.33)

Equation 6.33 holds good if

$$\det | Ax - \lambda x | = 0.$$ (6.34)

If Equation 6.34 is expanded, it will result in an Nth degree polynomial in λ. This polynomial is termed as the characteristic polynomial and the roots of the characteristic polynomial are called eigenvalues. The concepts are illustrated with a simple example.

Example 1

Find all the eigenvectors and eigenvalues of the matrix $A = \begin{bmatrix} 2 & 1 \\ 1 & 2 \end{bmatrix}$.

Solution: Let us first find the characteristic equation.

$$\det | Ax - \lambda I | = 0.$$

$$\begin{bmatrix} 2-\lambda & 1 \\ 1 & 2-\lambda \end{bmatrix} = 0. \tag{6.35}$$

$$\lambda^2 - 4\lambda + 3 = 0 \Rightarrow (\lambda - 1)(\lambda - 3) = 0.$$

Eigenvalues are $\lambda = 1$ and $\lambda = 3$.

The eigenvector equation reduces to

$$| Ax - \lambda I | x = 0.$$

$$\begin{bmatrix} 2-\lambda & 1 \\ 1 & 2-\lambda \end{bmatrix} \begin{bmatrix} x_1 \\ x_2 \end{bmatrix} = 0.$$

$$2x_1 + x_2 = \lambda x_1 \text{ and } x_1 + 2x_2 = \lambda x_2. \tag{6.36}$$

Put $\lambda = 1 \Rightarrow x_1 + x_2 = 0 \Rightarrow x_1 = -x_2$.

The eigenvectors are $\begin{bmatrix} -x_2 \\ x_2 \end{bmatrix}$.

Put $\lambda = 3 \Rightarrow -x_1 + x_2 = 0 \Rightarrow x_1 = x_2$.

$$\tag{6.37}$$

The eigenvectors are $\begin{bmatrix} x_2 \\ x_2 \end{bmatrix}$.

Let us now try to understand the eigen analysis algorithm for power spectrum estimation. It was indicated previously that if the signal is generated using AR model and the signal has additive noise, the overall system becomes ARMA model. Let us consider a special case of the signal, for example, the sinusoidal signal corrupted by additive white noise. The overall system will again be ARMA model. To generate a sinusoid, let us use a difference equation for a system

$$x(n)- = -a_1 x(n-1) - a_2 x(n-2),$$

where $a_1 = 2\cos(2\pi f_k)$, $a_2 = 1$ with initial conditions

$$\tag{6.38}$$

$$x(-1) = -1, x(-2) = 0.$$

The system has complex conjugate poles at f_k and $-f_k$.

The solution of the system is $x(n) = \cos(2\pi f_k n)$ for $n \geq 0$.

The generalized equation for generating P sinusoidal components can be written as

$$x(n) = -\sum_{m=1}^{2P} a_m x(n-m) \cdot \tag{6.39}$$

The transfer function of the system can be written as

$$H(Z) = \frac{1}{1 + \displaystyle\sum_{m=1}^{2P} a_m Z^{-m}}.$$

(6.40)

Let $A(Z) = 1 + \displaystyle\sum_{m=1}^{2P} a_m Z^{-m}$. It has $2P$ roots on the unit circle.

The roots of the denominator polynomial correspond to the frequencies of the sinusoids. Let the sinusoids be corrupted by white noise sequence, $w(n)$ with $E[\,|w(n)|^2] = \sigma_w^2$.

The signal with a number of sinusoids and additive white noise is represented as $y(n) = x(n) + w(n)$.

We can write x(n) as

$$x(n) - y(n) - w(n) = -\sum_{m=1}^{2P} [y(n-m) - w(n-m)] a_m.$$

(6.41)

Equivalently, we can write

$$\sum_{m=0}^{2P} a_m y(n-m) = \sum_{m=0}^{2P} a_m w(n-m) \text{ with } a_0 = 1.$$

(6.42)

We can easily observe that Equation 6.42 represents the difference equation for ARMA model. The coefficients of the AR and MA system are identical. This is found to be the characteristic feature of the ARMA system for sinusoids with white noise.

The difference equation in Equation 6.42 may be expressed in the matrix form as

$$Y^T A = W^T A, \; A = [a_0, a_1, ..., a_m].$$

(6.43)

If we multiply Equation 6.43 by Y and take the expected value, we obtain

$$E[YY^T A] = E[YW^T A] = E[(X + W)W^T A].$$

$$\Gamma_{YY} A = \sigma_w^2 A.$$

$$(\Gamma_{YY} - \sigma_w^2 I) A = 0.$$

(6.44)

$w(n)$ has zero mean and X is a deterministic sequence.

Equation 6.44 can be recognized as an eigen equation with σ_w^2 as the eigenvalue of the autocorrelation matrix Γ_{YY}. We will understand Pisarenko harmonic decomposition method with the help of the following simple example.

Example 2

Autocorrelation values $\gamma_{YY}(0) = 3$, $\gamma_{YY}(1) = 1$, and $\gamma_{YY}(2) = 0$ are given for a process of a single sinusoid with additive white noise. Find the frequency, its power, and variance of white noise.

Solution: Let us first write the correlation matrix.

$$\Gamma_{YY} = \begin{bmatrix} \gamma_{YY}(0) & \gamma_{YY}(1) & \gamma_{YY}(2) \\ \gamma_{YY}(-1) & \gamma_{YY}(0) & \gamma_{YY}(1) \\ \gamma_{YY}(-2) & \gamma_{YY}(-1) & \gamma_{YY}(0) \end{bmatrix} = \begin{bmatrix} 3 & 1 & 0 \\ 1 & 3 & 1 \\ 0 & 1 & 3 \end{bmatrix} \tag{6.45}$$

$\gamma_{YY}(-1) = \gamma_{YY}(1)$ and $\gamma_{YY}(-2) = \gamma_{YY}(2)$

The minimum eigenvalue is obtained as a root of the characteristic polynomial.

$$g(\lambda) = \begin{bmatrix} 3-\lambda & 1 & 0 \\ 1 & 3-\lambda & 1 \\ 0 & 1 & 3-\lambda \end{bmatrix} = (3-\lambda)(\lambda^2 - 6\lambda + 7) = 0.$$

The eigenvalues are $\lambda_1 = 3, \lambda_2 = 3+\sqrt{2}, \lambda_3 = 3-\sqrt{2}$.

$$\sigma_w^2 = \lambda_{\min} = 3 - \sqrt{2}. \tag{6.46}$$

The corresponding eigenvalue is nothing but the vector that satisfies Equation 6.44.

$$(\Gamma_{YY} - \sigma_w^2 I)A = 0.$$

$$\begin{bmatrix} \sqrt{2} & 1 & 0 \\ 1 & \sqrt{2} & 1 \\ 0 & 1 & \sqrt{2} \end{bmatrix} \begin{bmatrix} 1 \\ a_1 \\ a_2 \end{bmatrix} = \begin{bmatrix} 0 \\ 0 \\ 0 \end{bmatrix} \Rightarrow$$

$$a_1 = -\sqrt{2} \text{ and } a_2 = 1. \tag{6.47}$$

We will use the values of the coefficients to form a Z polynomial and find the roots to get the frequency.

$$a_0 + a_1 Z^{-1} + a_2 Z^{-2} = 0.$$

$$1 - \sqrt{2}Z^{-1} + Z^{-2} = 0.$$

$$Z^2 - \sqrt{2}Z + 1 = 0.$$

$$Z_{1,2} = \frac{1}{\sqrt{2}} \pm j\frac{1}{\sqrt{2}}, |Z_1| = |Z_2| = 1. \tag{6.48}$$

The roots lie on a unit code

We will now find the frequency using

$$Z_1 = e^{j2\pi f_1} = \frac{1}{\sqrt{2}} \pm j\frac{1}{\sqrt{2}} \Rightarrow f_1 = \frac{1}{8}. \tag{6.49}$$

The power of the sinusoid is obtained using the equation

$$P_1 \cos(2\pi f_1) = \gamma_{YY}(1) = 1 \Rightarrow$$

$$P_1 \times \frac{1}{\sqrt{2}} = 1 \Rightarrow P_1 = \sqrt{2} \quad (f_1 = 1/8, \cos(\pi/4) = 1/\sqrt{2}). \tag{6.50}$$

Concept Check

- What is the principle of minimum variance spectrum estimation?
- Define eigenvectors and eigenvalues.
- What is a characteristic equation?
- Write a difference equation to generate a sinusoid.
- State the characteristic feature of an ARMA system for sinusoids with white noise.
- How will you find the variance of noise from the eigenvalues?

6.5 Evaluation of Formants Using the Power Spectral Density Estimate

Let us discuss one application of power spectrum estimation, namely, evaluation of formants for a speech signal. But first let us understand the meaning of formants and their relation to LPCs using the vocal tract as an example.

The vocal tract can be considered as a circular waveguide with proper terminating conditions. It is closed at the larynx end and open at the mouth end. Excitation for voiced and unvoiced sound differs. It is a periodic pulse train for voiced speech generation. For unvoiced signals, excitation comes from the random noise generator or turbulence. When the excitation passes through the vocal tract, it resonates at certain resonating frequencies decided by the mode of excitation in the cavity, that is, TE10, TE20, TM11, mode, etc. These resonating frequencies are called formants.

Using these resonating frequencies, one can write the transfer function of the vocal tract, say in Z domain, as shown in Equation 6.51.

$$H(Z) = \frac{Z^M}{(Z - P_1)(Z - P_2), ..., (Z - P_M)}. \tag{6.51}$$

Here, we represent, say, M number of poles of the transfer function. This model of the vocal tract is called an all-pole model or AR model. In this model, it is assumed that there are zeros only at $Z = 0$, that is, the origin. If we multiply the factors of the denominator, we can obtain the denominator polynomial

$$H(Z) = \frac{Z^M}{Z^M + a_1 Z^{M-1} + a_2 Z^{M-2} + \cdots + a_{M-1} Z + a_M}, \tag{6.52}$$

where a_1, a_2, etc. are the coefficients of the polynomial called LPCs (as discussed in Chapter 5). If we factorize the denominator polynomial again, we will get the poles called formant frequency locations. This defines the relation between the formants and LPCs.

As we have observed, there are different methods to estimate power spectrum density. We have discussed the power spectrum density estimation using the method of periodograms in Section 6.2.1 and using the Welch method in Section 6.2.3.

As the offset used in the calculation of autocorrelation increases, the variance of the power spectrum density estimate also increases. There is a need to reduce the variance. This is where the Welch method comes in. Here, the method of periodograms is modified to reduce the variance of the estimate. The method is also called method of modified periodograms. Basically, this method divides the signal into small overlapping segments with 50% overlap between the successive segments. The second modification is that the data segments are first passed through a window, either Bartlett or Hamming, before calculation of the periodogram.

Similarly, for PSD estimates, the Welch method uses a window of small length, thereby decreasing the time window length and increasing the frequency window width, that is, decreasing the frequency resolution. The Heisenburg uncertainty principle states

$$\Delta t \times \Delta f \geq 1/2 . \tag{6.53}$$

When the time window length reduces, the variance of the estimate also reduces by the same factor. For example, if we are using a window of length 2048 for the periodogram method and a window of length 64 for the Welch method, the window length reduces by a factor of 2048/64, that is, 32. Hence, the variance of the estimate will reduce by a factor of 32. It must be remembered that improving the variance of the estimate is at the cost of frequency resolution. Some of the formants may be lost if the frequency resolution is not in a position to detect it.

We will now write a MATLAB program to estimate the power spectrum using the method of periodograms and the Welch method. We will use a direct command available in MATLAB, namely, spectrum.welch and spectrum.periodogram. The program listing given here is for a voiced speech segment using the Welch method. Figure 6.13 shows a plot of the signal and its periodogram. If we change the command in the program specrum.welch to spectrum.periodogram, we will get the periodogram estimate. Figure 6.14 shows the plot of the signal and its periodogram estimate.

```
%to find formants from the voiced signal using power spectral
estimation of speech signal using Welch method.
clear all;
fp=fopen('watermark.wav');
fseek(fp,224000,-1);
a=fread(fp,1024);
a=a-128;
subplot(2,1,1);plot(a);title('plot of voiced part of a signal');
xlabel('sample no.');ylabel('amplitude');
%1024 samples of the voiced signal are displayed
subplot(2,1,2);
Fs = 22100;   t = 0:1/Fs:.296;
h=spectrum.welch
d = psd(h,a,'Fs',Fs);
%Calculate the power spectrum estimate using Welch method
plot(d);
```

6.5.1 Interpretation of the Results

If we compare the plots of the power spectrum using the Welch method and the periodogram method, we find that the periodogram estimate almost tallies with the log spectrum graph. The peaks tracked in the power spectrum using the periodogram method tally with that obtained using log spectrum. The peaks occur at about 800, 1200, 1900, 3000, and

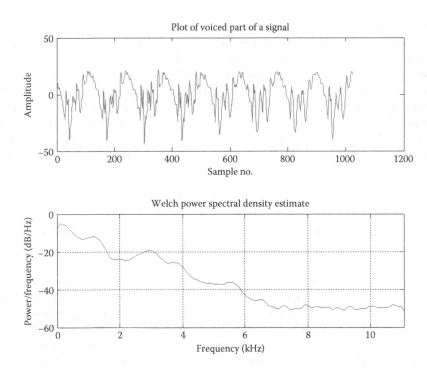

FIGURE 6.13
Plot of signal and its power spectrum estimate using the Welch method.

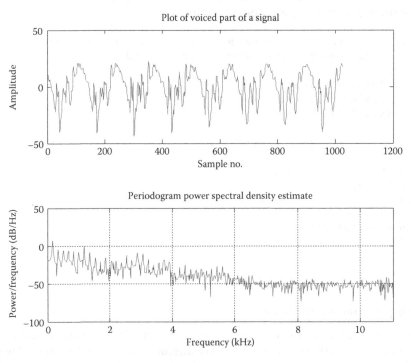

FIGURE 6.14
Plot of signal and its power spectrum estimate using the method of periodiogram.

3800 Hz. The peaks obtained using the Welch method are less in number; several peaks in the periodogram are missed when we use the Welch method. The Welch method used in the program uses a segment size of 64 allowing a frequency resolution of 22100/64, that is, 3431 Hz. The frequency points plotted are for 3431, 690.62, 10393, 1381.24, 1726.55, 2071.66, 2417.11, 2762.48, 3107.79 Hz, etc. The peaks for power spectrum density using the Welch method seem to occur at 1300, 3000, and 3800 Hz. The peaks at 800 and 1900 Hz are lost; we can say that the peaks near the frequency points resolved by the window are detected and other peaks are lost. To capture these peaks, one will have to use a longer data segment and a larger window size of, say, 2048 samples. If we do this, we will get almost the same power spectrum density graph as that for the periodogram method. We have to sacrifice for the variance of the estimate in this case. The reader is encouraged to verify this.

The same experiment can be repeated for the unvoiced segment of a signal. The program listing is given here for the Welch method. Figure 6.15 shows a plot of the signal and PSD estimate using the Welch method. Figure 6.16 shows the plot of the signal and its periodogram estimate for PSD.

```
%to find formants from the unvoiced signal using cepstrum of speech
signal
clear all;
fp=fopen('watermark.wav');
fseek(fp,23044,-1);
a=fread(fp,1024);
a=a-128;
subplot(2,1,1);plot(a);title('plot of unvoiced part of a signal');
xlabel('sample no.');ylabel('amplitude');
%1024 samples of the voiced signal are displayed
a1=window(@Hamming,1024);
for i=1:1024,
    a11(i)=a(i)*a1(i);
end
%the signal is passed via Hamming window and displayed.
Fs = 22100;    t = 0:1/Fs:.296;
h=spectrum.periodogram
d = psd(h,a11,'Fs',Fs);
figure;% Calculate the PSD
plot(d);
```

The observations made for the voiced segment can also be verified for the unvoiced segment. We see peaks in the PSD estimate using the Welch method at 1200, 2000, and 4000 Hz. The peaks using the periodogram method are found to occur at 800, 1200, 1500, 2200 Hz, etc. Hence, it is concluded that some of these peaks are lost using the Welch method.

Concept Check

- What are formants?
- What is the relation between formants and LPC?
- Why is it required to estimate the power spectrum density?

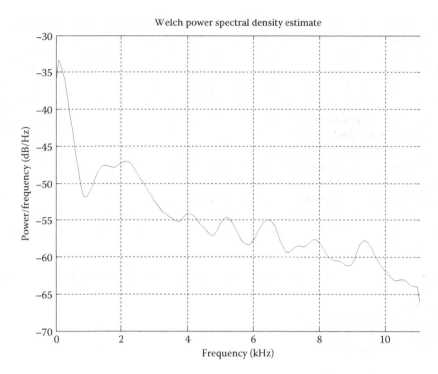

FIGURE 6.15
Plot of signal and its PSD estimate using Welch method.

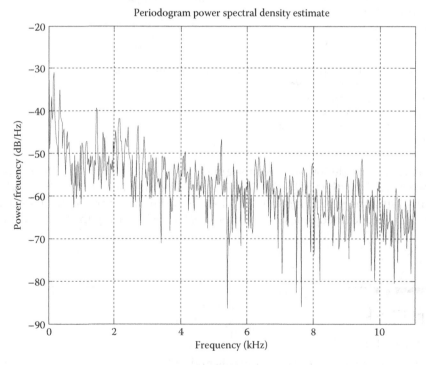

FIGURE 6.16
Periodogram power spectral density estimate.

- Why is an overlap of 50% recommended between the segments?
- Why do we call the Welch method a modified periodogram method?
- What is the drawback of the periodogram method?
- What is the drawback of the Welch method?

6.6 Evaluation of Cepstrum

This section is devoted to cepstral analysis. Let us first define a cepstrum. A cepstrum is obtained when we take an FT of the log spectrum. The term cepstrum is derived by reversing the first four letters of "spectrum." We can have a power cepstrum or a complex cepstrum.

Let us define a power cepstrum. It is a squared magnitude of the FT of the log of the squared magnitude of the FT of a signal. We can algorithmically define it as

signal $\rightarrow$ FT $\rightarrow$ abs() $\rightarrow$ square $\rightarrow$ log $\rightarrow$ FT $\rightarrow$ abs() $\rightarrow$ square $\rightarrow$ power cepstrum.

Let us also define a complex cepstrum. The complex cepstrum was defined by Oppenheim in his development of the homomorphic system theory as the FT of the logarithm obtained from the unwrapped phase of the FT of a signal. Sometimes, it is also called the spectrum of a spectrum. We will define it algorithmically as

signal $\rightarrow$ FT $\rightarrow$ abs() $\rightarrow$ log $\rightarrow$ phase unwrapping $\rightarrow$ FT $\rightarrow$ cepstrum.

The real cepstrum will use the log function defined for real values. The real cepstrum is related to the power cepstrum and the relation can be specified as

$$4(\text{Real spectrum})^2 = \text{Power spectrum}. \tag{6.54}$$

In some texts, the process of taking cepstrum is defined algorithmically as

FT $\rightarrow$ abs() $\rightarrow$ log $\rightarrow$ IFT.

This equation involves Inverse Fourier Transform (IFT) instead of FT. This hardly matters as when FT and IFT are applied on complex signals, the difference is only of a phase sign. The magnitude may change; but in this case, they define FT $X(k)$ and IFT $x(n)$ as given by Equations 6.54 and 6.55, respectively.

$$X(k) = \frac{1}{\sqrt{N}} \sum_{n=0}^{N-1} x(n) W^{nk}. \tag{6.55}$$

$$x(n) = \frac{1}{\sqrt{N}} \sum_{k=0}^{N-1} X(k) W^{-nk}. \tag{6.56}$$

The algorithm takes FT of a signal; which means we enter the spectrum domain. Taking absolute value and log, we remain in the spectrum domain, that is, the x-axis is frequency. When we take inverse fast fourier transform (IFFT) further, we actually come back to the time domain. But here there is a log block in between which says that the output domain is neither a frequency domain nor a time domain. It is generally called a quefrency domain—quefrency obtained by reversing the first four letters of "frequency." Hence, a cepstrum graph of a signal will be an amplitude vs quefrency graph.

When a signal is analyzed in the cepstral domain, the analysis is termed as a cepstral analysis. A short-time cepstrum analysis was proposed by Schroeder and Noll for pitch determination of human speech.

Basically, speech signal is a convolved signal resulting from convolution of impulse response of the vocal tract with either a periodic pulse train or a random signal or turbulence for voiced and unvoiced speech, respectively. When one tries to find pitch frequency as a first peak in the spectrum, there is a possibility that the first formant and the fundamental frequency may overlap. In such a situation, it is difficult to separate the first formant from the pitch frequency. The isolation of these two components can be achieved in the cepstral domain. Hence, the cepstral domain is preferred for pitch measurement. Let us see how this is made possible in the next section.

6.7 Evaluation of Higher Order Spectra

Let us discuss higher order spectral analysis, namely, the bispectrum. We have studied power spectral estimation. It gives the distribution of power over the frequency components. Here, the phase information is suppressed. Power spectrum can be obtained as an FT of autocorrelation sequence. Hence, it gives the statistical information about a Gaussian process with its mean. Often, it is required to find information related to the deviation from the assumption of Gaussian processes. However, the process is nonlinear. Higher order spectra (HOS) helps in this regard. HOS are defined in terms of the higher order cumulants. Higher order moment spectra for deterministic signals are also defined. Higher order cumulant spectra are defined for random signals. The study of HOS is useful in three respects. First, it suppresses the Gaussian noise of unknown parameters. Second, it preserves the true phase and magnitude information. Third, it can detect nonlinearity in the signal.

We will first define moments, cumulants, and HOS of random signals. HOS are also called cumulant spectra and are defined in terms of the cumulants. Let us define cumulants. Consider a set of real random variables $x_1, x_2, ..., x_n$. Their joint cumulants of order $r = k_1 + k_2 + \cdots + k_n$ are defined as

$$C_{k_1 k_2, ..., k_n} = (-j)^r \frac{\partial \ln \varphi(\omega_1, \omega_2, ..., \omega_n)}{\partial \omega_1^{k_1} \partial \omega_2^{k_2}, ..., \partial \omega_n^{k_n}} \Big|_{\omega_1 = \omega_2 = ... = \omega_n = 0} \qquad (6.57)$$

where $\varphi(\omega_1, \omega_2, ..., \omega_n) = E\{\exp[j\omega_1 x_1 + ... + \omega_n x_n]\}$ is their joint characteristic function.

The joint moments of the same set of random variables is given by

$$m_{k_1 k_2 ... k_n} = (-j)^r \frac{\partial \varphi(\omega_1, \omega_2, ..., \omega_n)}{\partial \omega_1^{k_1} \partial \omega_2^{k_2}, ..., \partial \omega_n^{k_n}} \Big|_{\omega_1 = \omega_2 = \cdots = \omega_n = 0}$$

$$= E\{x_1^{k_1} x_2^{k_2} ... x_n^{k_n}\}. \qquad (6.58)$$

The first four moments are given by

$$m_1 = E[x_1], \, m_2 = E[x_1^2],$$

$$m_3 = E[x_1^3], \, m_4 = E[x_1^4]. \tag{6.59}$$

The cumulants can be written in terms of moments as

$$C_1 = m_1, \, C_2 = m_2 - m_1^2,$$

$$C_3 = m_3 - 3m_2 m_1 + 2m_1^2,$$

$$C_4 = m_4 - 4m_3 m_1 - 3m_2^2 + 12m_2 m_1^2 - 6m_1^4. \tag{6.60}$$

The computation of the nth cumulant requires the knowledge of all its previous moments. If the process is zero mean, that is, $m_1^x = 0$, then the second and third order cumulants are the same as their moments by putting $m_1^x = 0$ in Equation 6.60 for all x.

6.7.1 Cumulant Spectra

The HOS are called cumulant spectra as they are defined in terms of their cumulants. If the cumulant sequence is absolutely summable, then the nth order cumulant spectra of $x(k)$ exists. The nth order cumulant spectra of $x(k)$ is defined as the $(n-1)$ dimensional FT of the nth order cumulant sequence.

Case 1: Consider the power spectrum, that is, $n = 2$.

$$C_2^x(\omega) = \frac{1}{2\pi} \sum_{\tau=-x}^{x} c_2^x(\tau) \exp(-j\omega\tau), \tag{6.61}$$

where $|\omega| \leq \pi$.

The second order spectrum is the FT of the autocorrelation sequence. This is the well-known result.

Case 2: Consider the trispectrum for $n = 3$. It is defined on similar lines.

$$C_3^x(\omega_1, \omega_2) = \frac{1}{(2\pi)^2} \sum_{\tau=-x}^{x} c_3^x(\tau_1, \tau_2) \exp[-j(\omega_1 \tau_1 + \omega_2 \tau_2)], \tag{6.62}$$

where $|\omega_1| \leq \pi, |\omega_2| \leq \pi, |\omega_1 + \omega_2| \leq \pi$. There are two methods to find the HOS—the direct method and the indirect method. The definitions of cumulants depend on expectation evaluation that assumes the data to be of infinite length. In practice, the data available are of finite length. Hence, the cumulants can only be estimated. We will discuss the indirect method.

6.7.1.1 Indirect Method

Let $X(k)$ for $k = 1, ..., N$ be the available data. The algorithm can be stated as follows.

Step 1: Segment the data in K records of M samples each. Consider the ith segment or record of the data.

Step 2: Subtract the mean of each record from each sample of the record. This makes the data segment to have a zero mean. The third order cumulant will now be the same as the third order moment.

Step 3: Estimate the third order moment for each segment as follows.

$$m_3^{x_i}(\tau_1, \tau_2) = \frac{1}{M} \sum_{l=-l_1}^{l_2} x^i(l) x^i(l + \tau_1) x^i(l + \tau_2). \tag{6.63}$$

Step 4: Compute the average third order cumulant as

$$c_3^{x_i}(\tau_1, \tau_2) = \frac{1}{K} m_3^{x_i}(\tau_1, \tau_2). \tag{6.64}$$

Step 5: Obtain the bispectrum as

$$C_3^x(\omega_1, \omega_2) = \frac{1}{(2\pi)^2 K} \sum_{\tau=-x}^{x} m_3^x(\tau_1, \tau_2) \exp[-j(\omega_1 \tau_1 + \omega_2 \tau_2)] w(\tau_1, \tau_2),$$

where $|\omega_1| \le \pi, |\omega_2| \le \pi, |\omega_1 + \omega_2| \le \pi.$ \hfill (6.65)

$w(\tau_1, \tau_2)$ is a two-dimensional window.

Let us do it practically. Let us use a speech signal and apply all the steps to find the bispectrum. A MATLAB program to do it is given here..

```
clear all;
f=fopen('watermark.wav','r');
fseek(f,80000,-1);
a=fread(f,2048);
plot(a);title('speech segment'); xlabel('sample no.');
ylabel('amplitude');
y = cum3est (a, 100, 128, 50, 'b', 32);
figure;
plot(y);title('third order cumulant estimated'); xlabel('sample no.');
ylabel('amplitude');
[Bspec,waxis] = bispeci (a, 100, 128, 50,'b',128, 0);
figure;
plot(abs(Bspec));title('bispectrum'); xlabel('sample no.');
ylabel('amplitude');
plot(angle(Bspec));title('phase of bispectrum'); xlabel('sample no.');
ylabel('angle in radians');
```

We have used the MATLAB commands, namely, cum3est and bispeci. The command cum3est estimates the third order cumulant for the data vector a. We have to specify the data vector a, the maximum lag to be computed (100), length of the data subsegments (128), and % overlap of the segments (50) have to be computed. The flag-if biased estimation is also to be computed; specify it as b or otherwise an unbiased estimation is done and the last parameter is fixed lag (32). The next command, namely, bispeci computes the bispectrum indirectly. It uses the parameters: data vector; maximum lag (100), length of the data sub-segment (128), % overlap of the segments (50), which are all to be computed; flag-if biased estimation is also to be computed—specify it as b or otherwise unbiased estimation is done. The last parameter is nfft—FFT length is to be used [default = 128] and wind—window function is to be applied. If wind = 0, the Parzen window is applied (default), otherwise the hexagonal window with unity values is applied. Figure 6.17 shows a speech segment plot. Figure 6.18 depicts its third order cumulant and Figure 6.19 shows the plot of the bispectrum.

6.7.1.2 Direct Method

The algorithm for the direct method can be stated as follows.

Let $X(k)$ for $k = 1, \ldots, N$ be the available data. The algorithm can be stated as follows.

Step 1: Segment the data in K records of M samples each. Consider the ith segment or record of the data.

Step 2: Subtract the mean of each record from each sample of the record. This forces the data segment to have a zero mean. The third order cumulant will now be the same as the third order moment.

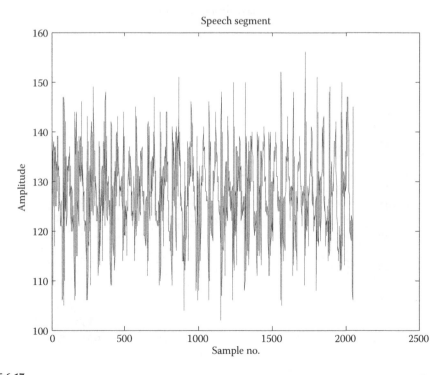

FIGURE 6.17
Plot of speech segment.

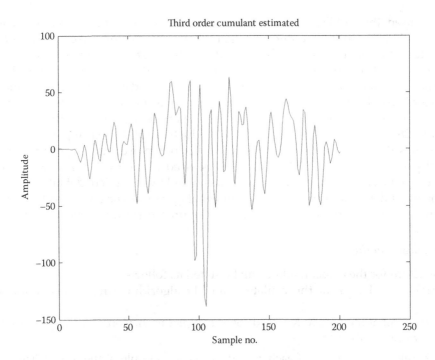

FIGURE 6.18
Third order cumulant estimated.

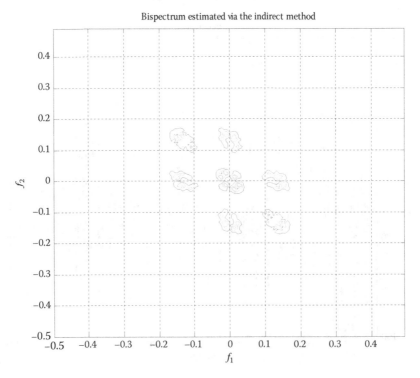

FIGURE 6.19
Plot of bispectrum.

Step 3: Find DFT of each segment.

$$F_X^i(k) = \sum_{n=0}^{M-1} x^i(n) e^{-j\frac{2\pi}{M}nk}, \quad k = 0, 1, \ldots, M-1, i = 1, 2, \ldots, k. \tag{6.66}$$

Step 4: The third order spectrum of each segment is obtained as

$$C_3^{Xi}(k_1, k_2) = \frac{1}{M} F_X^i(k_1) F_X^{i*}(k_2) F_X^{i*}(k_1 + k_2), \, i = 1, 2, \ldots, K. \tag{6.67}$$

Step 5: The variance of the estimate can be reduced by smoothing over a rectangular window given by

$$C_3^{Xi}(k_1, k_2) = \frac{1}{M_3^2} \sum_{n_1=-M_3/2}^{M_3/2-1} \sum_{n_2=-M_3/2}^{M_3/2-1} C_3^{Xi}(k_1+n_1, k_2+n_2), \, i = 1, 2, \ldots, K. \tag{6.68}$$

Step 6: Finally, find the average of all third order spectra.

$$C_3^{Xi}(\omega_1, \omega_2) = \frac{1}{K} \sum_{i=1}^{K} C_3^{Xi}(\omega_1, \omega_2), \, \omega_i = \frac{2\pi}{M} k_i, i = 1, 2, \ldots, K. \tag{6.69}$$

A MATLAB program to find the bispectrum using direct method is given here.

```
clear all;
f=fopen('watermark.wav','r');
fseek(f,80000,-1);
a=fread(f,2048);
a=a-128;
plot(a);title('speech signal'); xlabel('sample no.');
ylabel('amplitude');
y = cum3est (a, 100, 128, 50, 'b', 32);
figure;
plot(y);title('third order cumulant estimated'); xlabel('sample no.');
ylabel('amplitude');
[Bspec1,waxis1] = bispecd (a,  128, 0, 128, 50);
figure;
subplot(2,1,1);
plot(abs(Bspec1));title('magnitude of bispectrum using direct
method'); xlabel('sample no.'); ylabel('amplitude');
subplot(2,1,2);plot(angle(Bspec1)); title('phase of bispectrum using
direct method'); xlabel('sample no.'); ylabel('amplitude');
```

Figure 6.20 shows the bispectrum plot using the direct method and Figure 6.21 shows a magnitude plot for the bispectrum for the same speech segment.

Let us find the bispectrum of a musical instrument signal, say, a flute. A MATLAB program to find the third order cumulant, bispectrum, bispectrum magnitude, and phase using the indirect method, and bispectrum, bispectrum magnitude, and phase using the direct method is given here.

Figure 6.22 shows a signal and its third order cumulant. Figure 6.23 plots the bispectrum using the indirect method.

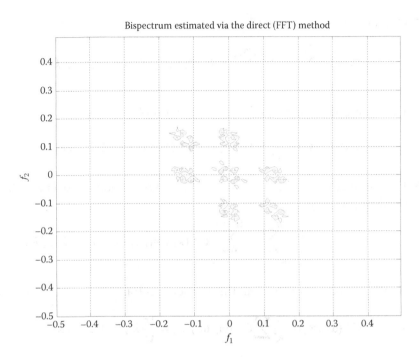

FIGURE 6.20
Bispectrum plot using the direct method.

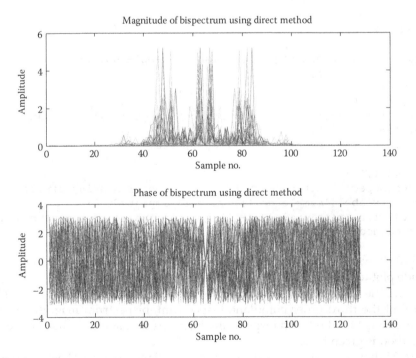

FIGURE 6.21
Magnitude and phase plot for a bispectrum using the direct method.

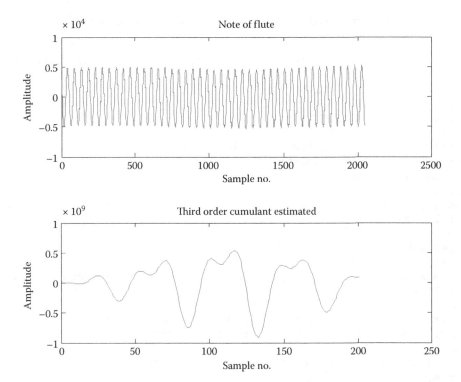

FIGURE 6.22
Plot of signal and its third order cumulant.

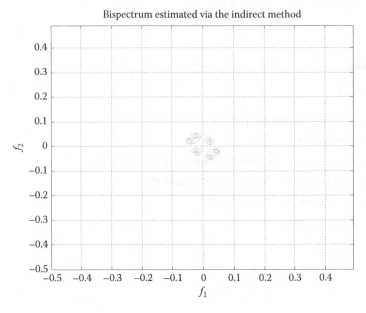

FIGURE 6.23
Plot of bispectrum using indirect method.

Trispectrum and further HOS are also computed on similar lines.

```
clear all;
%fp=fopen('Guitar_A_52.wav','r');% use this statement for Guitar bi
spectrum
fp=fopen('Flute_A_51.wav','r');
fseek(fp,90000,-1);
a=fread(fp,4096,'int16');
a=a-128;subplot(2,1,1);
plot(a);title('Note of flute');xlabel('Sample
number');ylabel('Amplitude');
y = cum3est (a, 100, 128, 50, 'b', 32);
subplot(2,1,2)
plot(y);title('third order cumulant estimated'); xlabel('sample no.');
ylabel('amplitude');
figure;
[Bspec,waxis] = bispeci (a,100,128,50,'b',128, 0);
figure;subplot(2,1,1);
plot(abs(Bspec));title('bispectrum magnitude using indirect method');
xlabel('sample no.'); ylabel('amplitude');
subplot(2,1,2);
plot(angle(Bspec));title('phase of bispectrum using indirect method');
xlabel('sample no.'); ylabel('angle in radians');
figure;
[Bspec1,waxis1] = bispecd (a,  128, 0, 128, 50);
figure;
subplot(2,1,1);
plot(abs(Bspec1));title('bispectrum magnitude using direct method');
xlabel('sample no.'); ylabel('amplitude');
subplot(2,1,2);
plot(angle(Bspec1));title('phase of bispectrum using direct method');
xlabel('sample no.'); ylabel('angle in radians');
```

Concept Check

- What are higher order cumulants?
- What are moments?
- What is the difference between moment and cumulant?
- How will you find the third order cumulant?
- Why do we use higher order spectra.
- Define a bispectrum.

Summary

This chapter mainly focuses on PSD estimation using nonparametric and parametric methods. Methods such as the minimum variance method and the eigen analysis method are also discussed. Lastly, HOS is introduced.

1. At first, energy signals and power signals are defined. Energy signals have finite energy and zero average power. Power signals have infinite energy and finite average power. The problems involved in calculation of energy density spectrum for deterministic signals are discussed. Energy density spectrum is calculated as the FT of autocorrelation. Power density spectrum calculation also has two problems, namely, the data record available is finite and the pdf is not known. We assume that the random process is ergodic so that ensemble autocorrelation is calculated as the time average autocorrelation.

2. Next, different nonparametric methods for power spectrum computation are discussed. Use of DFT to find the periodogram is explained. The Bartlett method uses average value of power for each DFT coefficient. Padding extra zeros to the signal extrapolates the periodogram. The Welch method is also termed as the method of modified periodograms. It allows the segments to overlap. The second modification suggested by Welch is to window the segments prior to computing the periodograms. The Blackman–Tukey method uses autocorrelation to find the power spectrum. It passes the autocorrelation via a smooth window to compensate for error in the calculation of the autocorrelation. The measures used for comparison of different nonparametric methods are quality and number of computations. The Welch method requires a little more computational power than the other methods. Quality is highest for the Blackman–Tukey method. MATLAB programs illustrate the different methods.

3. The next section concentrated on parametric methods using AR, ARMA, and MA model. The procedure for computation of power spectrum by finding AR parameters using the Levinson–Durbin algorithm is illustrated. Burg's method for power spectrum estimation is discussed. Burg uses forward and backward prediction error and minimizes the same in the algorithm. Burg postulated that the interpolation of autocorrelation that gives flattest spectrum will maximize the entropy. Hence, Burg's method is called as the MEM. The minimum variance method is described. If the variance of the output is minimized, the filter passes that frequency, say f_1, undistorted and attenuates other frequencies away from f_1. Changing f_1 over the entire range, we can find the power spectrum estimate with minimum variance. Eigenvalues and eigenvectors are defined. The eigen analysis method for power spectrum estimation is described. The sinusoidal signal corrupted by additive white noise is used. In this case, the overall system will again be an ARMA model. The coefficients of AR and MA system are identical.

4. One application of power spectrum estimation, namely, evaluation of formants in case of a speech signal, is described. We have briefly discussed the meaning of formants and its relation to LPC. We have illustrated the use of the periodogram method and the Welch method for evaluation of formants. The advantage of the Welch method and its drawback are discussed in detail with a MATLAB program.

5. Cepstrum domain is defined, and the pitch period measurement using cepstrum is illustrated using a block schematic. The method of entering a cepstrum domain is discussed, and the correct way to use a cepstrum window to eliminate the vocal tract information is pointed out. The peaks can be seen in the cepstrum in case of the voiced segment.

6. Finally, the higher order cumulant, namely, the third order cumulant, and HOS, that is, bispectrum, are discussed. Higher order cumulant spectra are defined for random signals. HOS suppresses the Gaussian noise of unknown parameters, preserves the true phase and magnitude information, and detects the nonlinearity in the signal. Higher order cumulants and spectra are defined. The indirect method and direct method for computation of bispectrum are explained. MALAB programs illustrate the computation of bispectrum for speech and music signals.

Key Terms

Energy signal	Time resolution
Power signal	Frequency resolution
Energy density spectrum	Heisenberg uncertainty principle
Power density spectrum	Levinson–Durbin algorithm
Periodogram	Berg's method
Modified periodogram	Maximum entropy method (MEM)
Bartlett method	Lattice structure
Welch method	Pitch period
Blackman–Tukey method	Cepstrum
AR model	Cepstrum domain
ARMA model	Cumulants
MA model	Moments
Eigen analysis method	Higher order spectra (HOS)
Minimum variance method	Bispectrum
Discrete Fourier transform (DFT)	Trispectrum
Autocorrelation	Magnitude plot
Formants	Phase plot
LPC	

Multiple-Choice Questions

1. Energy signal is one for which
   a. Energy is highest.
   b. Energy is finite.
   c. Energy is infinite.
   d. Energy is finite and average power is zero.

2. Power signal is one for which
   a. Average power is finite and energy is infinite.
   b. Power is infinite.
   c. Power is zero.
   d. Power is infinite and energy is infinite.

3. Power density spectrum calculation is difficult because
   a. Data record is infinite.
   b. Finite data record is available.
   c. Probability density function is not known.
   d. Finite data record is available and probability density function is not known.

4. The Bartlett method of finding the periodogram is
   a. The method of averaging periodogram.
   b. The method of modified periodogram.
   c. The method of finding periodogram.
   d. The method of finding average of each DFT output.

5. Welch method is
   a. The method of averaging periodogram.
   b. The method of modified periodogram.
   c. The method of finding periodogram.
   d. The method of finding average of each DFT output.

6. Blackman–Tukey method
   a. Finds the average periodogram.
   b. Finds the periodogram.
   c. Finds autocorrelation and then FT of autocorrelation.
   d. Finds the autocorrelation.

7. Computational power required is more for
   a. Welch method.
   b. Blackman–Tukey method.
   c. Bartlett method.
   d. Periodogram method.

8. Parametric methods extrapolate
   a. The autocorrelation.
   b. The periodogram.
   c. The parameters.
   d. The Fourier transform

9. Levinson–Durbin algorithm
   a. Finds AR parameters.
   b. Finds MA parameters.

    c. Finds ARMA parameters.

    d. Finds lattice parameters.

10. The ARMA model is suitable when

    a. The signal is a clean signal.

    b. The signal is corrupted by additive noise.

    c. The signal is corrupted by multiplicative noise.

    d. The signal is corrupted by convoluting noise.

11. Matrix A is said to have an eigenvector x and eigenvalue λ if

    a. $Ax = \lambda$.

    b. $Ax = \lambda x$.

    c. $Ax = \lambda x^2$.

    d. $Ax = \lambda^2 x$.

12. The characteristic feature of ARMA system for sinusoids with white noise is that the

    a. Parameters of AR and MA model are the same.

    b. Parameters of AR model are different from MA model.

    c. Number of parameters of AR model is less than that of MA model.

    d. Number of parameters of AR model is more than that of MA model.

13. The formants are

    a. Resonating frequencies of vocal cords.

    b. Resonating frequencies of vocal tract.

    c. Resonating frequencies of both vocal tract and vocal cords.

    d. Resonating frequencies of mouth cavity.

14. The Welch method reduces the variance of the power spectrum estimate but

    a. Reduces frequency resolution.

    b. Reduces time resolution and frequency resolution.

    c. Increases frequency resolution.

    d. Increases time resolution.

15. The cepstrum is

    a. IFFT of the log spectrum.

    b. FFT of the spectrum.

    c. Windowing of the spectrum.

    d. Windowed FFT of the log spectrum.

16. If the speech segment is voiced,

    a. We observe repetition in the cepstrum.

    b. We observe peaks at regular intervals in the cepstrum.

    c. We observe the random nature in the cepstrum.

    d. We observe no peaks in the cepstrum.

17. Higher order cumulant spectra are defined for
    a. Deterministic signals.
    b. Random signals.
    c. Stationary signals.
    d. Strict sense stationary signals.
18. Higher order spectra are defined in terms of
    a. Cumulants.
    b. Moments.
    c. Autocorrelation.
    d. Fourier transform.
19. Burg postulated that the interpolation that gives flattest spectrum
    a. Will maximize the interpolation.
    b. Will maximize the entropy.
    c. Will maximize the error.
    d. Will maximize the spectrum.

Review Questions

1. Define energy signal and power signal. Define energy spectral density and power spectral density estimation. Why is it called an estimation? How can one estimate the pdf?
2. Describe estimation of power density spectrum of random signals using direct method and indirect method.
3. Describe the use of DFT for power spectrum estimation.
4. Draw a block schematic for the Bartlett method for power spectrum estimation and explain the algorithm.
5. Draw a block schematic for the Welch method for power spectrum estimation and explain the algorithm. What is the advantage of the Welch algorithm? Why is the Welch method also called the method of modified periodogram?
6. Compare the performance of different periodogram methods for power spectrum estimation. What are the performance measures?
7. Discuss the use of parametric methods for power spectrum estimation. Write the algorithm for estimation of power spectrum using the Yule–Walker method.
8. Describe Burg's method for power spectrum estimation.
9. Discuss the situation when Burg's method results in the maximum entropy method (MEM).
10. How will you use AR model for power spectrum estimation of voiced speech?
11. How will you use ARMA model for power spectrum estimation of signal corrupted with noise?

12. Explain minimum variance power spectrum estimation.

13. Explain the eigen analysis method for power spectrum estimation.

14. Define eigenvalues and eigenvectors. Explain the Pisarenko harmonic decomposition method with the help of simple example.

15. What are formants? How will you use periodogram for finding formants?

16. What is a cepstrum? How do we evaluate the cepstrum of a speech segment?

17. How will you take a V/UV decision from the cepstrum plot?

18. Define higher order cumulants. How will you find the bispectrum using indirect method?

19. How will you find the bispectrum using direct method?

20. What is the advantage of using higher order spectra?

Problems

1. Record a speech file of your voice and track the voiced speech. Read 1024 samples of speech and use DFT to compute the periodogram.

2. Record a speech file of your voice and track the voiced speech. Read 1024 samples of speech and use the Bartlett method to compute the periodogram. Compute the quality and number of complex multiplications.

3. Record a speech file of your voice and track the voiced speech. Read 1024 samples of speech and use the Welch method to compute the periodogram. Compute the quality and number of complex multiplications.

4. Record a speech file of your voice and track the voiced speech. Read 1024 samples of speech and use the Blackman–Tukey method to compute the periodogram. Compute the quality and number of complex multiplications.

5. Use the pyulear command in MATLAB to find the power spectrum for 1024 samples of speech file recorded of your voice. Use the Levinson–Durbin algorithm to find LPC.

6. Find all the eigenvectors and eigenvalues of the matrix $A = \begin{bmatrix} 4 & -1 \\ 1 & 2 \end{bmatrix}$.

7. The autocorrelation values

$\gamma_{YY}(0) = 4$, $\gamma_{YY}(1) = 2$, and $\gamma_{YY}(2) = 0$ are given for a process of a single sinusoid with additive white noise. Find the frequency, its power, and the variance of white noise.

8. Find the first three formants for 1024 samples of speech recorded in your voice using the periodogram method and the Welch method.

Suggested Projects

1. Record a speech file in your own voice and use the Bartlett method to plot a periodogram for a segment of size 1024 samples.

2. Record a speech file in your own voice and use the Welch method to plot a periodogram for a segment of size 1024 samples.

3. Record a speech file in your own voice and use the Blackman–Tukey method to plot a periodogram for a segment of size 1024 samples.

4. Record a speech file in your own voice and use the AR model to plot a power spectrum for a segment of size 1024 samples.

5. Record a speech file in your own voice and use the AR model by Burg to plot a power spectrum for a segment of size 1024 samples.

6. Record a speech file in your own voice and use power spectrum estimation using the Welch method to find the formants for a segment of size 1024 samples.

7. Record a speech file in your own voice and use the power spectrum estimation using the Blackman–Tukey method to find the formants for a segment of size 1024 samples.

8. Record a speech file in your own voice and use cepstrum analysis to find the pitch period for a segment of size 1024 samples.

9. Record a speech file in your own voice and use the indirect method for higher order spectral analysis to find third order cumulant, a bispectrum, magnitude, and phase plot of bispectrum. Check if the bispectrum is different for different spoken words.

10. Record a musical instrument note and use the indirect and direct method for higher order spectral analysis to find third order cumulant, a bispectrum, magnitude, and phase plot of bispectrum. Check if the bispectrum is different for different instruments.

Answers

Multiple Choice Questions

1. (d)
2. (a)
3. (d)
4. (a)
5. (b)
6. (c)
7. (a)
8. (a)
9. (a)
10. (b)
11. (b)
12. (a)
13. (b)
14. (a)

15. (a)
16. (b)
17. (b)
18. (a)
19. (b)

Problems

1. Refer to Section 6.2.1. The only modification to be done is as follows. You have to record and plot the entire wav file and see by observation where the voiced part is. Use fseek command to track the voiced part and use the program given in the text on this voiced part.

2. Refer to Section 6.2.2 and Section 6.2. The only modification to be done is as follows. You have to record and plot the entire wav file and see by observation where the voiced part is. Use fseek command to track the voiced part and use the program given in the text on this voiced part.

3. Refer to Section 6.2.3 and Section 6.2. The only modification to be done is as follows. You have to record and plot the entire wav file and see by observation where the voiced part is. Use fseek command to track the voiced part and use the program given in the text on this voiced part.

4. Refer to Section 6.2.4 and Section 6.2. The only modification to be done is as follows. You have to record and plot the entire wav file and see by observation where the voiced part is. Use fseek command to track the voiced part and use the program given in the text on this voiced part.

5. Refer to Section 6.3.1. The only modification to be done is as follows. You have to record and plot the entire wav file and see by observation where the voiced part is. Use fseek command to track the voiced part and use the program given in the text on this voiced part.

6. Eigenvalues are $\lambda_1 = 3$ and $\lambda_2 = \therefore$. The eigenvectors are $\begin{bmatrix} x_2 \\ x_2 \end{bmatrix}$.

$$P_1\cos(2\pi f_1) = \gamma_{YY}(1) = 2 \Rightarrow$$

7. $Z_1 = e^{j2\pi f_1} = -\frac{1}{2} \pm j\frac{\sqrt{3}}{2} \Rightarrow f_1 = \frac{1}{6}. \quad P_1 \times \frac{1}{2} = 2 \Rightarrow P_1 = 4 \quad (f_1 = 1/6, \cos(\pi/3) = 1/2).$

$$g(\lambda) = \begin{bmatrix} 4-\lambda & 2 & 0 \\ 2 & 4-\lambda & 2 \\ 0 & 2 & 4-\lambda \end{bmatrix} = (4-\lambda)(\lambda^2 - 8\lambda + 12) = 0.$$

The eigenvalues are $\lambda_1 = 4, \lambda_2 = 6, \lambda_3 = 2$.

$\sigma_w^2 = \lambda_{min} = 2$.

8. Refer to Section 6. The only modification to be done is as follows. You have to record and plot the entire wav file and observe where the voiced part is. Use fseek command to track the voiced part and use the program given in the text on this voiced part.

7

Statistical Speech Processing

LEARNING OBJECTIVES

In this chapter, we will learn about the following concepts.

- Evaluation of standard deviation
- Skew and kurtosis
- Dynamic time warping (DWT)
- Hidden Markov modeling
- Gaussian mixture models
- Statistical sequence recognition
- Statistical pattern recognition and parameter estimation
- Vector quantization and hidden Markov modeling (VQ-HMM) speech recognition and model training
- Discriminant acoustic probability estimation
- Parametric speech processing
- Pitch synchronous analysis of speech

This chapter will discuss statistical speech processing methods. The first section deals with statistical parameters such as mean, standard deviation, skew, and kurtosis. The dynamic time warping (DTW) required due to natural changes in time duration when same speech is uttered a number of times is discussed in the next section. Speech recognition methods use statistical modeling of speech such as the HMM and Gaussian mixture models. These are described in detail in Section 7.3. We then discuss statistical sequence recognition and pattern recognition. Speech recognition uses vector quantization (VQ) and HMM models. These are explained in brief. HMM models require acoustic probability estimation. We will also describe parametric speech processing. The pitch synchronous analysis is dealt with in detail.

7.1 Measurement of Statistical Parameters of Speech

The statistical parameters used for speech processing are the estimation of mean, standard deviation, variance, skew, and kurtosis. The probability density function (pdf) is also estimated. Let us discuss these measurements one by one.

Only finite segment of speech is available with us. The limitations of the computing capabilities of even our most state-of-the-art computers also limit the length of the speech segment that can be used for processing. To estimate the mean, we process each small segment of speech at a time and find its mean. The mean of means of all the segments is then the estimation of mean for the signal. Let us do it practically. We will use a segment of recorded speech and try to estimate the true mean using the following algorithm.

1. Divide the speech signal into a number of small segments of size, say, 128 samples. The segment size is so selected that the time for each segment is of the order of 10–20 ms.
2. Find the mean of each segment.
3. Find the mean of all these mean values.
4. Find the variance and standard deviation.

We will write a MATLAB program for calculation of mean, variance, and standard deviation. The plot of the speech signal is shown in Figure 7.1.

```
clear all;
clc;
fp=fopen('watermark.wav', 'r');
fseek(fp,100,-1);
a=fread(fp,10240);
a=a-128;
plot(a);
title('plot of speech segment');
xlabel('sample number');ylabel('amplitude');
for j=1:10,
    for i=1:1024,
    m(j)=mean(a);
    end
end
mn=mean(m);
disp(mn);
vr=0.0;
for i=1:10240,
    vr=vr+(a(i)-mn)^2;
    end
disp(vr/10240);
std=sqrt(vr/10240);
disp(std);
```

Mean = −0.4093; Variance = 115.6394; Standard deviation = 10.7536.

The second order center moment is called the variance of the random variable X. The standard deviation is the square root of the variance. The third second order center moment is a measure of asymmetry of the density function about the mean. Skewness can be considered as a measure of symmetry, or, in fact, the lack of symmetry. Kurtosis is defined as ratio of the fourth order center moment and fourth power of standard deviation. The reader can refer to Chapter 2, Section 2.1 for detailed definitions of statistical parameters like skew and kurtosis. We will now consider the measurement of skew and kurtosis. The graph of the speech signal and its histogram are shown in Figure 7.2.

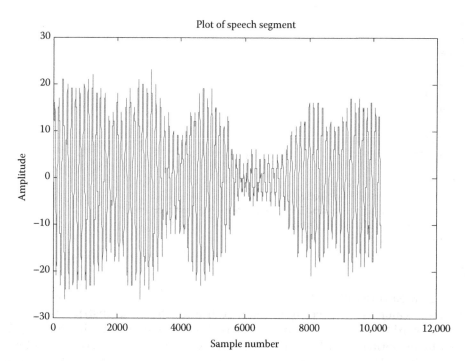

FIGURE 7.1
Plot of speech signal.

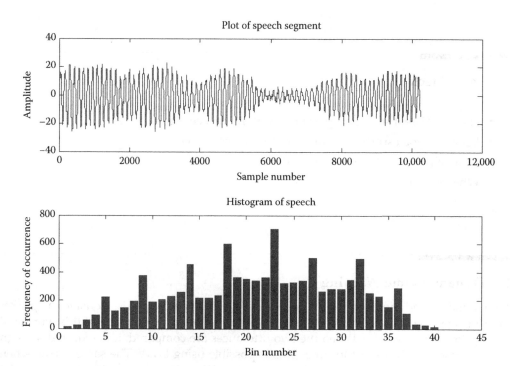

FIGURE 7.2
Plot of speech signal and its histogram.

```
clear all;
clc;
fp=fopen('watermark.wav', 'r');
fseek(fp,8000,-1);
a=fread(fp,10240);
subplot(2,1,1);
a=a-128;
plot(a);
title('plot of speech segment');
xlabel('sample number');ylabel('amplitude');
p=hist(a,40);
subplot(2,1,2);bar(p);
title('histogram of speech');
xlabel('bin number');ylabel('frequency of occurrence');
s=3*( mean(p)-median(p))/std(p);
disp(s);
z=kurtosis(p);
disp(z);
```

Skew=0.2378; Kurtosis=3.5079.

The skew value is positive and small, indicating that the distribution is slightly inclined toward the left. The kurtosis value is closer to 3 indicating that the distribution is closer to the normal distribution. We have divided the total range of parameter values in 40 small bins. The histogram is a plot of frequency of occurrence with respect to the bin number.

Concept Check

- Name different statistical parameters of the speech signal.
- How will you estimate the mean of the speech signal?
- How will you find the skew and kurtosis?
- What is the histogram?

7.2 Dynamic Time Warping

Let us understand the need for the time warping for speech signals. Consider a speaker uttering some word, say, "word," a number of times. The time duration of this word will change for each utterance. When the two utterances are compared, it is required to align the two words with respect to time. This is possible using DTW. The same is true when two different speakers are uttering the same word. We will understand word warping by considering linear time warping first.

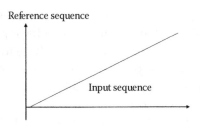

FIGURE 7.3
Plot illustrating linear warping.

7.2.1 Linear Time Warping

Consider two signal sequences of the same length. The problem of finding the distance between the two sequences is simple in this case. We have to find the local distances and sum it up to generate the global distance. In general, however the reference template and the input template will have different number of points in a vector. Let us consider a simple case wherein one template is twice as large as the other template. To make the two templates of the same length, you may decimate the longer template or interpolate the smaller one. This leads to a linear relation between the two templates as shown in Figure 7.3 and is called linear time warping. Linear time warping is a linear function used for warping the two templates.

The speech recognition process consists of matching the incoming speech template with the stored templates. The template with the lowest Euclidean distance from the input pattern is the recognized word. In case of speech signals, linear time warping will not help. The best match (lowest distance measure) is based on dynamic programming. We have to use a DTW word recognizer. Dynamic warping is required for the following reason. Generally, variations in speech duration are not spread evenly over different sounds. For example, stop consonants vary only slightly in length, whereas diphthongs and glides vary by a great amount over different utterances of the same word. The warping will be uneven in this case. A process for optimal warping must be found out. This is done by using a nonlinear or DTW.

7.2.2 Dynamic Time Warping

When the two sequences of feature templates of two different speech words are to be compared, the sequences must be warped dynamically. We have to find the best possible warping of the time axis for one or both sequences for optimal comparison. The criterion will be the minimization of global error. The problem can be formulated as a sequential optimization strategy in which the current estimate of the global error function is updated for each step.

Consider a problem of template matching for speech recognition. The warping function is required to minimize the overall cost function (A typical cost function is the square of the distances between the samples.) subject to the following constraints. The major optimizations to the DTW algorithm arise from observations on the nature of best paths through the grid. The optimization of DTW algorithm is based on different conditions and these can be summarized as:

- *Monotonic condition*: The function must be monotonically increasing. The path is not allowed to turn back on itself; both i and j indexes either stay the same or increase, they will never decrease.
- *Continuity condition*: The function must not skip any points in between. The path is to advance one step at a time.

- *Boundary condition*: The function must match the end points of the two templates. The path will start at the bottom left and ends at the top right.
- *Adjustment window condition*: A good path is not likely to wander very far from the diagonal. The distance that the path is allowed to wander is the window length, *r*.
- *Slope constraint condition*: The path must not be too steep or too shallow. This prevents or will not match very short sequences with very long ones. The condition is expressed as a ratio n/m, where m is the number of steps in the x direction and n is the number in the y direction. After m steps in x, we must make a step in y and vice versa. In other words, we can say that there is a global limit on the slope of a warping function. This can be achieved by restricting the domain of the process to the parallelogram as shown in Figure 7.4.

These observations can be applied and we can restrict the moves that can be made from any point in the path. This restricts the number of paths that need to be considered. For example, with a slope constraint of $P = 1$, if a path has already moved one square up, it must next move either diagonally or to the right.

Computing the cost function can be considered as a process of finding a minimum cost path from the point (1, 1) to (L, M), where L and M are the number of points in the two templates. This is a problem in dynamic programming. Thus, time warping algorithms can be implemented using dynamic programming techniques.

We can state the algorithm for DTW. Imagine that the rows represent the input template and the columns, the reference template. Each point in the template can be a set of 12-Mel frequency cepstral coefficients (MFCC) points for one section of the speech file. The search space is restricted to a parallelogram as shown in Figure 7.4.

1. Start with the first point in each template; compute the distance, that is, a cost function.
2. Now move on to next point in a shorter template and find its cost function with the next point in the longer template, and then move to the further next point in the longer template. Find the minimum cost. Starting from point (1, 1), the next point in the warping function is fixed at the minimum cost point.
3. Go on traversing in the forward direction for each point in the shorter template until both the end points meet.

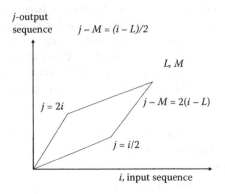

FIGURE 7.4
Parallelogram defined by the slope limits in time warping.

The global distance $D(i, j)$ is computed as

$$D(i, j) = \mathbf{min}[D(i+1, j+1), D(i+1, j), D(i, j+1)] + d(i, j), \qquad (7.1)$$

where $d(i, j)$ is the local distance of the point (i, j). Initial condition is $D(1, 1) = d(1, 1)$.

This process of DTW is illustrated in Figure 7.5. It shows the time alignment between the test and the training template. Here, the input "SsPEEhH" is a "noisy" version of the template "SPEECH." The input "SsPEEhH" can be matched against all templates in the system's training set. The best matching template is found that will have a smaller distance with the input template than any other templates. A simple global distance score for a path is the sum of the local distances that make up the path.

For DTW algorithm, local warping is constrained to skip no frames of either reference or input template, but is permitted to repeat either one. Thus, the minimum is computed only for three predecessors, namely, $(i+1, j)$, $(i, j+1)$, $(i+1, j+1)$. The features for the template must be selected as per human perception, for example, MFCC as discussed in Chapter 5.

The power of the DTW algorithm goes even beyond these observations. Instead of searching for all possible routes that satisfy these constraints in the grid, the DTW algorithm keeps track of the cost of the best path to each point in the grid. During the match process, one will have no idea as to which path is the lowest cost path; this can be traced back when we reach the end point.

The main source of inaccuracy in template matching is due to wrong end point detection in case of isolated word recognition. Hence, the end point match condition in the algorithm is removed. Normally, a pause between the words is required for correct recognition. Sometimes, the short-term spectrum of breath may be confused with fricative. Often, a click is observed preceding the spoken word. If this click gets attached to the word, it cannot be correctly recognized. Normally, the length of the time between the click and the word is checked to see if it can be treated as the same word.

The DTW algorithm is able to achieve the following goals.

1. The temporal integration of local distances between the acoustic frames can be effectively established.

2. It normalizes the time variations for speech sounds.

3. In case of continuous speech, the approach segments the speech and there is no need for any explicit segmentation.

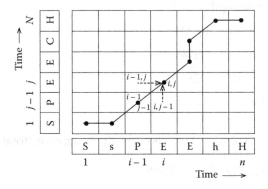

FIGURE 7.5
Process of DTW for matching two templates for the word "speeh."

There are a number of limitations to the DTW-based approach for sequence recognition in case of speech or any other signal.

1. Comparison of templates requires endpoint detection and this can be error-prone in real-world acoustic conditions.
2. A strong mathematical structure is always required for global distance computation and minimization.
3. Since continuous speech is not a concatenation of words, the mechanism to represent the context is required.

Statistical distributions can be a reasonable choice for representation of variability in real-speech samples. We will discuss the statistical sequence recognition method in the following section.

We have conducted innovative experiments for speech recognition. In these experiments, we gave emphasis on the visual features, that is, the speech waveform plot. We extracted a positive envelope of the utterance, which was a feature for speech recognition. The typical visual features captured are

1. Positive envelope of spoken word,
2. Word size,
3. Number of peaks of the envelope, and
4. Direct current (DC) value of the utterance.

The positive envelope of the utterance is tracked by finding the local peaks for a finite number of samples. The local peaks are joined together to get the envelope of the utterance. The result of envelope tracking is shown in Figure 7.6. The word size, number of peaks, and DC value are used for classification of the word so that the template is compared with the templates of the words in the same class. We compare the templates of the positive envelope using the DTW algorithm. The word is not recognized when the distance between the templates is less for the correct word but is recognized when it is less than the set threshold. The threshold is calculated for each word and stored along with the template.

Total words used for recognition = 250

Words falsely recognized using DTW = 12

Words not recognized using DTW = 10

Words correctly recognized = 228

Accuracy of the system:

Accuracy % = (Number of words correctly recognized/total words) × 100.

Accuracy % = (228 / 250) × 100 = 91.2%.

The experiments are also conducted for tracking the negative envelope.

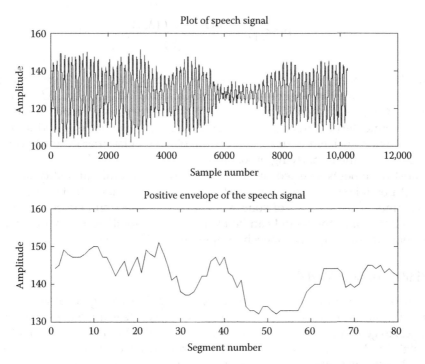

FIGURE 7.6
Positive envelope of the word detected is shown by a black line.

7.3 Statistical Sequence Recognition for Automatic Speech Recognition (ASR)

Statistical models are found to be useful to solve the problem of speech recognition. Statistical framework allows density estimation, alignment of training data, and silence detection for isolated or continuous word recognition. Speech signals can be assumed to be generated according to some probability distribution. As there are different speech sounds, it is necessary to assume that there is more than one distribution. We extract the statistical parameters and construct a model for every linguistic unit. During training, the parameters are estimated to approximate a minimum probability of error. We use Bayes' solution. In the recognition phase, we will search through the space of hypothesized utterances to find the hypothesis that has a maximum *a posterior* (MAP) probability. The probabilities are estimated so that a true minimum may not be reached.

7.3.1 Bayes Rule

The problem can also be formulated using Bayes' rule as follows. Let Y be the observed sequence. There will be a number of statistical models per sequence for every linguistic unit. Let M_j represent the jth statistical model for a sequence. The vector Y will typically consist of cepstral vectors or MFCC vectors computed for an acoustic window. The problem is to classify the vector Y and assign it to a class. According to Bayes' rule,

$$P(M_j \mid Y) = \frac{P(Y \mid M_j)P(M_j)}{P(Y)}. \tag{7.2}$$

If $P(M_j \mid Y)$ is maximized, then Y is assigned to class M_j. The Bayes' decision rule classification is shown in Figure 7.7. There are classes M_1, M_2,..., M_j. The product $P(Y \mid M_j)P(M_j)$ is calculated for each class. If the vector Y is assigned to class M_j for which the product term is maximized, then the error probability is found to be the lowest. The problem associated with this procedure of Bayes' classifier is how to compute the product terms and the underlying probabilities?

The statistical models selected must represent the temporal variability and temporal structure. A possible structure is a stochastic finite state automaton that is used as a speech model. The model will consist of some states and a connection between the states. The parameters describing the model can be transition probabilities. The parameters will be learned from the training set of speech samples and will be used for testing.

7.3.2 Hidden Markov Model

The automata most commonly used for speech recognition are generative models in the sense that the states have outputs that are the observed feature vectors. To concatenate such models to generate concatenated sequences, typically Markov assumptions are made use of. With the Markov assumptions, the general stochastic automata become Markov models. A Markov model is a finite state automaton for which each transition has an associated probability and a sequence of states is a Markov chain.

A strategy that uses a stochastic model of speech production is known as a HMM. It offers performance comparable to time warping at a fraction of the computational cost. HMM is explored by many researchers. In case of HMM, a model can be in only a finite number of different states. HMM for speech recognition has each state capable of generating a finite number of possible outputs. In generating a word, the system passes from one state to another. Each state continues to emit an output until the entire word is out. Such a model is illustrated in Figure 7.8. It is called a state transition diagram. The state is represented by a circle and the arrows represent the transition between the states. The transition probabilities are usually written on the arrows. The transitions between the states and the outputs are made random to cope with the variations in the pronunciations and timings. Hence, HMM is a statistical Markov model where a system is assumed to be a Markov process with unobserved or hidden states. An HMM can be considered the simplest dynamic Bayesian network.

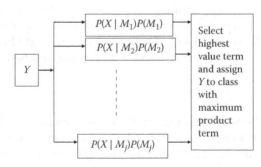

FIGURE 7.7
Bayes' criterion for class selection.

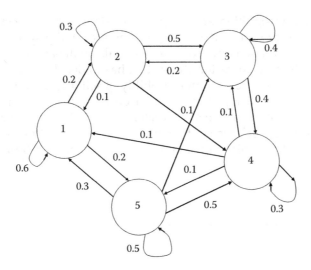

FIGURE 7.8
State diagram for an HMM.

In a regular Markov model, the state is directly visible to the observer, and therefore, the state transition probabilities are treated as the only parameters. In a *hidden* Markov model, the state is not visible, but the output that depends on the state is visible. Each state has a probability distribution over the possible output tokens. Therefore, the sequence of tokens generated by an HMM gives information about the sequence of states. Note that the adjective "hidden" refers to the state sequence through which the model passes. It does not refer to the parameters of the model; even if the model parameters are known exactly, the model is still "hidden." For example, each word in the recognition vocabulary is represented by a model. The recognizer will work with all the outputs and decide which model gave rise to the outputs. The model itself is not visible to the recognizer, but must be inferred from the available data. Hence, it is termed as hidden.

It must be clearly stated that HMM does not require any correspondence to articulatory organs. The number of states in the model does not have any bearing on the expected phonetic content of the word. It is assumed that the transitions from one state to the other occur at discrete times and each transition from state q_i to q_j has a probability which depends only on q_i. If there are N states, one can represent the transition probabilities using an N by N matrix A, where $A(i, j) = P$(transition from q_i to q_j). The transition matrix for Figure 7.10 can be written as follows.

$$A = \begin{bmatrix} 0.3 & 0.5 & 0.1 & 0 & 0.1 \\ 0.2 & 0.4 & 0.4 & 0 & 0 \\ 0 & 0.1 & 0.3 & 0.5 & 0.1 \\ 0 & 0.1 & 0.1 & 0.5 & 0.3 \\ 0.2 & 0 & 0 & 0.2 & 0.6 \end{bmatrix}. \tag{7.3}$$

Note that the sum of all outgoing transitions from any state is one. Hence, the sum of probabilities in a row is one. By making the transitions nondeterministic, we enable the model to handle omissions or repetitions of states. This will allow variability in pronunciation, which is a desirable feature.

HMMs for speech recognition are the group of states interconnected. The states are assumed to emit feature vectors depending on the probability specified for each transition. Consider one observation frame of speech signal. It will have a corresponding feature vector. Observing the feature vector, the system has to decide a state to which it can be associated. If the HMM indicates high probability of self looping for that state, it implies that the feature vector will repeat more than once. A directed connection with high probability indicates that the next feature vector associated with the next state is likely to follow. The models for speech recognition tend to be left to right models in the sense that the state transitions obey a certain temporal order. Normally, each phone is modeled by several states.

Concept Check

- How will you classify feature vector X using Bayes' rule?
- What is the problem with a Bayes' classifier?
- What is a Markov model?
- What is HMM?
- What is hidden in HMM?
- Why is the sum of probabilities in a row equal to one for a transition probability matrix of HMM?
- How will you allow variability in pronunciation using HMM?
- What does it mean when we say that the HMM indicates high probability of self looping for a state?

7.4 Statistical Pattern Recognition and Parameter Estimation

Consider a problem of speech recognition. The feature vectors are found from the speech frames. If there are K words to be recognized, there will be K classes. Let the feature vector be represented as X. The problem is to classify the vector X as belonging to the correct class ω_k, that is, the kth class. We can use *a posteriori* probability criterion or maximum likelihood criterion. Let us discuss the meaning of these criteria.

1. *MAP criterion*: Given that the feature vector X has occurred, the probability that the correct class ω_k is identified, represented as $P(\omega_k \mid X)$, is called the *a posteriori* probability as it is calculated after the data have occurred. The class is identified *a posteriori* probability is found to be maximum. According to Bayes' rule, $P(\omega_k \mid X) = \dfrac{P(X \mid \omega_k)P(\omega_k)}{P(X)}$. The term $P(\omega_k)$ represents the probability of class k, which is *a priori* probability of class k—it is normally evaluated before the data X occurs.

2. Likelihood-based MAP criterion: The probability $P(X|\omega_k)$ is the conditional probability. It is also called the likelihood function. $P(X)$ is constant for all classes. Finding MAP probability is equivalent to determining $P(X|\omega_k)P(\omega_k)$ over all classes and evaluating the maximum to identify the class. The decision for a class is taken by finding the ratio of posterior probabilities given by

$$\frac{P(\omega_k|X)}{P(\omega_j|X)} = \frac{P(X|\omega_k)|P(\omega_k)}{P(X|\omega_j)P(\omega_j)}. \tag{7.4}$$

Bayes' rule says that if the ratio is greater than one, the decision will be taken for class k over class j. This condition to select class k over all j other than k is equivalent to

$$\frac{P(X|\omega_k)}{P(X|\omega_j)} > \frac{P(\omega_j)}{P(\omega_k)}. \tag{7.5}$$

The left-hand side of Equation 7.6 is often termed as a likelihood ratio. If we take the log of the likelihood ratio, we can write the condition as

$$\log P(x|\omega_k) + \log P(\omega_k) > \log P(x|\omega_j) + \log P(\omega_j). \tag{7.6}$$

MAP classification based on likelihood ratio is the same as choosing a class to maximize a statistical discriminant (which can discriminate between the classes) function given by

$$\arg\max_k[\log P(x|\omega_k) + \log P(\omega_k)]. \tag{7.7}$$

7.4.1 Statistical Parameter Estimation

The difficulty in applying Bayes' rule or other such criteria lies in the estimation of probability densities. The main requirement for statistical estimation will be a large training set. Usually, some parameters are learned from the training data and are used for estimation of probabilities. Sometimes, a parametric form of density, say, Gaussian density, is assumed, and parameters are then extracted using the training data. When we assume Gaussian density, its formula is used to find the densities for the problem. The exact Gaussian density may not be acceptable in certain cases. In such cases, the density is assumed as the weighted sum of or a mixture of Gaussian densities. Mean and variance are learned from the training data through some iterative algorithm such as expectation maximization (EM) algorithm. The algorithm can be stated as follows.

1. Select some probability density associated with each class.
2. Choose initial set of parameters for the density function.
3. Compute the posterior estimate for a hidden variable.
4. Find the values of parameters that maximize the joint density for the data and the hidden variable.
5. Evaluate the log likelihood and use MAP criteria. If the stopping criterion is not met, continue with the next iteration.

Consider the problem of estimation of the mean. We assume Gaussian density function for a particular class, that is, a word in case of word recognition. We will assume some value of mean found using some training data. This is termed as initialization of variables. If the density is assumed as the weighted sum of or a mixture of Gaussian densities, find the values of the weights that will maximize the probability for mean. Continue with the new iteration until a true maxima is reached.

7.4.2 Acoustic Probability Estimation for ASR

In this section, we will learn how acoustic probabilities are estimated. The training set data is used to divide each spoken word into either phones or syllables. A model is to be generated for each subword unit. The subword unit will be divided into a number of frames of size 10 ms each. The feature vectors such as MFCC are calculated for each frame. These feature vectors are the outputs of the states of the given model. The feature vectors of the frames are vector quantized using any standard VQ algorithm such as the k-means algorithm.

Clusters are formed and the centroid of each cluster is found using k-means algorithm. The centroid template is called the code word and it represents an index in the code book. Counting the number of templates falling in each cluster gives the probability for the centroid template, that is, the state probability. Each state in an HMM of a word is visible as its output; in other words, the centroid template is available. The transition probability is found as the number of times the state changes from i to j in a training data. If the training data set is very large, a probability estimate is made close to the actual estimate. The radius of each cluster accounts for the variation of features resulting from style of utterance, changes in health conditions of the speaker, and so on. This method of probability estimation is called discrete HMM or VQ-based HMM. It is attractive because of its simplicity and low computational requirements. Neural networks can also be used to estimate the probabilities.

Concept Check

- What is the meaning of MAP criterion?
- What is a likelihood function?
- How will you estimate the probability density?
- How will you find the mean using EM algorithm?
- How will you estimate the state probability using the VQ based approach?

7.5 VQ-HMM-Based Speech Recognition

The block schematic for isolated word recognition using VQ-HMM technique is shown in Figure 7.9. The essential blocks in the speech recognition system are feature vector

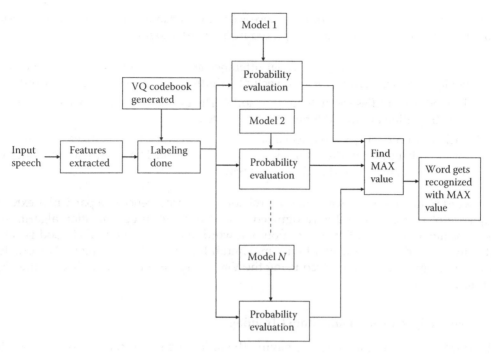

FIGURE 7.9
Block schematic for speech recognition using VQ-HMM.

extraction, VQ codebook, labeling of the external vector, probability calculation for each word model, and the word recognition block. The system has two modes of operation—the training mode and the recognition mode. The feature extraction model will extract the features for the speech segment of, say, 20 ms. The feature vector may consist of MFCC, delta MFCC, and double-delta MFCC.

Training mode: The code book is prepared for all possible vectors for a word using a large data set. The code words in the code book of the word will generate states in a model for each word. The possible transitions from one code word to another will generate the model for each word. The code book is prepared using a process of pruning, the steps of which are listed here.

1. The first data vector is put in the code book.

2. The next vector is compared with the first one. If the distance between the two vectors is less than threshold, the vector is not entered; otherwise, it is entered in the code book. This is how the code book grows. The size of the code book will be equal to the number of possible states.

3. The next word utterance is then used for coding the feature vectors in it. The vectors are put in a cluster of the code word to which it closely matches. The pointer index for each code word generates a label for each feature vector pointing to which code word cluster the external vector falls. After all the feature vectors are coded, the centroid of the cluster is found. This forms the updated new code word.

4. The feature vectors for all utterance of the same word are again coded using these updated code words. The process listed in Steps 2 and 3 is repeated until the centroids do not change.

Recognition mode: The external word utterance is first divided into a number of frames. The features are extracted for each frame. The procedure for word recognition can be listed as follows.

1. The feature vectors in the external utterance are vector quantized and the code book index is generated for each vector by finding a close matching code word.
2. The index identifies the state for the model. The probability for each state and the state transition are found for each word model.
3. Each word model has a sequence of code word vectors, that is, states. The maximum probability for a word model is to be found out. The word with maximum likelihood is recognized as the correct word.

If in word number 2 there is maximum probability for occurrence of a particular external sequence of vectors, word 2 is recognized. The Baum–Welch optimization algorithm is used for model training. For recognition of a word, the word model likelihood is calculated for all models and the word with maximum likelihood is recognized. The problem of word recognition can be solved using the Viterbi algorithm. We will discuss the algorithms in the next section.

7.5.1 HMM Specification and Model Training

The HMM is a finite state machine having the following specifications. A set of hidden states is denoted by Q, an output that is visible from the state is denoted by O, transition probabilities between the states are denoted by A, the emission probabilities (to generate the output from the states) are denoted by B, and initial state probabilities are written as Π (Pi). The current state is not observable. HMM is defined as a triplet (A, B, Π) as shown in Figure 7.10. The parameters for HMM are initial state probabilities, transition probabilities, and emission probabilities. These can be estimated using VQ-HMM as follows. Count the number of feature vectors falling in the cluster representing the state. The ratio of this count to the total number of feature vectors in a word will give the state probability. The number of times the transition from one cluster to another is made is counted to estimate the transition probability. The emission probability is the number of times the output vector is obtained, that is, a feature vector when the word is in a cluster.

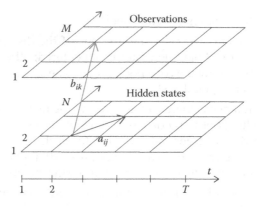

FIGURE 7.10
Model characterized by the set of parameters: $\Lambda = \{A, B, \Pi\}$.

There are three problems to be solved in case of HMMs for speech recognition:

1. Given the model parameters A, B, and Π, compute the probability of a particular output sequence O. This problem can be solved by the forward and backward algorithms.

2. Given the model parameters, find the most likely sequence of (hidden) states that have generated a given observed output sequence. This problem can be solved by the Viterbi algorithm.

3. Given an output observed sequence, find the most likely set of state transition and output or emission probabilities. This is solved by the Baum–Welch algorithm, which is used in the training mode of the system.

Let us discuss these algorithms namely forward and backward algorithm, Viterbi algorithm and Baum Welch algorithm one by one

7.5.1.1 Forward Algorithm

Let us consider a particular phone or a phoneme of a word. Each phone/phoneme has a set of output sequences appearing serially as time progresses. We need to calculate the probability for every output sequence. This output sequence O_t will be reached after passing through state q_t at time t. The particular output sequence is likely to be a sequence obtained from a number of possible output sequences. We have to add the probabilities of all these paths coming from different output sequences. The problem is to find the probability for a particular output sequence at time t denoted by O_t. Let $\alpha_t(i)$ be the probability of the observation sequence $O_t = \{\sigma(1), \sigma(2), \ldots, \sigma(t)\}$ to be produced by all possible state sequences that end at the ith state, namely, q_i.

$$\alpha_t(i) = P(\sigma(1), \sigma(2), \ldots, \sigma(t) \mid q(t) = q_i). \tag{7.8}$$

Then the unconditional probability of the partial observation sequence is given by the sum of $\alpha_t(i)$ over all N states.

The forward algorithm is a recursive algorithm for calculating $\alpha_t(i)$ for the observation sequence of increasing length t. First, the probabilities for the single-symbol sequence are calculated as a product of the initial ith state probability and emission probability of the given symbol $\sigma(1)$ in the ith state. This is called the initialization step. The probability is calculated as follows.

$$\alpha_1(i) = p_i b_i(\sigma(1)), \quad i = 1, \ldots, N. \tag{7.9}$$

Then the recursive formula is to be applied. Assume that we have calculated $\alpha_t(i)$ for some t. To calculate $\alpha_{t+1}(j)$, multiply every $\alpha_t(i)$ by the corresponding transition probability from the ith state to the jth state, sum the products over all states, and then multiply the result by the emission probability of the symbol $\sigma(t+1)$. This is called the recursion process represented as

$$\alpha_{t+1}(i) = \left[\sum_{j=1}^{N} \alpha_t(j) a_{ji} \right] b_i(O(t+1)). \tag{7.10}$$

Here $i = 1, \ldots, N; t = 1, \ldots, T-1$.

Iterating the process, we can eventually calculate $\alpha_T(i)$, which is the total probability to get the observation sequence symbols $\sigma(1)$, $\sigma(2)$,..., $\sigma(t)$ from all possible states to state i. Summing these symbols over all states, we can obtain the required probability as given by

$$P(\sigma(1)\sigma(2),...,\sigma(T)) = \sum_{j=1}^{N} \alpha_T(j). \tag{7.11}$$

7.5.1.2 Backward Algorithm

This algorithm works on lines similar to the forward algorithm. It uses a symmetrical backward variable $\beta_t(i)$ as the conditional probability of the partial observation sequence from $\sigma(t+1)$ to the end to be produced by all state sequences that start at the ith state. The initialization is done as follows.

$$\beta_t(i) = P\left(\sigma(t+1),\sigma(t+2),...,\sigma(T) \,|\, q(t) = q_i\right).$$

$$\beta_T(i) = 1, i = 1,...,N. \tag{7.12}$$

The backward algorithm calculates recursively backward variables going backward along the observation sequence. The forward algorithm is typically used for evaluating the probability of an observation sequence to be emitted by an HMM. Both algorithms are used for finding the optimal state sequence and estimating the HMM parameters. The recursion can be described as

$$\beta_t(i) = \sum_{j=1}^{N} a_{ij} b_j(\sigma(t+1)) \beta_{t+1}(j). \tag{7.13}$$

Here $i = 1,...,N; t = T-1, T-2,...,1.$

Iterating the process, we can calculate the total probability to get the observation sequence symbols $\sigma(1)$, $\sigma(2)$, ..., $\sigma(t)$ from all possible states to state j. Then summing the symbols over all states, we can obtain the required probability as

$$P(\sigma(1)\sigma(2),...,\sigma(T)) = \sum_{j=1}^{N} p_j b_j(\sigma(1)) \beta_1(j). \tag{7.14}$$

Both forward and backward algorithms give the same results for total probabilities $P(O) = P\left(\sigma(1),\sigma(2),...,\sigma(T)\right)$.

7.5.1.3 Viterbi Algorithm

The likelihood for the model can be found out by evaluating the likelihood of the most likely sequence of states. This is called the Viterbi approximation. In Viterbi training, efforts are made to optimize the parameters to maximize the likelihood of the best path in the model. The Viterbi algorithm chooses the best state sequence that maximizes the likelihood of the state sequence for the given observation sequence in a signal. This is used for recognition of the unknown word.

The Viterbi algorithm is a dynamic programming algorithm similar to the forward algorithm except for two differences:

1. It uses maximization in place of summation at both the recursion and termination steps.
2. It keeps track of the arguments that maximize the likelihood for each t and i. These are stored in N by T matrix ψ. This matrix is used to retrieve the optimal state sequence at the backtracking step.

The EM algorithm steps are as follows.

1. Assume an initial set of parameters for the density functions.
2. Determine the most likely state sequence.
3. Update the parameters.
4. Assess the solution and repeat Steps 2 and 3 if required.

Viterbi training procedure finds the likelihood of the best path (state sequence) for each model corresponding to every word to be recognized. The most probable transition is used at each step, so backtracking generates the state sequence. Let us illustrate the Viterbi alignment procedure. It uses dynamic programming. Figure 7.11 illustrates the procedure.

Each utterance Y is segmented into an initial estimate of state occupancies for some model N. The linear match is used. The parameters are selected to maximize the likelihood of the data. Using the new parameters, a new segmentation is found that is shown as a solid curve in Figure 7.11. The arrows drawn on the x-axis show the state transition time for new segmentation.

In Viterbi training, the best path through each model is enforced. Hence, during training, the best path is likely to correspond to the correct model.

7.5.1.4 Baum–Welch Algorithm

Summing up, it can be stated that forward and backward recursions are used to find the estimates of the probabilities of the hidden states and the state transition probabilities. The probability of the output sequence can be expressed as

$$P(O \mid A) = \sum_{i=1}^{N} \alpha_t(i)\beta_t(i). \tag{7.15}$$

The probability of being in state q_i at time t is given by

$$\gamma_t(i) = \frac{\alpha_t(i)\beta_t(i)}{P(O \mid A)}. \tag{7.16}$$

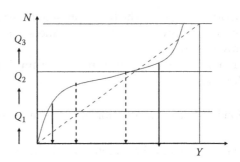

FIGURE 7.11
Viterbi alignment.

For the estimation of all these parameters, estimation andmaximization steps are repeated until some stopping criterion is met. The overall procedure is referred to as the forward backward or the Baum–Welch algorithm.

7.5.1.5 Posterior Decoding

There are several possible criteria for finding the most likely sequence of hidden states. One such criterion is to choose states that are most likely at the time when a symbol is emitted. This approach is called posterior decoding.

Let $\lambda_t(i)$ be the probability of the model to emit the symbol $\sigma(t)$ in the ith state for the observation sequence O.

$$\lambda_t(i) = P(q(t) = q_{i|} \,|\, O). \tag{7.17}$$

It can be shown that

$$\lambda_t(i) = \alpha_t(i)\beta_t(i) / P(O), \quad i = 1, \dots, N;\ t = 1, \dots, T. \tag{7.18}$$

We can select the state $q(t)$ that will maximize the value of $\lambda_t(i)$.

$$q(t) = \arg \max\{\lambda_t(i)\}. \tag{7.19}$$

Posterior decoding is found to work well when the HMM is observed to be an ergodic process, that is, there is transition from any state to any other state. If applied to an HMM of different architecture, the approach may arrive at a sequence that may not be a legitimate path because some transitions are not permitted.

Concept Check

- What are the essential blocks of ASR using VQ-HMM?
- Describe the block VQ code book.
- What is represented by a centroid of the cluster?

- How will you generate a model for a word?
- How will you calculate the state probability for a word using VQ?
- Define a triplet representing HMM.
- What is the use of forward and backward algorithm?
- Which algorithm will you use for word recognition?
- Name the algorithm used for the training mode of the system?

7.6 Discriminant Acoustic Probability Estimation

Two different approaches for estimation of acoustic probabilities, namely, the VQ based approach and use of Gaussians or Gaussian mixtures, are considered. When the parameters are estimated, the process must increase the likelihood of the correct model and reduce the likelihood of the incorrect one. This is essentially the discriminant estimation. We must focus on discrimination during training.

7.6.1 Discriminant Training

The statistical method for recognition uses Bayes' rule

$$P(\omega_k \mid X) = \frac{P(X \mid \omega_k)P(\omega_k)}{P(X)}. \tag{7.20}$$

Given the observation sequence X, one has to try to maximize the probability that it falls in a class k. The main difficulty lies in the fact that the actual probabilities are not known and the estimates of the probabilities depend on the parameters obtained during training. The parameter may be mean or variance. We must incorporate the dependence on these parameters in Bayes' equation. During the training phase, the mean changes and typically, we must change the mean to maximize $P(X \mid \omega_k, \text{mean})$. However this will also change the value of $P(X \mid \text{mean})$, and hence, the quotient may not increase.

Bayes' rule is changed to the rule given by

$$P(\omega_k \mid X, \text{mean}) = \frac{P(X \mid \omega_k, \text{mean})P(\omega_k \mid \text{mean})}{\sum_{k=1}^{K} P(X \mid \omega_k, \text{mean})P(\omega_k \mid \text{mean})}. \tag{7.21}$$

This rule means that the probability of a particular model is related to the ratio of the likelihood of that model weighted by its prior probability to the sum of such products for all other models. If the likelihood for correct class increases, the likelihood for incorrect models will decrease and the criterion will be discriminant.

Another measure used for discrimination of classes is the maximum mutual information (MMI). It is the mutual information between the model and acoustics and given by

$$I(\omega_k, X \mid \text{mean}) = E\left[\log \frac{P(\omega_k, X \mid \text{mean})}{P(\omega_k \mid \text{mean})P(X \mid \text{mean})}\right]. \tag{7.22}$$

E stands for the expectation operator.

The formula is modified to make it discriminant. It takes the form

$$I(\omega_k, X \mid \text{mean}) = \log \frac{P(X \mid \omega_k, \text{mean})}{\displaystyle\sum_{k=1}^{K} P(X \mid \omega_k, \text{mean})P(\omega_k \mid \text{mean})}. \tag{7.23}$$

If we compare Equations 7.21 and 7.23, we find that in the numerator in Equation 7.23, the prior probability term is missing and there is a log function. This is also a discriminant function. The MMI method is used in work at IBM.

7.6.2 Use of Neural Networks

Model probabilities are also estimated using neural networks, typically multilayer perceptrons. Neural networks under very general conditions estimate the HMM state probabilities. These are used to estimate model probabilities. It has been proved by a number of researchers that the outputs of the gradient trained classification systems may be interpreted as posterior probabilities of the output classes. The system must be trained in the classification mode, that is, for k classes, the target is one for correct class and zero for all others. The use of a softmax function is more common than a sigmoid. The function can be written as

$$f(y_i) = \frac{e^{y_i}}{\displaystyle\sum_{k=1}^{K} e^{y}}. \tag{7.24}$$

The neural networks can be trained to produce state probabilities for HMM assuming that each output is trained to correspond to a state. HMM emission probabilities are estimated if Bayes' rule is applied to artificial neural network (ANN) outputs. Let X represent the feature vector, which is the output emitted from the state q. We can write the expression for the emission probability as

$$\frac{P(X_n \mid q_k)}{P(X_n)} = \frac{P(q_k \mid X_n)}{P(q_k)}. \tag{7.25}$$

The probability $P(q_k \mid X_n)$ is obtained from the ANN outputs and $P(q_k)$ is a prior probability for a class obtained using the VQ-HMM approach. The process using ANN uses discriminant analysis. It uses the back propagation process that includes the training from the targets associated with the incorrect states.

Concept Check

- What is the meaning of discriminant?
- Write Bayes' rule for discriminant analysis.
- Define maximum mutual information.
- Write the function used for neural networks for discriminant analysis.

Summary

The chapter began with an introduction to the statistical properties of random variables, namely, mean, standard deviation, skew, kurtosis, and so on. We then explained DTW, HMM, statistical parameter estimation, discriminant acoustic probability estimation, etc. The chapter highlights the following concepts.

1. The statistical properties include mean, variance, skew, kurtosis, and higher order moments. All these properties are evaluated for segment of speech signal. A MATLAB program is written to find these statistical parameters. The calculation of skew and kurtosis for speech signal is described using a MATLAB program.

2. The need for time warping due to utterance variability was emphasized. We discussed linear time warping and DTW in detail. The optimization of the DTW algorithm is based on different conditions such as monotonic condition, continuity condition, boundary condition, and adjustment window condition. It was indicated that there are a number of limitations of the DTW-based approach for sequence recognition. Statistical sequence recognition was then discussed. It was indicated that we have to extract the statistical parameters and construct a model for every linguistic unit. We described Bayes' rule and HMM in detail. HMMs for speech recognition are the group of states interconnected. The states are assumed to emit the feature vectors according to the probability specified for each transition.

3. The problem of statistical pattern recognition and parameter estimation was considered. We then focused on the statistical parameter estimation. Acoustic probability estimation was discussed for ASR using the VQ approach. VQ-HMM-based speech recognition problem is considered. The generation of code book using pruning algorithm is described. The calculation of centroid for a cluster generating the centroid as the new updated code word is illustrated. A block schematic for speech recognition using VQ-HMM is explained. The method for specifying HMM for each word is illustrated. The estimation of model parameters such as initial state probability A, transition probability B, and emission probability H is illustrated. It was indicated that there are three problems to be solved in case of HMMs for speech recognition. Given the model parameters A, B, and H, compute the probability of a particular output sequence O. This problem is solved by the forward and backward algorithms.

Given the model parameters, find the most likely sequence of (hidden) states, which must have generated a given observed output sequence. This problem can be solved by the Viterbi algorithm. Given an output observed sequence, find the most likely set of state transition and output or emission probabilities. This is solved by the Baum–Welch algorithm. All these algorithms are described.

4. We have considered two different approaches for estimation of acoustic probabilities, namely, the VQ-based approach and the use of Gaussians or Gaussian mixtures. The discriminant analysis is discussed. We illustrated how Bayes' rule can be modified to use it for discriminant analysis. When the likelihood for correct class increases, the likelihood for the incorrect model will decrease and the criterion will be discriminant. Use of neural networks is also considered. It is shown by a number of researchers that the outputs of the gradient trained classification systems may be interpreted as posterior probabilities of the output classes.

Key Terms

Estimation of random variables	Posterior decoding
Mean	Forward algorithm
Standard deviation	Backward algorithm
Variance	Baum–Welch algorithm
Skew	Viterbi algorithm
Kurtosis	VQ-HMM
Hidden Markov modeling	k-means algorithm
Transition probability	Pruning algorithm
Emission probability	Clustering
State probability	Discriminant analysis
State transition diagram	Bayes rule
State variables	Gaussian mixtures
Transition matrix	Posterior probability

Multiple-Choice Questions

1. Time normalization of a template can be done using
   a. Division of time by its maximum value.
   b. Dynamic time warping.
   c. Time-scale modification.
   d. Time doubling.

2. In case of HMM for speech recognition,
   a. The state output is hidden.
   b. The transition is hidden.
   c. The probabilities are hidden.
   d. The states are hidden.

3. *A posteriori* probability for a class is evaluated
   a. Before the observation vector occurs.
   b. After the observation vector occurs.
   c. When the observation vector is known.
   d. When the probability value is known.

4. HMM is represented as a triplet consisting of
   a. State probability, emission probability, and transition probability.
   b. State probability, output probability, and input probability.
   c. Emission probability, input probability, and transition probability.
   d. State probability, transition probability, and input probability.

5. Given an observation sequence, the Baum–Welch algorithm finds the
   a. Most likely set of state transition and emission probabilities.
   b. Most likely set of input and output probabilities.
   c. Most likely set of emission probabilities.
   d. Most likely set of transition probabilities.

6. The discriminant estimation
   a. Increases the likelihood of the correct model and decreases the likelihood of the incorrect one.
   b. Decreases the likelihood of the correct model and increases the likelihood of the incorrect one.
   c. Increases the likelihood of both the correct model and the incorrect one.
   d. Decreases the likelihood of both the correct model and the likelihood of the incorrect one.

7. The mean of the speech signal can be estimated as
   a. The mean of the means of all small segments of speech.
   b. The mean of the entire speech signal.
   c. The mean of any one segment of speech.
   d. The mean of the middle segment of speech.

8. The kurtosis of the distribution is close to 3 means
   a. The distribution is Gaussian.
   b. The distribution is closer to uniform distribution.
   c. The distribution is closer to normal distribution.
   d. The distribution is closer to Rayleigh distribution.

Review Questions

1. Name the different statistical properties of any random signal.
2. How will you estimate the mean of a speech signal given to you?
3. Write the algorithm to find the skew and the kurtosis of a segment of speech signal.
4. Why is time normalization required for sequence recognition in case of speech recognition?
5. Explain DTW with respect to speech recognition.
6. What are the limitations of DTW?
7. Explain the use of Bayes' rule for class selection.
8. What is HMM? What is hidden? Draw a state diagram for HMM as a general case and show how you will write a transition matrix.
9. Explain MAP criterion and the likelihood-based MAP criterion.
10. State the expectation maximization algorithm for, say, estimation of mean.
11. How will you obtain the acoustic probability estimation in ASR using VQ?
12. Draw the block schematic for speech recognition using VQ-HMM and explain the function of each block.
13. How will you specify HMM using a triplet?
14. State the three problems to be solved in case of HMM for speech recognition with the algorithm to be used for each problem solution.
15. How will you find the probability of any observation sequence using forward or backward algorithm?
16. How will you use Viterbi algorithm to recognize a word?
17. How will you use the Baum–Welch algorithm for training?
18. Explain the meaning of discriminant training. How will you modify Bayes rule for discriminant analysis?
19. Explain the use of neural network for discriminant analysis.

Problems

1. Record a speech file for the word "Hello" and measure the skewness and kurtosis for the signal. Use a MATLAB program.
2. Record five different words in your voice by uttering each word 10 times. Find the pitch contour for all utterances using the method of average magnitude difference function (AMDF) and use DTW for template matching. Find the recognition rate.
3. Record the same word spoken by 20 different speakers (use 10 utterances for each speaker). Find the pitch contour for all utterances using the method of AMDF and use DTW for template matching. Find the recognition rate. Comment on your

results. Is pitch contour a useful parameter for speech recognition or speaker identification?

4. Record five different words each uttered 10 times. Try to generate HMM for each word. Try to find the speech recognition rate.

Answers

Multiple-Choice Questions

1. (b)
2. (d)
3. (b)
4. (a)
5. (a)
6. (a)
7. (a)
8. (c)

8

Transform Domain Speech Processing

LEARNING OBJECTIVES

- Small segment analysis of speech
- Pitch synchronous analysis of speech
- Use of transforms for speech processing
- Discrete cosine transform (DCT)
- Short-time Fourier transform (FT)
- Wavelet transform (WT)
- Processing of speech signal using different transforms
- Wavelet based time varying (WTV) model for speech
- Speech compression
- Sampling rate conversion
- Speech enhancement
- Speech recognition
- Speech synthesis
- Emotional speech synthesis
- Emotion recognition from speech

This chapter deals with small segment analysis of speech and pitch synchronous analysis of speech. It discusses transform domain speech processing, namely, processing using short-time Fourier transform (FT), wavelet transform (WT), etc.

The statistical analysis of speech has limitations due to the length of the speech segment available. Practically, we have only a limited duration of speech available for analysis. The statistical estimate of parameters approaches the actual value only when the length of the speech segment is very large. Again, the probability density function is not known. We have to remain satisfied with only the estimate of statistical parameters. We may extract the speech parameters such as pitch frequency, formants, linear prediction coefficients (LPC), and mel frequency cepstral coefficients (MFCC) and use these parameters for speech recognition, speaker recognition, etc. This is termed as the parametric approach. This parametric approach is not directly valid, as the speech is time varying. The speech parameters also vary with time. The only possibility of using parametric approach is to use short segment analysis of speech. We will first discuss short segment analysis of speech and then describe the pitch synchronous analysis.

8.1 Short Segment Analysis of Speech

Let us again consider the linear time varying (LTV) model for speech. Even though the parameters of speech, namely, pitch frequency for the excitation and the impulse response of the vocal tract, vary with time, they can be assumed to remain constant over a small segment of speech of the order of 10–20 ms. This is named quasi-periodic nature of speech. So, when we analyze a small segment of speech, we can assume that it is an linear time invariant (LTI) model. For every small segment, the parameters will remain constant, but they will vary from one segment to the other. The time variations of speech parameters from one segment to the other can be captured using a short time analysis of speech. What should be the length of the segment? Should it remain constant or it should vary as the pitch period varies? The answer to this question is the segment duration must be selected as equal to the number of samples equal to pitch period in terms of number of samples. This is termed as pitch synchronous analysis of speech.

8.1.1 Pitch Synchronous Analysis of Speech

Let us now understand pitch synchronous analysis of speech. Recall the definition of pitch frequency "f." The inverse of "f" is the pitch period, which can be measured in terms of number of samples.

The pitch synchronous analysis improves the results of speech recognition, speaker verification, speech synthesis, etc. Speech recognition or speaker verification systems extract certain features from speech segment, for example, pitch frequency, formants, LPC, and MFCC. If the segment size is made equal to the pitch period in terms of the number of samples, it is called pitch synchronous measurement. The values of these parameters vary with time. Hence, the parameters measured will vary form one pitch segment to the next pitch segment. The variation can be tracked properly when the pitch synchronous analysis is done. The experiments carried out by the author revealed that the results of all such experiments are improved using pitch synchronous analysis. The block schematic for pitch synchronous analysis can be drawn as shown in Figure 8.1.

It is observed that the speech parameters vary from one pitch to the next as the excitation is periodic for voiced segment. Hence, it is recommended to track the speech parameters for each pitch period. The author has proved that pitch synchronous WT improves the results for speaker verification. It is also experimentally proved that pitch synchronous analysis improves speech synthesis quality.

A MATLAB program illustrates the use of pitch synchronous analysis. The input speech signal is taken. We have used fseek to directly enter the voiced segment. The reader may use any algorithm for voiced/unvoiced detection discussed in Chapter 5, Section 5.2.

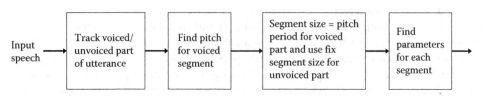

FIGURE 8.1
Block schematic for pitch synchronous analysis.

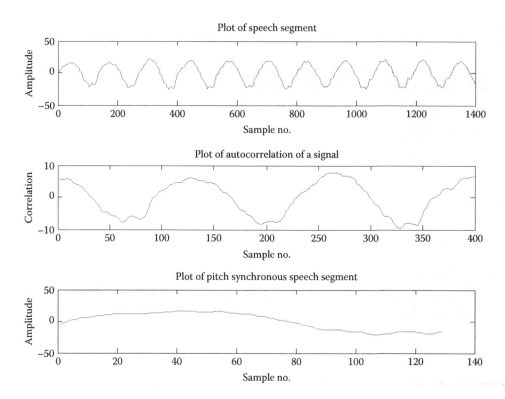

FIGURE 8.2
Plot of speech signal, autocorrelation, and pitch synchronous segment.

The voiced segment is processed further to find autocorrelation. The distance between two successive maximum of the autocorrelation will give the pitch period in terms of the number of samples (refer to Figure 8.2 for plot of autocorrelation). The number of samples equal to pitch period is taken from the speech segment to get a pitch synchronous segment of speech. Figure 8.2 shows a lot of speech segment, plot of autocorrelation, and plot of pitch synchronous speech segment. Compare pitch synchronous segment with the first 129 samples of speech segment. (Here, pitch period is 129 samples.)

```
Clear all;
clc;
fp=fopen('watermark.wav', 'r');
fseek(fp,98,-1);
a=fread(fp,1400);
a=a-128;
subplot(3,1,1);
plot(a);
title('plot of speech segment');
xlabel('sample number');ylabel('amplitude');
for k=1:400,
sum(k)=0;
end
for k=1:400,
for i=1:45,
sum(k)=sum(k)+(a(i)*a(i+k));
```

```
sum(k)=sum(k)/45;
end
end
subplot(3,1,2);plot(sum);title(' plot of autocorrelation of a signal');
xlabel('sample no.');ylabel('correlation');
max=2;
for i=1:150,
    if sum(i)>max,
        max=sum(i);
        pitch=i;
    end
end
disp(max);
disp(pitch);
for i=1:pitch,
    b(i)=a(i);
end
subplot(3,1,3);
plot(b);
title('plot of pitch synchronous speech segment');
xlabel('sample number');ylabel('amplitude');

max=6.1818; Pitch period=129 samples;
```

Concept Check

- What is a short segment analysis of speech?
- What is a pitch synchronous segment?
- How will you track a pitch synchronous segment?

8.2 Use of Transforms

We can extract the speech parameters using transform domain analysis. Transform domain analysis is a frequency domain analysis. Let us go through the short-time FT, WT, and discrete cosine transform (DCT) analysis of speech. We will discuss all these transforms and illustrate their use in speech processing. We will first go through the basics of these transforms and then describe their different applications for one-dimensional signal-like speech. When naturally occurring signals such as speech are to be processed, it is required to extract the signal parameters or features. A transform is advantageous in four ways. First, it is observed that the spectral properties of the signal are better revealed in the transform domain. Second, the denominator polynomial of system transfer function in a transform domain will indicate the locations of poles and zeros for analyzing the stability of the system. Third, when it comes to transmitting a signal over a communication channel, it is required to compress the signal. The compression requires energy compaction and it can be achieved in a transform domain. Last, the transform reduces the correlation of the signal. Transform domain processing assumes that there is high correlation between the signal samples in time domain. Such a signal will exhibit considerably reduced correlation when it is transformed. A transformation maps the time domain (correlated) data into transformed (uncorrelated) coefficients.

We will study the following parameters for the transforms, namely, basis functions, basis matrix of the transform, orthogonality of a transform, and the reversibility of a transform.

8.2.1 Discrete Cosine Transform

DCT is a transform related to FT and DCT coefficients are real and will have double the resolution as that of DFT. One-dimensional DCT of a sequence $u(n)\ 0 \leq n \leq N-1$ is defined as

$$v(k) = \alpha(k) \sum_{n=0}^{N-1} u(n) \cos\left[\frac{\pi(2n+1)k}{2N}\right] \quad 0 \leq k \leq N-1 \tag{8.1}$$

where

$$\alpha(0) = \sqrt{\frac{1}{N}} \text{ and } \alpha(k) = \sqrt{\frac{2}{N}} \text{ for } 1 \leq k \leq N-1 \tag{8.2}$$

Inverse DCT (IDCT) is defined as

$$u(n) = \sum_{n=0}^{N-1} \alpha(k) v(k) \cos\left[\frac{\pi(2n+1)k}{2N}\right] \quad 0 \leq n \leq N-1 \tag{8.3}$$

Properties of DCT are listed as follows:

1. It is real and orthogonal
2. It is not the real part of DFT
3. It can be used as a fast transform as we can make use of fast DCT for its computation.
4. It exhibits excellent energy compaction for highly correlated data sequence.

The basis vectors of DCT are the cosine functions. If we take DCT of any sequence and take its inverse, we will get the same sequence back, indicating that it is revertible. The reader is encouraged to verify revertibility by taking 4-point DCT of any 4-point sequence. Let us understand the meaning of these properties. To understand orthogonality, we have to find the basis matrix for DCT and verify that matrix multiplied by its transpose is equal to 1. This will prove that the basis vectors are orthogonal to each other. To verify the revertibility of a transform, we will take DCT of the 4-point signal and then take IDCT and verify that the 4-point IDCT result is same as the 4-point signal.

We will define the term energy compaction. Then we take a transform of a signal and verify the following. Signal energy, which is distributed over all samples of a signal, gets concentrated in few transform coefficients. If the energy gets compacted in smaller number of coefficients, we can conclude that the energy compaction is high for a transform. When the signal has high correlation, the next samples can be predicted from previous samples. We can inspect that the signal shows some amount of repetition. The dictionary definition of correlation can be stated as follows. Two random variables are positively correlated if high values of one signal are likely to be associated with high values of the other signal. They are negatively correlated if high values of one signal are likely to be associated with low values of the other signal. We will compute DFT of the highly correlated data. We will show that the DFT coefficients of the signal are decorrelated (reduced correlation) when the time domain signal has high correlation.

Let us write a MATLAB program to plot spectrum of the signal using DCT and find the correlation of DCT transformed signal.

```
%to find frequency contents of speech signal using DCT
clear all;
fp=fopen('watermark.wav','r');
fseek(fp,44,-1);
a=fread(fp,512);
plot(a);title('plot of speech signal');xlabel('sample number');
ylabel('Amplitude')
for i=1:128,
    a1(i)=a(i);
end
for i=129:256,
    a2(i-128)=a(i);
end
disp(corrcoef(a1,a2));
c=dct(a);
figure;
cc=idct(c);
for i=1:511,
    d(i)=c(i+1);
end
subplot(2,1,1);
plot(d);title('plot of frequency contents of speech
signal');xlabel('DCT coefficient number'); ylabel('Amplitude');
subplot(2,1,2); plot(cc);title('plot of idct same as signal');
xlabel('sample no.'); ylabel('amplitude');
for i=1:128,
    d1(i)=d(i);
end
for i=129:256,
    d2(i-128)=d(i);
end
disp(corrcoef(d1,d2));
```

A plot of signal and its spectrum using DCT is shown in Figures 8.3 and 8.4, respectively. Figure 8.4 shows a plot of IDCT, which is found to be same as the original signal, thus confirming the revertibility property. The correlation for input speech signal is 0.8304 and that for transformed signal is −0.1031. We say that the correlation of the signal is reduced in transform domain.

A plot of spectrum of the signal indicates that the first 200 DCT coefficients are finite with some non-zero value. We find that the total energy of a signal that was distributed over almost infinite samples in time domain gets compacted or concentrated only in the first 200 DCT coefficients.

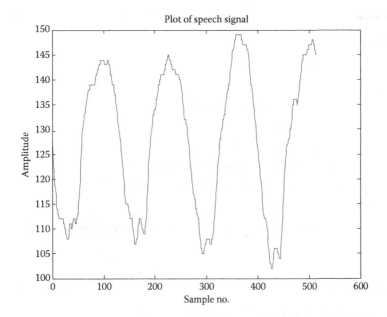

FIGURE 8.3
Plot of speech signal.

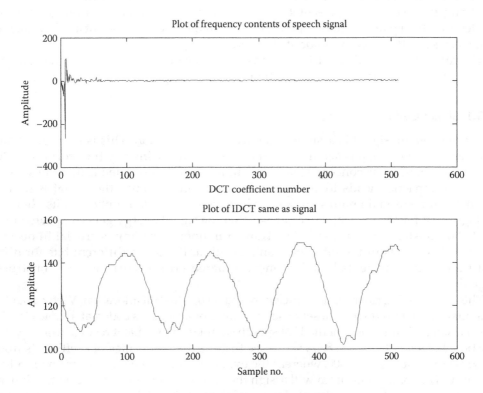

FIGURE 8.4
Plot of spectrum of speech signal using DCT and plot of IDCT.

Concept Check

- Define DCT of a sequence $x(n)$.
- What is energy compaction?
- Are we getting energy compaction in DCT?
- Is DCT a revertible transform?
- What is a basis function?
- What is the condition for a matrix to be orthogonal?
- When will the transform be revertible or invertible?
- What is the definition of correlation?
- What do you mean by saying that the signal is decorrelated in transform domain?

8.3 Applications of DCT for Speech Processing

DCT uses fixed kernels of infinite duration varying from $-\infty$ to $+\infty$ (exponential function, cosine and sine function, respectively). As these kernels (basis functions) are not localized in time domain, time variations in the signal cannot be tracked using these kernels. Basically, the frequency representation using these transforms is the average frequency contents of the signal. It cannot represent which frequencies are present at what time. The transforms are, therefore, appropriate for stationary signals.

The different applications of these transforms include signal coding, signal compression, filtering, identification, or recognition of features, etc.

8.3.1 Signal Coding

The time domain signal is assumed to have high correlation. This is true for all naturally occurring signals. When the signal is transformed using any transform, say, DCT, the signal energy gets concentrated in less number of decorrelated transform samples. Energy compaction leads to signal compression, and because the signal is decorrelated, the transform domain samples can be coded using less number of bits. Referring to Figure 8.4 for the DCT plot, we can see that the signal energy gets compacted in the first 200 transform domain samples. Transform domain coding is crucial in deciding the number of bits in the bit allocation algorithm. It assigns different bits for different transform coefficients based on mean squared error occurring for reconstructed signal.

The signal coding can be made efficient based on optimal bit allocation. Vector quantization may also be used where a set of scalar values or a vector is coded at a time. Consider speech signal coding. DCT of the 1024 speech samples shows that it results in energy compaction. Referring to Figure 8.4, we can see that the dynamic range of coefficients from 0 to 50 is large and is above 500, whereas the dynamic range of coefficients from 51 to 100 is about 50. The coding algorithm will assign the number of bits depending on the dynamic range of values, and the bits allocated for coefficients 0–50 will be large as compared to bits used for coefficients from 51 to 100. If the number of bits allocated for coding is large, the quantization step size becomes low, and if the number of bits allocated for coding is less, the quantization step size is found to be large.

$$\text{Quantization step size } q = \frac{\text{Dynamic range}}{\text{No. of steps} = 2^B}$$

(8.4)

B is the number of bits allocated

Consider the coding of coefficients from 51 to 100 that have a value less than the step size. The coefficients with value less than the step size will, of course, be coded with zero. The reader can verify that the coefficients from coefficient number 51 onward will all be coded with a zero value. Hence, the signal samples from 1 to 1024 are compressed to coefficients from 0 to 50. We thus achieve the compression ratio of 1024/50, i.e., >20=20.48 for speech signal using DCT. Remember that transforming the signal in DCT domain does not give compression. Compression is achieved when the transform domain coefficients are coded using appropriate step size.

8.3.2 Signal Filtering

Filtering is also simplified in frequency domain or transform domain. In time domain, consider examples of a low-pass filter (LPF). Here, we have to convolve the signal samples with the impulse response of the filter or the filter kernel. Convolution is computationally costly.

When it is required to low-pass filter the signal in DCT domain, we have to design a window corresponding to the ideal LPF response and pass the DCT output via this window. Consider a rectangular window as shown in Figure 8.5. The DCT output is shown in Figure 8.6, and the DCT output after passing via a window is shown in Figure 8.7. The windowed output passed via the inverse transformed to get a filtered signal in time domain. The filtered signal is shown in Figure 8.8. Compare the speech output with original speech signal shown in Figure 8.3. Note that high frequencies are removed from the signal.

A MATLAB program for filtering in DCT domain can be written as

```
%to filter speech signal in DCT domain
clear all;
fp=fopen('watermark.wav','r');
fseek(fp,44,-1);
a=fread(fp,512);
c=dct(a);
```

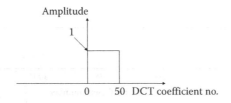

FIGURE 8.5
Rectangular window: Filter response for a LPF passing DCT coefficients from 0 to 50.

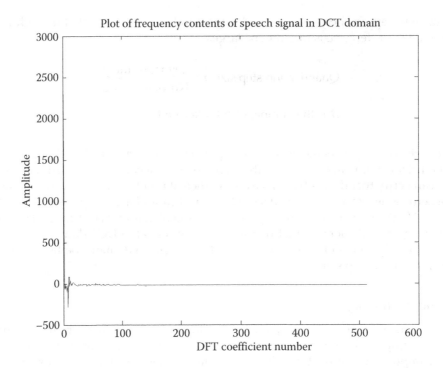

FIGURE 8.6
Plot of DCT output. (We have zoomed onto first 300 samples of DCT.)

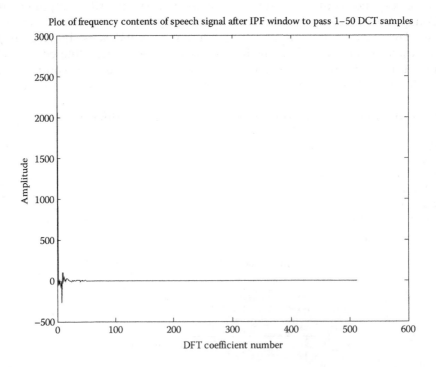

FIGURE 8.7
Plot of DCT output after passing via a window of length 64. (We have zoomed onto first 300 samples of DCT.)

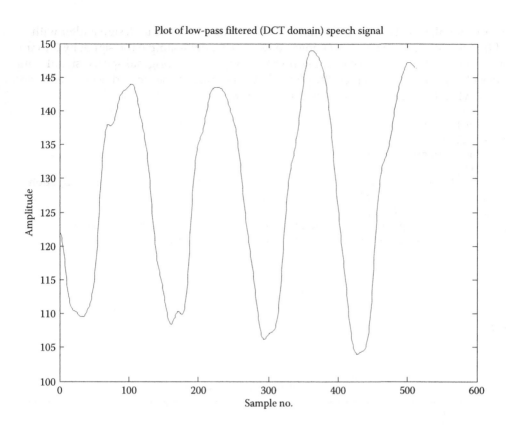

FIGURE 8.8
Low-pass filtered speech signal (filtering in DCT domain).

```
plot(c);title('plot of frequency contents of speech signal in DCT
domain');xlabel('DFT coefficient number'); ylabel('Amplitude');
for i=1:50,
    d1(i)=c(i);
end
for i=51:512,
    d1(i)=0;
end
figure;
plot(d1);title('plot of frequency contents of speech signal after lPF
window to pass 1 to 50 DCT samples');xlabel('DFT coefficient number');
ylabel('Amplitude');
d3=idct(d1);
figure;
plot(d3);title('plot of low pass filtered (DCT domain) speech
signal');xlabel('sample number'); ylabel('Amplitude');
```

8.3.3 Sampling Rate Conversion and Resizing

DCT can be used to change the sampling rate of the speech signal. Consider a simple example of decimation by a factor of 2. The antialiasing filter is to be used as the first stage of the decimator. It can be designed as a window in direct current (DC) domain. Let us consider the speech signal, take its DCT, and retain only half the coefficients to design a filter with a cut-off frequency of $\pi/2$. We will take inverse DCT and down-sample the signal by 2 to implement a decimator by a factor of 2. Figure 8.9 shows a plot of original speech signal samples and its DCT. Figure 8.10 depicts the plot of low-pass filtered speech and decimated speech.

A MATLAB program to implement is as follows.

```
%to implement a decimator by a factor of 2 using DCT
clear all;
fp=fopen('watermark.wav','r');
fseek(fp,44,-1);
a=fread(fp,512);subplot(2,1,1);plot(a);title('plot of original speech
signal');xlabel('sample no.'); ylabel('Amplitude');
c=dct(a);subplot(2,1,2);
plot(c);title('plot of frequency contents of speech signal in DCT
domain');xlabel('DFT coefficient number'); ylabel('Amplitude');
for i=1:256,
    d1(i)=c(i);
end
for i=257:512,
    d1(i)=0;
end
figure;
subplot(2,1,2);
```

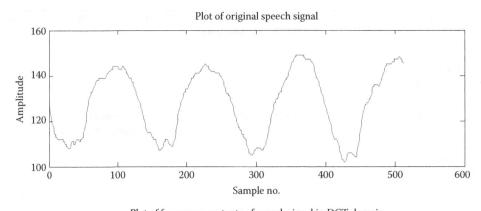

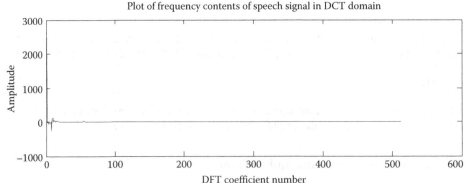

FIGURE 8.9
Original speech signal and its DCT.

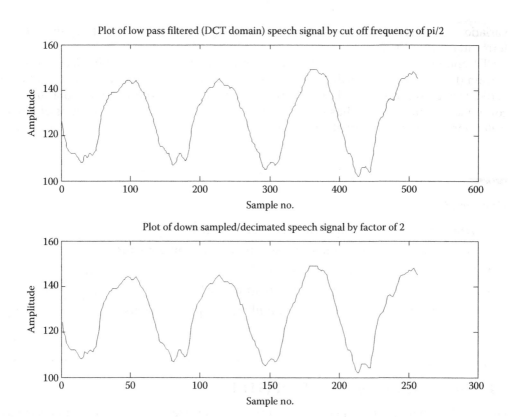

FIGURE 8.10
Low-pass filtered and decimated speech.

```
d3=idct(d1);subplot(2,1,1);
plot(d3);title('plot of low pass filtered (DCT domain) speech signal by
cut off frequency of pi/2');xlabel('sample number'); ylabel('Amplitude');
for i=2:2:512,
    e(i/2)=d3(i);
end
subplot(2,1,2);
plot(e);title('plot of down sampled/decimated speech signal by factor of
2');xlabel('sample number'); ylabel('Amplitude');
```

8.3.4 Feature Extraction and Recognition

There are different applications for transforms. Consider applications in the area of speech processing. For speech recognition, we can extract DCT coefficients for different spoken words and prepare a template for each spoken word. This is done in the training mode. In the recognition mode, the external word is used to find the parameters and the template is compared against the stored template. It will show a smaller distance with the particular word and the word will be recognized. For speaker verification, similar features can be extracted and used to form templates. Some transform coefficients will be more sensitive to spoken word and some will be more sensitive to the speaker. The main problem with these transforms lies in the fact that they cannot capture time variations in the selected parameters as explained in the earlier section. The transforms capable of capturing time variations are short-time Fourier transform (STFT) and WT. We will study these transforms in the next sections.

DFT and other transforms are also used for extracting the features in electroencephalogram (EEG) and electrocardiogram (ECG) signal processing. Now, the authentication of audio or images is done by hiding the watermark in the transform domain output of the cover audio or image. DCT has proved its usefulness, as it provides robustness of hidden watermark even after different attacks on the watermarked audio or watermarked image.

Concept Check

- What are the different applications of transforms?
- How will you implement filtering action in transform domain?
- How will use an LPF in the DCT domain?
- Can you use transforms for signal compression?
- How will you use transform domain filters for speech signals?

8.4 Short-Time Fourier Transform (STFT)

DFT and DCT have fixed kernel, and the kernel extends over infinity from $-\infty$ to $+\infty$. Hence, these transforms cannot capture time variations in the signal. The naturally occurring signals are found to be time varying, and they have some random component. We will introduce STFT and multiresolution analysis (MRA).

The most intuitive way for the analysis of a nonstationary signal is to perform a time-dependent spectral analysis. A non-stationary signal is divided into a number of time segments, over which the signal may be assumed to be stationary. Then, the FT is applied to each of the segment. The STFT is associated with a window of fixed length. Gabor, in 1940, first introduced the STFT, which is known as the sliding FT. When a signal is analyzed, the window is slid over the entire signal with, or sometimes without, the overlap between the successive windows. The size of the window is small and decided as per the required resolution. The transform is defined as

$$S(\omega, \tau) = \int f(t) g^*(t - \tau) \exp(-j\omega t) dt \qquad (8.5)$$

Here, the function $g(t)$ is a square integrable short time window. The window has a fixed width, and it is shifted along the time axis by a factor of τ. Thus, STFT uses the kernels as short time windowed sines and cosines. They are also called basis functions. STFT can thus be considered as a Fourier analysis of the short time windowed signal. According to the time-varying signal, a speech signal is divided into segments of fixed size, say, 128. The length of the segment is decided according to the resolution required. We find out the time interval of speech over which the parameters of the vocal tract filter remain almost constant. It is observed that for a speech signal, the parameters of vocal tract can be assumed to remain constant over the duration of 10 ms of speech. So, STFT is useful for analysis of speech signals. The duration of the segment decides the number of samples in a segment. We can equivalently say that we are viewing a signal via a short duration window of fixed length. The length of the window decides the resolution of the signal frequencies. The time frequency trade-off for STFT is shown in Figure 8.11. We then take fast fourier transform (FFT) of the windowed signal. In the windowing process, sample to sample multiplication of signal samples and window coefficients is carried out and FFT of the multiplication result is taken. The FFT output for different successive windowed signals can be plotted. Overlapping or nonoverlapping windows are used. Multiplying the signal with window localizes the signal in time domain, but it will result in a convolution of signal spectrum and the spectrum of the window. The signal is thus blurred in the frequency domain. The narrower the window, the better we localize the signal in time domain and the poorer is the localization of its spectrum. If we want to improve the time resolution of the signal, we must reduce the size of the window, and if we want to improve the frequency resolution of the signal, we will need to increase the size of the window. Heisenberg Uncertainty Principle applies here, which says that $\Delta t \times \Delta f \geq 1/2$. This means, if we improve the resolution in time domain by reducing the window size, we have to sacrifice for frequency domain resolution, and vice versa.

Figure 8.12 shows a three-dimensional (3D) spectrogram of the speech signal. This plot is called as 3D spectrogram as it has x axis as time, y axis as frequency, and z axis as magnitude. This plot can show the time variations of the signal frequencies. A 3D plot can be converted to a 2D plot by representing the amplitude of the FFT component using a gray value or color.

Procedure for plotting the 3D spectrogram can be listed as follows:

1. Divide a speech signal into segments of size, say, 128.
2. Pass the signal segment via a short time window of the same length. We may use any smooth window such as Gaussian window or Hamming window. The use of window can avoid the abrupt transition, and hence the Gibbs phenomenon.

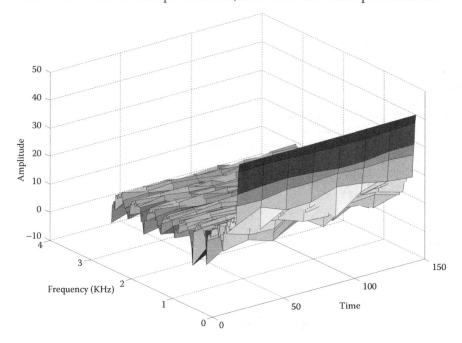

FIGURE 8.11
Plot of 3D spectrogram.

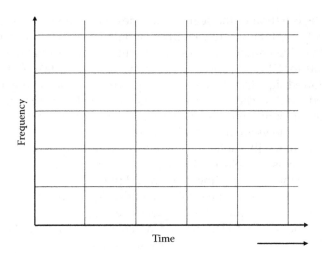

FIGURE 8.12
Time frequency resolution for STFT.

3. Take FFT of each segment.

4. Plot FFT magnitude with respect to frequency. This is the result for one segment of data. Using a 3D spectrogram, the magnitude of FFT is plotted with respect to frequency and time.

A MATLAB program to plot a 3D spectrogram is given as follows.

```
%plot of 3-D spectrogram
clear all;
fp=fopen('watermark.wav','r');
t=0:0.016:2;  % 2 secs @ 16 kHz sample rate 256 samples at a time=0.016
seconds data taken at a time.
fseek(fp,44,-1);
x=fread(fp,1024);
[y,f,t]=spectrogram(x,window,128);
%length of window is 128 samples with y as spectrogram
%amplitude varying with time and frequency
surf(t,f,10*log10(abs(y)));
xlabel('Time');ylabel('Frequency (KHz)');
zlabel('amplitude');
```

Concept Check

- What is an STFT?
- Why is STFT suitable for speech signals?
- What is a 3D spectrogram?
- What is a 2D spectrogram?

- Why is the tile size fixed for STFT?
- Can you observe frequency variations from one speech segment to the other in a 2D spectrogram?

8.5 Wavelet Transform

The WT provides the time–frequency representation of the signal. WT can represent the time and frequency information simultaneously. Hence, we say that it gives a time–frequency representation of the signal. WT is a new mathematical tool used for the local representation of non-stationary signals. WT is in a way mapping of a signal to time frequency joint representation. It preserves the temporal aspects of signal and provides MRA using dilated window functions. The high-frequency analysis is possible using narrow windows and low-frequency analysis is possible using wide windows. When we use a narrow window, the time resolution is better and we can resolve the high frequencies in the time domain. When the window is wide, the time resolution is poor, but frequency resolution is better, and we can resolve low frequencies in the frequency domain. We can say that in a joint time and frequency representation, we can resolve low frequencies in frequency domain and high frequencies in time domain simultaneously. The time frequency resolution trade-off for WT is represented in Figure 8.13. We can observe that wherever the time resolution is good (at high-frequency region), the frequency resolution is poor. Wherever the time resolution is poor (at low-frequency region), the frequency resolution is fine. The tile size is uneven. When time side is long, the frequency side of the tile is short, and vice versa. Note that the relative bandwidth (Q) is constant. The relative bandwidth is evaluated as the ratio of bandwidth and center frequency of the band. We can observe that the relative bandwidth, Q, is constant for wavelet decomposition. This property is useful in the analysis of natural signals such as speech signal that also exhibits a constant Q distribution of energy.

The basis functions of WT are generated from a basic wavelet function by using the operation of dilations and translations. The basic wavelet or mother wavelet, i.e., a kernel, is not fixed. You may select the kernel from standard kernels available, or it can be generated from the signal itself. This is the main merit of the WT. The child wavelets are generated by translation and scaling of main mother wavelet (kernel). Kernel or mother wavelet is used to generate the further basis functions. Hence, it is called mother wavelet. The mother wavelet must be localized in the time domain as well as in the frequency domain, so that it will resolve the time domain signals that are separated by duration greater than the window length and will resolve frequency domain signals that are separated by duration greater than the frequency domain window length. When time domain window is narrow, it will resolve high-frequency signals in time domain, and when frequency domain window is narrow, it will resolve low-frequency signals in the frequency domain. The set of basis functions thus formed contains the time domain windows of different lengths, so the wavelet is capable of resolving the high-frequency and low-frequency signals simultaneously.

The signal is expressed as the weighted sum of the basis functions. The WT coefficient at a scale "a" and translation "b" is given by

$$W(a,b) = \sum_{t=-\infty}^{\infty} s(t) h_{a,b}^{*}(t) \mathrm{d}t \qquad (8.6)$$

Here,

$$h_{a,b}(t) = \left(\frac{1}{a}\right)^{1/2} g[(t-b)/a] \tag{8.7}$$

The wavelet coefficient represents the degree to which the signal $s(t)$ is similar to the scaled and translated mother wavelet (kernel). To reduce the time bandwidth product of WT output, the discrete WT with wavelets of discrete dilation and translation can be used with $a = 2^k$ and $b = 2^k \times l$. These are called dyadic wavelets. The discrete wavelets can be made orthogonal to their dilations and translation by a special choice of basis wavelets having the following property.

$$\sum h_{ik}(l)h_{mn}^*(l) = 1 \quad \text{if } i = m, k = n \tag{8.8}$$

$$= 0 \quad \text{otherwise}$$

The orthogonal wavelet decomposition does not have a redundancy in the signal representation. An arbitrary signal is represented as a sum of orthogonal basis functions (scaled and translated mother wavelet) weighted by WT coefficients.

Restrictions on mother wavelets (kernel):

If $g(t)$ is a mother wavelet, then

1. $g(t)$ has finite energy,

$$\int_{-\infty}^{\infty} |g(t)|^2 \, dt < \infty \tag{8.9}$$

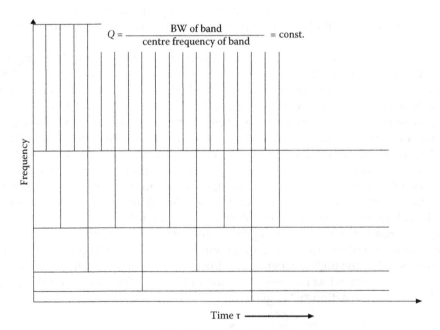

FIGURE 8.13
Time frequency resolution selection for WT.

This condition is termed as the regularity condition.

2. FT of $g(t)$ is finite. (FT) of $g(t)$ must have a zero value at zero frequency. This means that the time average of the wavelet function is zero. So, we can say that WT removes the DC component of the signal. Hence, wavelets are band pass filters in frequency domain and they are oscillatory functions like a wave having zero mean value. This is called as the admissibility condition.

3. If p(n) represents the LPF filter, then HPF coefficients $q(n)$ are given by

$$q(n) = (-1)^n p(N - 1 - n) \tag{8.10}$$

If orthogonal basis function is used, $p(n)$ and $q(n)$ form mirror filters or perfect reconstruction filters or a mother wavelet. The filter pairs $P(\omega)$ and $Q(\omega)$ are called mirror filters because

$$Q(\pi/2 - \omega) = p(\pi/2 + \omega) \tag{8.11}$$

This is the mirror image property about $\pi/2$.

Using property 1 and 2, the mother wavelet is localized in time domain and in frequency domain too. Localization in time domain allows it to get time resolution, depending on the duration of the wavelet in time domain. Localization in frequency domain allows it to get frequency resolution proportional to the duration of the wavelet in frequency domain. This is why it is called as a small wave or a wavelet.

Concept Check

- How can WT resolve low as well as high frequencies in a signal simultaneously?
- Comment on time frequency trade-off for WT.
- What is a mother wavelet?
- Why must a mother wavelet be localized in time domain?
- Why must a mother wavelet be localized in frequency domain?
- Why is a kernel called mother wavelet?
- Why are the filters $p(n)$ and $q(n)$ called mirror filters?

8.6 Haar Wavelet and Multiresolution Analysis

We will distinguish wavelets in two classes.

1. Continuous WT
2. Discrete WT

We will discuss one example of continuous wavelet, which is called Haar wavelet. The Haar wavelet was historically introduced by Haar in 1910. Haar wavelet is a bipolar function given by

$$\psi(t) = \begin{cases} 1 & 0 < t < 1/2 \\ -1 & 1/2 < t < 1 \\ 0 & \text{otherwise} \end{cases} \tag{8.12}$$

The scaling function is defined as

$$\phi(t) = \begin{cases} 1 & 0 < t < 1 \\ 0 & \text{otherwise} \end{cases} \tag{8.13}$$

The WT coefficient at a scale "a" and translation "b" of a signal $s(t)$ can be written as

$$w(a,b) = \int_{-\infty}^{\infty} s(t) h_{a,b}^*(t) \mathrm{d}t \tag{8.14}$$

where

$$h_{a,b}(t) = \frac{1}{\sqrt{a}} g[(t-b)/a], \tag{8.15}$$

where $g(t)$ represents a mother wavelet. The basis functions for WT are dilated and shifted versions of $g(t)$ as given in Equation 8.51. The signal $s(t)$ can be represented as the weighted sum of the basis functions. The wavelet coefficient at a particular scale "a" and translation "b" represents the similarity between the signal $s(t)$ and the scaled and translated mother wavelet. Note that scale represents the inverse of frequency.

Properties of Haar wavelet are as follows.

1. Haar transform is real and orthogonal.
2. Haar is a very fast transform. On $N \times 1$ vector, it can be implemented in $O(N)$ operations.
3. Haar transform has poor energy compaction.

Using a Haar transform,

1. Any function can be written as linear combinations of $\phi_1(t), \phi_2(2t), \phi(4t), \dots$, etc. and their shifted functions.
2. Any function can be approximated by linear combinations of the constant function, $\psi_1(t), \psi_2(2t), \psi(4t), \dots$, etc. and their shifted functions.
3. Wavelet/scaling functions with different scale m have a functional relationship:

$$\varphi(t) = \varphi(2t) + \varphi(2t - 1) \tag{8.16}$$

$$\psi(t) = \varphi(2t) - \varphi(2t-1) \tag{8.17}$$

The 2×2 Haar matrix associated with the Haar wavelet is given by

$$H = \frac{1}{\sqrt{2}}\begin{bmatrix} 1 & 1 \\ 1 & -1 \end{bmatrix} \tag{8.18}$$

Haar decomposition is implemented in the following examples:

Example 1

Consider a sequence $f = (2, 5, 8, 9, 7, 4, -1, 1)$. We wish to expand this signal in the Haar basis. In practice, we do this by performing the following steps.

Step 1:

$$\begin{aligned} f &= (2+5, 8+9, 7+4, -1+1, 2-5, 8-, 7-4, -1-1)/\sqrt{2} \\ &= (7, 17, 11, 0, -3, -1, 3, -2)/\sqrt{2} \end{aligned} \tag{8.19}$$

Step 2:

$$\begin{aligned} f &= \left(\frac{7+17}{\sqrt{2}}, \frac{11+0}{\sqrt{2}}, \frac{7-17}{\sqrt{2}}, \frac{11-0}{\sqrt{2}}, -3, -1, 3, -2\right)\Big/\sqrt{2} \\ &= \left(\frac{24}{\sqrt{2}}, \frac{11}{\sqrt{2}}, \frac{-10}{\sqrt{2}}, \frac{11}{\sqrt{2}}, -3, -1, 3, -2\right)\Big/\sqrt{2} \end{aligned} \tag{8.20}$$

Step 3:

$$\begin{aligned} f &= \left(\frac{24+11}{(\sqrt{2})^2}, \frac{24-11}{(\sqrt{2})^2}, \frac{-10}{\sqrt{2}}, \frac{11}{\sqrt{2}}, -3, -1, 3, -2\right)\Big/\sqrt{2} \\ &= \left(\frac{35}{2}, \frac{13}{2}, \frac{-10}{\sqrt{2}}, \frac{11}{\sqrt{2}}, -3, -1, 3, -2\right)\Big/\sqrt{2} \end{aligned} \tag{8.21}$$

$$\approx (12.4, 4.60, -5.00, 5.50, -2.12, -0.707, 2.12, -1.41)$$

The numbers in the final vector are the Haar coefficients obtained in the expansion. We add/subtract a pair of numbers, divide by the normalization factor $\sqrt{2}$, and between every step we keep the last half number of elements of the vector unchanged. This procedure is called "averaging and differencing." We can verify the result by using a function db1 in MATLAB. The notation db1 used in the MATLAB program stands for Haar wavelet. We use discrete WT (dwt command). The notation "ca" represents the low-resolution components and "cd" are detail function components. The component ca is decomposed further to go to higher resolution in the next level of decomposition.

```
clear all;
f=[2,5,8,9,7,4,-1,1];
[ca,cd]=dwt(f,'db1');
disp(ca);
disp(cd);
[ca1,cd1]=dwt(ca,'db1');
disp(ca1);
disp(cd1);
[ca2,cd2]=dwt(ca1,'db1');
disp(ca2);
disp(cd2);
```

Result of the first decomposition, i.e., step 1, is

$$
\begin{array}{cccc}
4.9497 & 12.0208 & 7.7782 & 0 \\
-2.1213 & -0.7071 & 2.1213 & -1.4142
\end{array}
$$

We have to retain the detail coefficients, i.e., last four coefficients, and decompose a vector of first four coefficients using a second decomposition. The result of second decomposition can be written as step 2.

$$
\begin{array}{cc}
12.0000 & 5.5000 \\
-5.0000 & 5.5000
\end{array}
$$

Now, retain the detail coefficients, i.e., the last two coefficients, and decompose a vector of the first two coefficients using a third-level decomposition. The result of third decomposition, i.e., step 3, is

$$
12.3744 \quad 4.5962
$$

The overall result using the low-resolution components for third decomposition and the detail coefficients for both first and second decomposition can be written as

$$
12.3745, \quad 4.5962, \quad -5.00, \quad 5.500, \quad -2.1213, \quad -0.7071, \quad 2.1213, \quad -1.4142.
$$

8.6.1 Multiresolution Analysis

The time and frequency resolution problems result due to a physical phenomenon (the Heisenberg uncertainty principle). Though the problems exist regardless of the transform used, it is possible to analyze any signal using an alternative approach called the *multiresolution analysis*. MRA, as implied by its name, analyzes the signal at multiple or different frequencies with different resolutions. Consider the decomposition of the signal using Haar wavelet. In the first level of decomposition, the signal is passed via two filters, one low pass and the other high pass, which form mirror filters or perfect reconstruction filters (PRFs), $\frac{1}{\sqrt{2}}[1\,1]$ and $\frac{1}{\sqrt{2}}[1-1]$, respectively, as shown in Equation 8.18.

Mirror filters or PRFs represent one low-pass filter and the other high-pass filter (HPF); they have the same cut-off frequency and their frequency response curves are the mirror image of the other as shown in Figure 8.14. The wavelet decomposition for the first stage of decomposition is shown in Figure 8.15. The result of each filter is down-sampled by a factor of 2 as shown in Figure 8.15.

In Example 1, we get the output of first level of decomposition as

$$f = (7, -3, 17, -1, 11, 3, 0, -2) / \sqrt{2}$$

after down sampling by 2 we get   (8.22)

$$= (7, 17, 11, 0, -3, -1, 3, -2) / \sqrt{2}$$

The output of the first-level decomposition is down-sampled by a factor of 2 to get the signal used for the second decomposition. This means that for the second level of decomposition, we are operating on a smaller length of the signal. Reducing the window length of the signal is equivalent to increasing the time domain window length of filters or dilating the filters. This will increase the frequency resolution for the second stage. Here, a vector $f_1 = (7, 17, 11, 0) / \sqrt{2}$ containing the first four samples obtained after down-sampling by a factor of 2 is a low-resolution component and a vector $f_2 = (-3, -1, 3, -2) / \sqrt{2}$ containing last four samples obtained after down-sampling by a factor of 2 is a detail function. The length of the time domain window for the second decomposition is now 4.

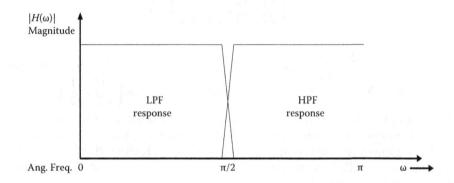

FIGURE 8.14
Mirror image filter (PRFs) responses.

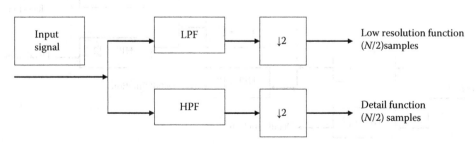

FIGURE 8.15
Wavelet decomposition using PRFs.

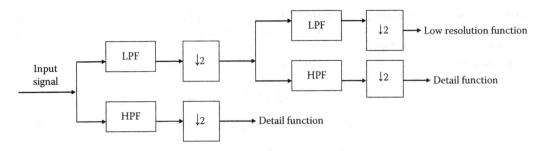

FIGURE 8.16
Second level of decomposition using WT.

The detail function is kept as it is and not decomposed further. Only low-resolution component is decomposed further (second level of decomposition) using same PRFs as shown in Figure 8.16.

In Example 1, we get the output of second level of decomposition as

$$f = \left(\frac{24}{\sqrt{2}}, \frac{-10}{\sqrt{2}}, \frac{11}{\sqrt{2}}, \frac{11}{\sqrt{2}} \right) \Big/ \sqrt{2}$$

after decimation by 2, (8.23)

$$f = \left(\frac{24}{\sqrt{2}}, \frac{11}{\sqrt{2}}, \frac{-10}{\sqrt{2}}, \frac{11}{\sqrt{2}} \right) \Big/ \sqrt{2}$$

where a vector $f_1 = \left(\frac{24}{\sqrt{2}}, \frac{11}{\sqrt{2}} \right) / \sqrt{2}$ consisting of first two samples after down-sampling by a factor of 2 is a low-resolution component and a vector $f_1 = \left(\frac{-10}{\sqrt{2}}, \frac{11}{\sqrt{2}} \right) / \sqrt{2}$ consisting of last two samples after down-sampling by a factor of 2 is a detail function.

The detail function is again kept as it is, as discussed earlier. It is not decomposed further. Only a low-resolution component is decomposed further (third level of decomposition) using the same PRFs as shown in Figure 8.17. Here, the size of time window for the signal is reduced to a factor of 2.

In Example 1, we get the output of the third level of decomposition as

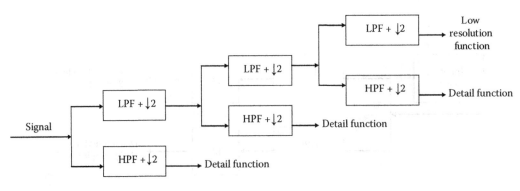

FIGURE 8.17
Third level of decomposition using wt.

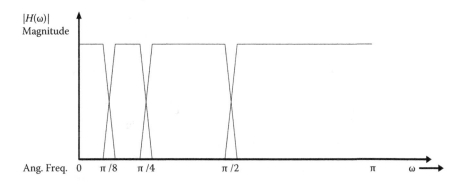

FIGURE 8.18
Frequency bands for the WT analysis tree.

$$f_1 = \left(\frac{35}{2}, \frac{13}{2}\right)\Big/\sqrt{2} = 12.3745, \, 4.5962,$$

here, 12.3745 is a low-resolution component and 4.5962 is a detail function component obtained after down-sampling by a factor of 2. Here, down-sampling is not required as there are only two samples.

The analysis tree for the frequency bands is drawn as depicted in Figure 8.18. During the first level of decomposition, the total range of angular frequency equal to π is divided into two bands, one from 0 to $\pi/2$ and the other from $\pi/2$ to π. During the second level of decomposition, low-resolution range of angular frequency between 0 and $\pi/2$ is again divided into two bands, one from 0 to $\pi/4$ and the other from $\pi/4$ to $\pi/2$. During the third level of decomposition, we divide the low-resolution range of angular frequency between 0 and $\pi/4$ into two bands, one from 0 to $\pi/8$ and the other from $\pi/8$ to $\pi/4$. From the frequency band, it is observed that LPF and HPF with different cutoff frequencies specifically equal to half the cutoff frequency for the previous decomposition are to be used. We decimate the signal itself, thereby reducing its sampling frequency. So, we can use the same LPF and HPF for all decompositions. (We can recall that the spectrum is stretched when we decimate.)

Let us understand the meaning of MRA. The LPF and HPF form the wavelet basis or the kernel or the mother wavelet. Let us consider LPF and HPF contain N number of impulse response coefficients. N is equal to 2 in the case of Haar wavelet. Once, N is fixed, the time duration of the filter window is fixed. This also fixes the frequency resolution for the first level of decomposition. In the next level of decomposition, dilate the filter window length by a factor of 2, i.e., double the length of the time window. When time domain filter window length is increased, the frequency resolution increases (frequency domain filter window reduces by a factor of 2). For the next level of decomposition, further dilate the filter window by a factor of 2, thereby reducing the frequency domain filter window by a factor of 2. This gives the famous multiresolution decomposition. We also observe from the time frequency tiling of WT that the tile size, if reduced in time domain, its frequency domain size increases and vice versa. The resolution for every stage is different. At low frequencies, frequency resolution is better, and at high frequencies, time resolution is better. The example indicates that we are not dilating the filter window but are shrinking the signal window by down-sampling by a factor of 2. These two operations, namely, shrinking a

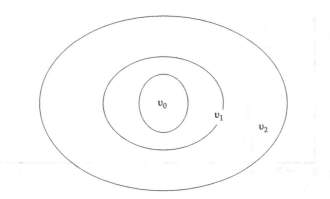

FIGURE 8.19
Nested vector spaces spanned by scaling functions.

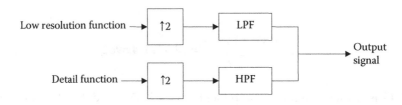

FIGURE 8.20
Reconstruction of a signal using PRFs.

signal window and dilating a filter window, are treated as equivalent. So, by shrinking a signal window, we can get MRA.

Using our intuitive idea of scale or resolution, we can say that the requirement of multi-resolution is the nesting of spanned spaces, namely, $v_0 \subset v_1 \subset v_2 \subset, \ldots, \subset L^2$, as shown in Figure 8.19.

The definition of scaling function spaces indicates that if

$$f(t) \in v_j \text{ then } f(2t) \in v_{j+1} \tag{8.24}$$

Equation 8.24 states that if a function is in a space v_j, then its frequency scaled function is in its next higher frequency resolution space v_j. This confirms that the space that contains high-resolution signals will definitely contain those of low resolution.

The signal can be reconstructed from the same orthogonal basis functions by exactly the reverse operation. For reconstruction, we interpolate and then use the same PRFs for filtering as shown in Figure 8.20.

Concept Check

- What is averaging and differencing?
- What are the properties of Haar wavelet?
- What are the characteristics of image filters or PRFs?

- Why are we down-sampling the wavelet decomposed signal by a factor of 2?
- Why can we use the same PRFs for all levels when we use wavelet decomposition?
- What is multiresolution analysis?
- How will you reconstruct the signal using wavelet coefficients?

8.7 Daubechies Wavelets

The Daubechies wavelet basis are a family of orthonormal, compactly supported scaling and wavelet functions having maximum regularity for a given length of the mirror filters. Daubechies family contains the wavelet basis of different lengths, 4, 6, 8, ..., etc. Daubechies wavelet basis for wavelet written as "D2" contains four filter coefficients given as a set of low-pass filter coefficients $P(n)$ (called as a wavelet function) and high-pass filter coefficients $q(n)$ (called as a scaling function) given by

$$p(n) = \left(\frac{1+\sqrt{3}}{4\sqrt{2}}, \frac{3+\sqrt{3}}{4\sqrt{2}}, \frac{3-\sqrt{3}}{4\sqrt{2}}, \frac{1-\sqrt{3}}{4\sqrt{2}} \right) = (c_0, c_1, c_2, c_3) \tag{8.25}$$

$$q(n) = \left(\frac{1-\sqrt{3}}{4\sqrt{2}}, -\frac{3-\sqrt{3}}{4\sqrt{2}}, \frac{3+\sqrt{3}}{4\sqrt{2}}, -\frac{1+\sqrt{3}}{4\sqrt{2}} \right) = (c_3, -c_2, c_1, -c_0) \tag{8.26}$$

WT using Daubechies wavelet is computed using these filters to filter a signal. The factor $1/\sqrt{2}$ is used as a normalization factor. We can observe that the filters are neither symmetric nor antisymmetric.

Let us consider a signal consisting of some eight data points. The basis matrix using $p(n)$ and $q(n)$ filter, as given by Equations 8.25 and 8.26, is constructed as follows.

$$p(n) = (c_0, c_1, c_2, c_3)$$

$$= (0.482962913,\ 0.636516303,\ 0.224143868,\ -0.129409522) \tag{8.27}$$

$$M = \begin{bmatrix} c_0 & c_1 & c_2 & c_3 & 0 & 0 & 0 & 0 \\ c_3 & -c_2 & c_1 & -c_0 & 0 & 0 & 0 & 0 \\ 0 & 0 & c_0 & c_1 & c_2 & c_3 & 0 & 0 \\ 0 & 0 & c_3 & -c_2 & c_1 & -c_0 & 0 & 0 \\ 0 & 0 & 0 & 0 & c_0 & c_1 & c_2 & c_3 \\ 0 & 0 & 0 & 0 & c_3 & -c_2 & c_1 & -c_0 \\ c_2 & c_3 & 0 & 0 & 0 & 0 & c_0 & c_1 \\ c_1 & -c_0 & 0 & 0 & 0 & 0 & c_3 & -c_2 \end{bmatrix} \tag{8.28}$$

We can observe that the first row contains LPF coefficients with extra zeros padded. The second row contains HPF coefficients again with zeros padded. The third row is shifted with respect to first row by two positions. The fourth row is shifted with respect to second row by two positions. This will be continued until wrap-around occurs, i.e., further rotation will result in a first and a second row. The reader is encouraged to verify that the matrix is orthogonal by calculating $M \times M^T$ and showing that $M \times M^T = 1$.

8.7.1 Matrix Multiplication Method for Computation of WT

Step 1: Consider N point signal. Generate $N \times N$ point matrix M using the aforementioned procedure. To carry out the first level of decomposition using matrix M, we multiply the signal column vector by matrix M and find the result. It is then downsampled by a factor of 2 to get low-resolution component and a detailed component. We will calculate

$$MS = \begin{bmatrix} c_0 & c_1 & c_2 & c_3 & 0 & 0 & 0 & 0 \\ c_3 & -c_2 & c_1 & -c_0 & 0 & 0 & 0 & 0 \\ 0 & 0 & c_0 & c_1 & c_2 & c_3 & 0 & 0 \\ 0 & 0 & c_3 & -c_2 & c_1 & -c_0 & 0 & 0 \\ 0 & 0 & 0 & 0 & c_0 & c_1 & c_2 & c_3 \\ 0 & 0 & 0 & 0 & c_3 & -c_2 & c_1 & -c_0 \\ c_2 & c_3 & 0 & 0 & 0 & 0 & c_0 & c_1 \\ c_1 & -c_0 & 0 & 0 & 0 & 0 & c_3 & -c_2 \end{bmatrix} \begin{bmatrix} s_1 \\ s_2 \\ s_3 \\ s_4 \\ s_5 \\ s_6 \\ s_7 \\ s_8 \end{bmatrix} \begin{bmatrix} a_1 \\ d_1 \\ a_2 \\ d_2 \\ s_3 \\ d_3 \\ a_4 \\ d_4 \end{bmatrix} \quad (8.29)$$

The output vector is then down-sampled by a factor of 2 to generate a 4-point low-resolution component containing $\begin{bmatrix} a_0 \\ a_1 \\ a_2 \\ a_3 \end{bmatrix}$ and a detail function component $\begin{bmatrix} d_0 \\ d_1 \\ d_2 \\ d_3 \end{bmatrix}$. The

reason for down-sampling by a factor of 2 is to use the same filters for second-level decomposition also. We are not dilating the filter window but are shrinking the signal window by down-sampling by a factor of 2. These two operations, shrinking a signal window and dilating a filter window, are equivalent. So, by shrinking a signal window, we can keep the filter length as it is.

Step 2: Consider $N/2$ point low-resolution component given above signal. Generate $N/2 \times N/2$ point matrix M using the aforementioned procedure. To carry out second level of decomposition using matrix M, we multiply the signal column vector of length 4 by matrix M and find the result. It is then down-sampled by a factor of 2 to get low-resolution component and a detailed component. We will calculate

$$MS = \begin{bmatrix} c_0 & c_1 & c_2 & c_3 \\ c_3 & -c_2 & c_1 & -c_0 \\ c_2 & c_3 & c_0 & c_1 \\ c_1 & -c_0 & c_3 & -c_2 \end{bmatrix} \begin{bmatrix} a_0 \\ a_1 \\ a_2 \\ a_3 \end{bmatrix} = \begin{bmatrix} a_0' \\ d_0' \\ a_1' \\ d_1' \end{bmatrix} \qquad (8.30)$$

The output vector is then down-sampled by 2 to generate a 2-point low-resolution component containing $\begin{bmatrix} a_0' \\ a_1' \end{bmatrix}$ and a detail function component $\begin{bmatrix} d_0' \\ d_1' \end{bmatrix}$.

Example 2

Consider a signal given by

$$f = [2, 5, 8, 9, 7, 4, -1, 1].$$

We want to use the wavelet decomposition for a signal using Daubechies wavelet and execute second-level decomposition. We will do this by performing the following steps.

Step 1: We will use the filter coefficients stated in Equation 8.27.

$$X = \begin{bmatrix} c_0 & c_1 & c_2 & c_3 & 0 & 0 & 0 & 0 \\ c_3 & -c_2 & c_1 & -c_0 & 0 & 0 & 0 & 0 \\ 0 & 0 & c_0 & c_1 & c_2 & c_3 & 0 & 0 \\ 0 & 0 & c_3 & -c_2 & c_1 & -c_0 & 0 & 0 \\ 0 & 0 & 0 & 0 & c_0 & c_1 & c_2 & c_3 \\ 0 & 0 & 0 & 0 & c_3 & -c_2 & c_1 & -c_0 \\ c_2 & c_3 & 0 & 0 & 0 & 0 & c_0 & c_1 \\ c_1 & -c_0 & 0 & 0 & 0 & 0 & c_3 & -c_2 \end{bmatrix} \begin{bmatrix} 2 \\ 5 \\ 8 \\ 9 \\ 7 \\ 4 \\ -1 \\ 1 \end{bmatrix} = \begin{bmatrix} 4.7770 \\ -0.6341 \\ 10.6437 \\ -0.5288 \\ 5.5733 \\ -2.9219 \\ -0.0452 \\ -1.2365 \end{bmatrix} \qquad (8.31)$$

After down-sampling by a factor of 2 to get the low-resolution component containing four points as

$$\begin{bmatrix} 4.7770 \\ 10.6437 \\ 5.5733 \\ -0.0452 \end{bmatrix} \text{ and a detail function component } \begin{bmatrix} -0.6341 \\ -0.5288 \\ -2.9219 \\ -1.2365 \end{bmatrix}. \qquad (8.32)$$

Step 2: Consider a 4-point low-resolution component obtained in step 1. Generate 4×4 point matrix M using the aforementioned procedure. To carry out the second level of decomposition using matrix M, multiply the signal column vector of length 4 by matrix M and evaluate the result. It is then down-sampled by a factor of 2 to get low-resolution component and a detailed component. The result of computation is given.

$$MS = \begin{bmatrix} c_0 & c_1 & c_2 & c_3 \\ c_3 & -c_2 & c_1 & -c_0 \\ c_2 & c_3 & c_0 & c_1 \\ c_1 & -c_0 & c_3 & -c_2 \end{bmatrix} \begin{bmatrix} 4.7770 \\ 10.6437 \\ 5.5733 \\ -0.0452 \end{bmatrix} = \begin{bmatrix} 6.2567 \\ 3.4050 \\ 6.5898 \\ -4.7206 \end{bmatrix} \tag{8.33}$$

The output vector will be down-sampled by a factor of 2 to generate a 2-point low-resolution component containing $\begin{bmatrix} 6.2567 \\ 6.5898 \end{bmatrix}$ and a detail function component $\begin{bmatrix} 3.4050 \\ -4.7206 \end{bmatrix}$

Output of dwt obtained using second-level decomposition is

$$[6.2567, \ 6.5898, \ 3.4050, \ -4.7206, \ -0.6341, \ -0.5288, \ -2.9219, \ -1.2365]$$

The detail function is propagated as it is to the next level of decomposition.

All Daubechies family of wavelets are said to have compact support, in the sense that they are localized in time and frequency domain with a small number of wavelet and scaling function coefficients. An important property of Daubechies wavelet and scaling functions is the lack of symmetry or antisymmetry. Daubechies has shown that it is impossible to obtain an orthonormal and compactly supported wavelet that is either symmetric or antisymmetric about any axis.

Characteristics of orthogonal wavelets:

1. The wavelet filter and a scaling filter are always of same length. The length of each filter is found to be even. Daubechies family of wavelets has length equal to 4, 6, 8, 10, etc.
2. The wavelet function and scaling function satisfy the orthogonality condition.
3. The transform is revertible.

8.7.2 Number of Operations

Consider the number of operations required for the discrete orthonormal WT of a vector of signal data. Let L represent the length of the data vector and N the length of finite impulse response filters $P(n)$, $Q(n)$. Usually, N is very small compared to L. At the highest frequency band, the first stage of decomposition will require $2NL$ multiplications and additions. The next stage of decomposition will require NL multiplications and additions, and so on. Full wavelet decomposition will require $2NL \times \left(1 + \dfrac{1}{2} + \dfrac{1}{4} + \cdots\right) \approx 4NL$ computations. The orthonormal WT requires only $O(L)$ computations. This is seen to be even faster than FT, which requires $O(L \log L)$ computations. The computational complexity is as low as the order of $O(L)$.

8.7.3 Time Band Width Product

The WT is a mapping between function of time in one-dimensional case to two-dimensional time-scale joint representation. At the first glance, the time bandwidth product of the WT output is found to be squared of that of the signal. However, in MRA, the size of the data vector is seen to reduce by a factor of 2 when one moves from one frequency band to the next coarser resolution frequency band. This reduces the time bandwidth product by a factor of 2. Considering that the original data vector has L samples, the first stage wavelet coefficient output has $L/2$ samples and the second stage has $L/4$ samples, and so on. So the total bandwidth product of the wavelet decomposition$=L(1/2+1/4+\ldots)=L$.

Concept Check

- State the LPF and HPF for Daubechies wavelet.
- In matrix multiplication method, why are we down-sampling by a factor of 2?
- What is the meaning of compact support?
- What are the characteristics of orthogonal transforms?
- What is the computational complexity for wavelet decomposition?
- What is the time band width product of the wavelet decomposition?

8.8 Some Other Standard Wavelets

There are some standard functions that are used as wavelets. We will describe these standard wavelet functions.

8.8.1 Mexican Hat Function

It is a second derivative of the Gaussian $e^{-x^2/2}$. If we normalize it, we obtain

$$\psi(x) = \frac{2}{\sqrt{3}}\pi^{-1/4}(1-x^2)e^{-x^2/2} \tag{8.34}$$

If we plot this function and imagine it rotated around its symmetry axis, we obtain a shape similar to the Mexican hat. This function is most popular in the computer vision analysis.

8.8.2 A Modulated Gaussian

This function is used by J. Morlet and R. Kronland-Martinet. Its FT is a shifted Gaussian. It is modified and adjusted slightly so that $\psi(0)=0$.

8.8.3 Spline and Battle–Lemarie Wavelets

Spline wavelets use triangular scaling function and are symmetrical. The function is a first-order spline given by

$$h(n) = \left\{\frac{1}{2\sqrt{2}}, \frac{1}{\sqrt{2}}, \frac{1}{2\sqrt{2}}, 0\right\} \tag{8.35}$$

A quadratic spline is generated as follows:

$$h(n) = \left\{ \frac{1}{4}, \frac{3}{4}, \frac{3}{4}, \frac{1}{4} \right\} / \sqrt{2} \tag{8.36}$$

Three sections of second-order polynomials are connected to give continuous first-order derivatives at the junctions. The cubic spline is generated from

$$h(n) = \left\{ \frac{1}{16}, \frac{1}{4}, \frac{3}{8}, \frac{1}{4}, \frac{1}{16} \right\} / \sqrt{2} \tag{8.37}$$

It is not orthogonal over integer translation. If they are orthogonal, their support becomes infinite. They generate Battle–Lemarie wavelet system.

8.8.4 Biorthogonal Wavelets

The properties of biorthogonal wavelets are as follows.

1. The restriction of having wavelet filer and scaling filter of same and even length does not apply to biorthogonal wavelet systems.
2. Symmetric wavelets and scaling functions are used in biorthogonal wavelets.
3. Parseval's theorem does not hold good, i.e., energy of the samples of the wavelet and scaling function in time domain is not the same as the energy of the frequency domain samples.

8.8.5 Cohen–Daubechies–Feauveau Family of Biorthogonal Spline Wavelets

Spline scaling functions are symmetric, smooth, and have dyadic filter coefficients. If these scaling functions in orthogonal wavelets are used, the wavelets have infinite support. It is easy to use splines in biorthogonal wavelets. In the reconstruction from biorthogonal WT, the discrete synthesis filters are different from the analysis filters, and these analysis filters do not have the same length. The perfect reconstruction condition requires orthogonality for the analysis and synthesis filters.

DFT and DCT have fixed kernel. Once, kernel characteristics are fixed, the resolution and localization characteristics are also fixed. WT does not use a fixed kernel. The user can select a kernel for specific application or generate his/her own kernel for WT decomposition. Note that we cannot use any standard wavelet for any type of signal. We have a great flexibility in wavelet decomposition. We have to select a standard wavelet having matching characteristics with the signal or we can generate a special wavelet from the signal itself so that it will have characteristics same as the signal itself. If the wavelet function shape matches that of the signal, the gain of the decomposition will be the highest. This is because the wavelet coefficient indicates the similarity between the mother wavelet and the signal. If the wavelet is derived from the signal, it will have high matching with the signal. Such a wavelet will naturally give high gain of decomposition, and the values of wavelet coefficients will be very high, closer to 1. Else, we get very small gain of decomposition and cannot extract the parameters from the signal satisfactorily. There is a possibility to design a wavelet form the signal itself. The wavelet so designed must satisfy the regularity and admissibility condition in order that it can be used for WT decomposition.

Different standard wavelets are designed. For a signal under consideration, any wavelet selected may not give good results. We have to select a wavelet that has a shape matching with a signal, so that a gain of decomposition will be high. Some standard wavelets fail in the case of speech processing applications. The author of this book has designed a wavelet from the speech signal and has shown that it works better for speech analysis, recognition, synthesis, speaker verification, etc. The idea of generation of mother wavelet is patented in India, and the patent is granted.

8.8.6 Wavelet Packets

When we refer to time frequency tiling of WT, low frequencies have smaller bandwidths and high frequencies have larger bandwidths. This is called constant Q property of WT. The wavelet packet system was proposed by Ronald Coifman to allow finer and adjustable resolution of frequencies at high frequencies. Wavelet packets are required to find the parameters of the signal hidden in some specific bands.

8.8.6 1 Full Wavelet Packet Decomposition

In order to generate a basis system that will give a higher resolution at high frequencies, the user will iterate the detail function branch along with the low-resolution branch. The resulting three-stage analysis tree will now also be present for the detail function component. When both the low passband and high passband are split at all stages, the resulting filter bank structure is like a full binary tree that requires $O(N\log N)$ calculations similar to DFT. The wavelet packet decomposition is shown in Figure 8.21. The use of wavelet packets is to extract the features hidden in particular higher frequency band. The frequency bands for the analysis tree of full wavelet packet decomposition are shown in Figure 8.22.

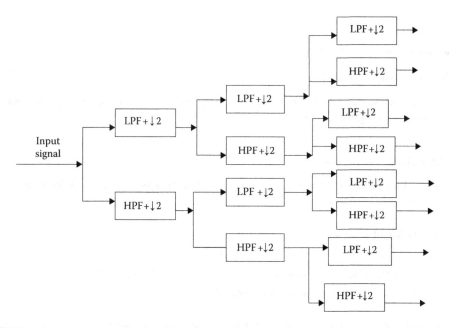

FIGURE 8.21
Wavelet packet decomposition.

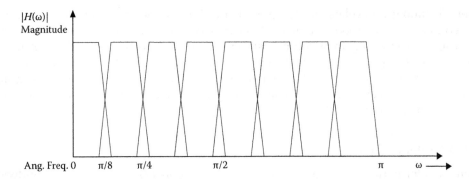

FIGURE 8.22
Frequency bands for full wavelet packet decomposition.

Concept Check

- What is the characteristic of Mexican hat wavelet?
- What are the characteristics of biorthogonal wavelets?
- Are the analysis and synthesis filters same for biorthogonal wavelets?
- What is the meaning of gain of decomposition?
- Why does a wavelet derived from the signal give high gain of decomposition?
- What is the flexibility in case of wavelet decomposition?
- What are wavelet packets?

8.9 Applications of WT

Diverse applications of WT such as signal compression, signal denoising, and sampling rate conversion are discussed in this section.

8.9.1 Denoising Using DWT

WT is a tool for removing noise from the signal. The denoising methods for noise removal follow these steps.

1. Take DWT of a signal.
2. Use a threshold for the transformed signal.
3. This removes the coefficients below a certain level (threshold).
4. Now take inverse discrete wavelet transform (IDWT).

This method can remove noise, and it achieves high compression ratios because of energy compaction in DWT. If a signal has its energy concentrated in less number of coefficients,

Plot of low resolution component of first-level decomposition of voiced part of a signal

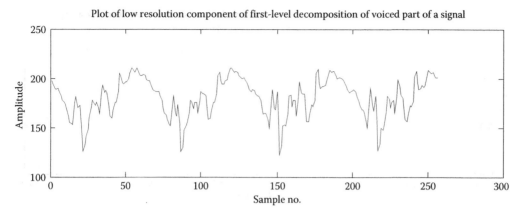

Detail function of first-level decomposition of voiced part of a signal made zero as is less than threshold

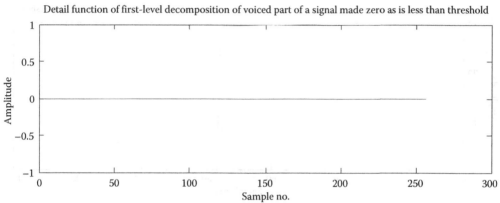

FIGURE 8.23
Plot of low-resolution and detail function for first-level decomposition.

these coefficients are found to be relatively large compared to noise component. The noise component has its energy spread over a large number of coefficients. This confirms that thresholding will remove low amplitude noise in wavelet domain. IDWT operation will retrieve the signal with little loss of details. There is little loss of details because the detail coefficients that contain some information are also removed. Let us consider a speech signal, take its DWT, and remove the noise. Figure 8.23 shows a plot of first-level wavelet decomposition of speech signal using Haar wavelet. The detail component is now made zero as it goes below the threshold. Figure 8.24 depicts a plot of second-level wavelet decomposition of speech signal using Haar wavelet. The detail component is again made zero as it goes below the threshold. Figure 8.25 shows a plot of original speech signal and the denoised speech signal. We can see that the speech waveform is made smooth.

A MATLAB program to execute the noise removal is as follows.

```
%program to achieve denoising using wavelet decomposition for 128 samples
of voiced speech using Haar wavelet
clear all;
fp=fopen('watermark.wav');
fseek(fp,224000,-1);
a=fread(fp,512);
```

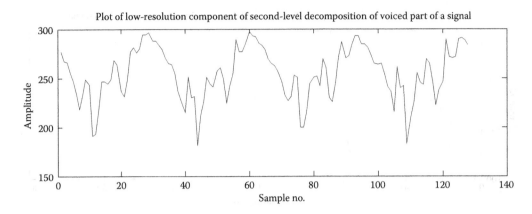

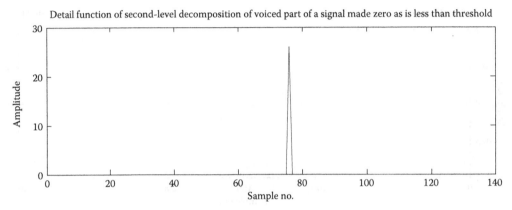

FIGURE 8.24
Plot of low-resolution and detail function for second-level decomposition.

```
%plot 256 points of voiced speech
[b,c]=dwt(a,'db1');
subplot(2,1,1);plot(b);title('plot of low resolution component of first
level decomposition of voiced part of a signal');
xlabel('sample no.');ylabel('amplitude');
  for i=1:256,
      if c(i)<20, c(i)=0;
      end
  end
subplot(2,1,2);plot(c);title('plot of detail function of first level
decomposition of voiced part of a signal made zero as is less than
threshold');
xlabel('sample no.');ylabel('amplitude');
figure;
[b1,c1]=dwt(b,'db1');
subplot(2,1,1);plot(b1);title('plot of low resolution component of second
level decomposition of voiced part of a signal');
xlabel('sample no.');ylabel('amplitude');
  for i=1:128,
      if c1(i)<20, c1(i)=0;
      end
  end
```

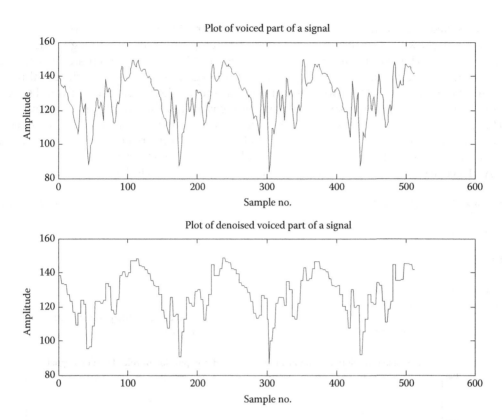

FIGURE 8.25
Plot of original and denoised speech signal.

```
subplot(2,1,2);plot(c1);title('plot of detail function of second level
decomposition of voiced part of a signal made zero as is less than
threshold'); xlabel('sample no.');ylabel('amplitude');
figure;
b2=idwt(b1,c1,'db1');
b3=idwt(b2,c,'db1');
subplot(2,1,1); plot(a);title('plot of voiced part of a
signal');xlabel('sample no.');ylabel('amplitude');
subplot(2,1,2);plot(b3);title('plot of denoised voiced part of a
signal');xlabel('sample no.');ylabel('amplitude');
```

8.9.2 Signal Compression

Energy compaction is the most important property for low bit rate transform coding. When we transform the signal using DWT, we get less number of significant coefficients. Threshold is selected. If the value of the coefficient is greater than the threshold, we transmit a "1" to indicate that the coefficient has a significant value, otherwise we transmit a "0" to indicate the insignificant coefficient. The significance of transform coefficients is represented in the form of significant map containing a "1" for significant coefficient and a "0" for insignificant coefficient. The significant map indicators are coded using N number of bits if there are N coefficients. We find that we need to actually code less number of significant coefficients only.

Let us take a speech signal, take its DWT, and try to obtain signal compression. A Figure 8.26 shows a plot of first-level wavelet decomposition of speech signal using the Haar wavelet. The detail component is made zero as it has a value less than the step size. Figure 8.27 depicts a plot of second-level wavelet decomposition of speech signal using Haar wavelet. Figure 8.28 shows a plot of original speech signal and the recovered speech signal. We can see that the speech waveform recovered is the same as the original one. This proves that for the detail function of first-level wavelet decomposition, we select a step size large enough so that all detail function coefficients are coded using zero value. This gives 50% signal compression as only half the coefficients are coded.

A MATLAB program to execute the signal compression is as follows.

```
%program to get compression using wavelet decomposition of a signal using
%using Haar wavelet
clear all;
fp=fopen('watermark.wav');
fseek(fp,224000,-1);
a=fread(fp,512);
%plot 512 points of voiced speech
[b,c]=dwt(a,'db1');
subplot(2,1,1);plot(b);title('plot of low resolution component of first
level decomposition of voiced part of a signal');
```

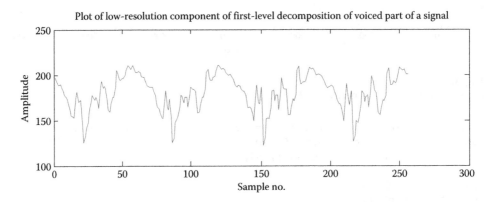

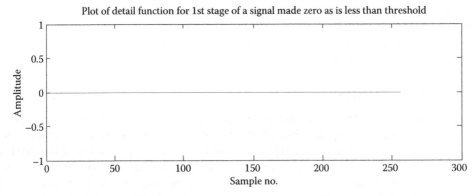

FIGURE 8.26
Plot of first-level decomposition, detail function is zero.

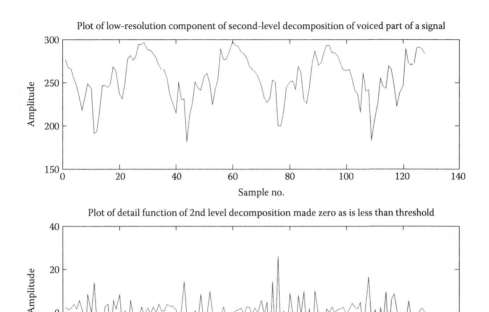

FIGURE 8.27
Plot of second-level decomposition.

```
xlabel('sample no.');ylabel('amplitude');
  for i=1:256,
      if c(i)<20, c(i)=0;
      end
  end
subplot(2,1,2);plot(c);title('plot of detail function for 1st stage of a
signal made zero as is less than threshold');
xlabel('sample no.');ylabel('amplitude');
figure;
[b1,c1]=dwt(b,'db1');
subplot(2,1,1);plot(b1);title('plot of low resolution component of second
level decomposition of voiced part of a signal');
xlabel('sample no.');ylabel('amplitude');
subplot(2,1,2);plot(c1);title('plot of detail function of 2nd level
decomposition made zero as is less than threshold'); xlabel('sample
no.');ylabel('amplitude');
figure;
b2=idwt(b1,c1,'db1');
b3=idwt(b2,c,'db1');
subplot(2,1,1); plot(a);title('plot of voiced part of a
signal');xlabel('sample no.');ylabel('amplitude');
subplot(2,1,2);plot(b3);title('plot of recovered voiced part of a
signal');xlabel('sample no.');ylabel('amplitude');
```

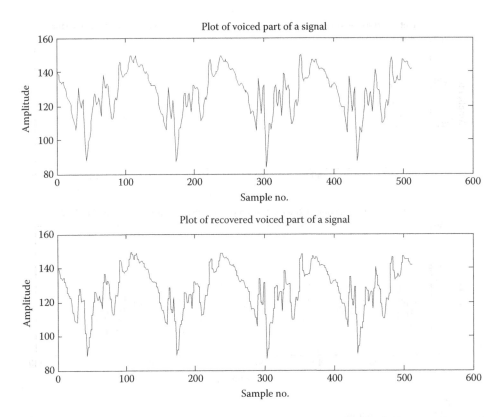

FIGURE 8.28
Plot of original signal and recovered signal.

8.9.3 Signal Filtering

Consider the filtering of speech signal using WT. Now, scan the speech signal samples and take DWT. The signal can be low-pass filtered if we retain only low-resolution component and make all detail coefficients zero. Then take IDWT to get the processed signal. To detect the detail function, we will make low-resolution component zero and will retain only the detail function to get high-pass filtered speech signal. After taking inverse wavelet transform (IWT), we will get the high-pass filtered speech signal. Figure 8.29 shows the original speech signal, low-pass filtered signal and high-pass filtered speech signal.

A MATLAB program to do it is as follows.

```
%program to get low pass filtered and high pass filtered speech signal.
%using Haar wavelet
clear all;
fp=fopen('watermark.wav');
fseek(fp,224000,-1);
a=fread(fp,512);subplot(3,1,1);plot(a);title('plot of voiced speech
signal');
xlabel('sample no.');ylabel('amplitude');
%plot 256 points of voiced speech
[b,c]=dwt(a,'db1');
```

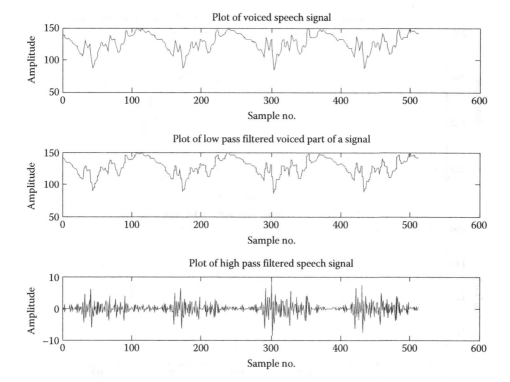

FIGURE 8.29
Original, low-pass filtered, and high-pass filtered speech.

```
for i=1:256,
    c(i)=0.0;
end
b1=idwt(b,c,'db1');
subplot(3,1,2); plot(b1);title('plot of low pass fitered voiced part of a
signal');xlabel('sample no.');ylabel('amplitude');
[b2,c2]=dwt(a,'db1');
for i=1:256,
    b2(i)=0.0;
end
b11=idwt(b2,c2,'db1');
subplot(3,1,3);plot(b11);title('plot of high pass filtered speech
signal');xlabel('sample no.');ylabel('amplitude');
```

8.9.4 Sampling Rate Conversion

Consider the case of decimation by a factor of 2 for a speech signal. We will read speech signal and take its WT using a Haar wavelet. Retain only the low-resolution component to implement a low-pass filter by a factor of $\pi/2$. The IWT output will be down-sampled by 2 to get the decimation by a factor of 2. Figure 8.30 depicts a plot of low-resolution function and recovered signal by making detail function zero. Figure 8.31 depicts the plots of original speech signal and the decimated speech signal.

A MATLAB program is as follows.

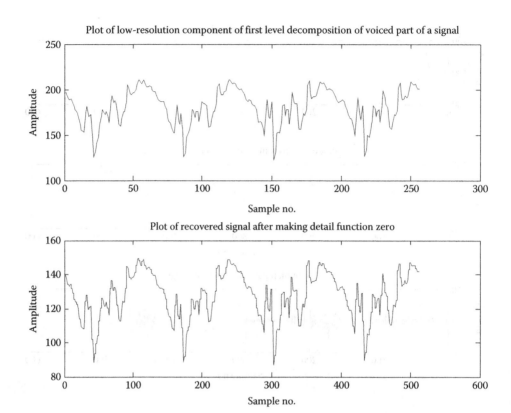

FIGURE 8.30
Plot of low-resolution function and recovered signal after making detail function zero.

```
%program to decimate speech signal by a factor of 2
clear all;
fp=fopen('watermark.wav');
fseek(fp,224000,-1);
a=fread(fp,512);
%plot 256 points of voiced speech
[b,c]=dwt(a,'db1');
subplot(2,1,1);plot(b);title('plot of low resolution component of first
level decomposition of voiced part of a signal');
xlabel('sample no.');ylabel('amplitude');
 for i=1:256,
    c(i)=0;
 end
 a1=idwt(b,c,'db1');
subplot(2,1,2);plot(a1);title('plot of recovered signal after making
detail function zero');
xlabel('sample no.');ylabel('amplitude');
 for i=2:2:512,
    a11(i/2)=a1(i);
 end
end
figure;subplot(2,1,1); plot(a);title('plot of original voiced part of a
signal');xlabel('sample no.');ylabel('amplitude');
subplot(2,1,2);plot(a11);title('plot of denoised voiced part of a
signal');xlabel('sample no.');ylabel('amplitude');
```

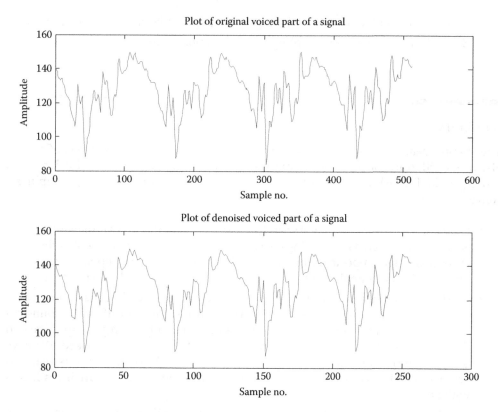

FIGURE 8.31
Plot of original speech signal and decimated speech signal.

Wavelet bases have the following useful properties.

1. They can represent smooth functions
2. They can represent singularities
3. We can select the bases for any application.
4. Wavelets are computationally less costly. The complexity is of the order of $O(N)$.

Wavelets have number of applications related to seismic and geographical signal processing. There are a number of applications in medical and biomedical signal and image processing field as well. These applications include enhancing the contrast for the signal such as image enhancement, extraction of some peculiar feature from the signal, etc.

Concept Check

- What are the different applications of WT?
- How will you achieve the compression by properly coding DWT coefficients?

- How will you achieve low-pass filtering and high-pass filtering for speech using WT?
- How will you implement sampling rate conversion using WT?

Summary

The chapter deals with short segment analysis, pitch synchronous analysis of speech. It also deals with transform domain analysis of speech using DCT, STFT, and WT, and applications of DCT, STFT, and WT for speech compression, coding, filtering, sampling rate conversion, etc.

1. We started with a shot segment speech analysis. It was shown that the speech parameters can be treated as remaining constant for a short duration segment of the order of 10–20 ms. The importance of pitch synchronous analysis is emphasized, and the pitch synchronous analysis of speech is illustrated with a MATLAB program.

2. We discussed the meaning of basis function, basis matrix, orthogonality of a transform, and revertibility of a transform. We then introduced different parameters for measuring the performance of the transforms, namely, energy compaction and decorrelation. The Fourier-related transform such as DCT is discussed. We have given definition and properties of DCT. We emphasized that DCT is real and orthogonal. The energy compaction and decorrelation properties are investigated with DCT.

3. We discussed different applications of transform domain processing such as compression and coding. We illustrated how to filter a signal in transform domains such as DCT. We have illustrated the use of DCT for speech coding and compression. We indicated the use of DCT for filtering of speech signal. Sampling rate converter for speech is implemented. We pointed out that these transforms cannot capture the time variations in the signals. To capture time variations, we have to use transforms such as STFT and WT.

4. We started with STFT. We discussed the meaning of sliding window transform and indicated how the frequency resolution is decided by the size of the window. STFT uses a short-time windowed sines and cosines as basis functions. STFT can be considered as a Fourier analysis of the short time windowed signal. Multiplying the signal with window localizes the signal in time domain, but results in a convolution of signal spectrum with the spectrum of the window. The narrower the window, the better we localize the signal in time domain and poorer we localize its spectrum. Heisenberg uncertainty principle applies here. $\Delta t \times \Delta f \geq 1/2$, i.e., if we improve the resolution in time domain, we have to sacrifice for frequency domain resolution, and vice versa.

5. We introduced the 2D and 3D spectrograms. 3D spectrogram is a plot of FFT amplitude with respect to frequency and time. 2D graph is a time–frequency plot with amplitude reduced to the color of the pixel. WT was introduced as a tool for analysis of time-varying signals. WT provides multiresolution analysis with dilated windows. The high-frequency analysis is done using narrow windows, and low-frequency analysis is done using wide windows. The relative bandwidth is calculated as a ratio of bandwidth and center frequency of the band.

We can observe that the relative bandwidth Q is constant for wavelet decomposition. It was pointed out that the constant Q property is useful in the analysis of natural signals such as speech signal, which also exhibits constant Q distribution of energy.

6. We defined WT. The wavelet coefficient represents how well the signal $s(t)$ is similar to the scaled and translated mother wavelet (kernel). The kernel must satisfy regularity and admissibility condition. If the orthogonal basis is used, the wavelet function and scaling function form PRFs. One example of continuous wavelet, namely, Haar wavelet, was considered for calculations and decomposition using Haar wavelet is illustrated. The concept of multiresolution was explained using complete decomposition of signal into frequency bands. At low frequencies, frequency resolution is better, and at high frequencies, time resolution is better. In case of the example explained, it was pointed out that we are not dilating the filter window, but we are shrinking the signal window by down-sampling by a factor of 2. These two operations are equivalent.

7. We considered one representative of orthogonal wavelets, namely, Daubechies wavelet. It was indicated that the wavelet and scaling functions are not symmetrical or are antisymmetrical. Matrix multiplication method for computation of wavelet decomposition is explained. The properties of orthogonal wavelet family are discussed. We calculated the number of operations and time bandwidth product. Other wavelets such as spline, Mexican hat, and Gaussian basis are introduced. We also considered biorthogonal wavelet basis and its properties. We introduced spline biorthogonal wavelets and wavelet packets.

8. Last, we indicated some applications of WT such as denoising, signal compression, signal filtering, and sampling rate conversion using WT. We also pointed out that there are several applications for speech.

Key Terms

Transform domain	Wavelet function
Optimality	Scaling function
Revertibility	Signal compression
Orthogonality	Daubechies wavelet
Basis functions	Compact support
Basis vectors	Wavelet packets
Basis matrix	Splines
Orthogonal matrix	Detail function
Energy compaction	Low-resolution function
Decorrelation	Orthogonal wavelets
Compression	Biorthogonal wavelets
Transform domain filtering	Time bandwidth product
Discrete cosine transform	Signal compression
Discrete sine transform	Signal coding
Wavelet transform	Signal filtering
Haar transform	Sampling rate conversion

Multiple-Choice Questions

1. The short time analysis assumes that the speech parameters
   a. Remain constant over a short time interval
   b. Vary over the short time interval
   c. Are quasi-constant over a short interval
   d. Are almost constant over a short interval

2. The pitch synchronous analysis uses the segment size equal to
   a. Pitch frequency
   b. Pitch period in terms of number of samples
   c. Pitch frequency in Hz.
   d. Two pitch periods

3. The following transform gives maximum energy compaction.
   a. DCT
   b. Discrete fourier transform (DFT)
   c. Discrete sine transform
   d. Wavelet transform

4. Filtering in time domain is done using following operation on signal and filter response
   a. Sample by sample multiplication
   b. Convolution
   c. Sample by sample addition
   d. Sample by sample division

5. The transform is revertible if
   a. The basis matrix is square matrix
   b. The basis matrix is unity matrix
   c. The basis matrix is orthogonal
   d. The basis matrix is same as its inverse

6. Decorrelation for transform domain coefficients means
   a. Transform coefficients have no relation to input signal
   b. Transform coefficients have reduced correlation among themselves
   c. Transform coefficients have maximum correlation with input signal
   d. Transform coefficients have maximum correlation with itself.

7. Transform domain coding relies on the fact that
   a. Input signal is decorrelated
   b. Input signal has high correlation
   c. Transform coefficients have high correlation
   d. Transform coefficients have reduced correlation

8. DCT is not useful for speech signal, because
   a. Speech signal is stationary
   b. Speech signal is time varying and DCT cannot capture time variations
   c. Speech signal is time varying and DCT can capture time variations
   d. Speech signal is a random process, but DCT is not

9. The basis functions for STFT are
   a. Sines and cosines
   b. Windowed sines
   c. Windowed cosines
   d. Windowed sines and cosines

10. Heisenberg uncertainty principle says that
    a. $\Delta t \times \Delta f \leq 1/2$
    b. $\Delta t \times \Delta f \geq 1/2$
    c. $\Delta t \times \Delta f \geq 1/3$
    d. $\Delta t \times \Delta f \leq 1/3$

11. A three-dimensional spectrogram is a plot of
    a. FFT amplitude with respect to frequency and time
    b. DCT amplitude with respect to time
    c. DCT amplitude with respect to frequency and time
    d. FFT amplitude with respect to frequency.

12. Wavelet transform provides multiresolution analysis using
    a. Windowed sines and cosines
    b. Shifted sines and cosines
    c. Dilated windows
    d. Shifted windows

13. The relative bandwidth is calculated as ratio of
    a. Bandwidth and maximum frequency of the band
    b. Maximum bandwidth and minimum frequency of the band
    c. Bandwidth and center frequency of the band
    d. Minimum bandwidth and maximum frequency of the band

14. For orthogonal basis, scaling and wavelet function for a wavelet are
    a. Orthogonal to each other
    b. Orthonormal to each other
    c. Symmetrical to each other
    d. Antisymmetrical to each other

15. At low frequencies, frequency resolution is better, and at high frequencies,
    a. Time resolution is better
    b. Time resolution is poor

    c. Time resolution is made half

    d. Time resolution is doubled

16. Biorthogonal wavelets have wavelet and scaling functions that are

    a. Doubly orthogonal

    b. Not orthogonal

    c. Symmetrical

    d. Not symmetrical

17. To implement low-pass filtering in WT domain, we have to retain only

    a. Low low (LL) component

    b. Low high (LH) component

    c. High low (HL) component

    d. High high (HH) component

18. To implement image resizing by a factor of 1/2 in WT domain, we will retain

    a. LL and LH component

    b. HL and LH component

    c. HH and LH component

    d. LL component

Review Questions

1. Explain the meaning of short-time speech analysis.
2. What is pitch synchronous analysis? Write the algorithm to track pitch synchronous segment.
3. Explain the term decorrelation of transform coefficients.
4. Explain the use of DCT for energy compaction and decorrelation.
5. State different applications of DCT.
6. Explain filtering using DCT.
7. How will you achieve compression by properly coding DCT output?
8. How will you implement sampling rate conversion using DCT?
9. Describe short-time FT analysis of signals. What are the basis functions used in STFT?
10. Explain how can STFT extract the variations in the signal? How will you increase the frequency resolution using STFT?
11. Explain how Heisenberg's uncertainty principle applies to STFT?
12. What is a 3D spectrogram? How can a 3D plot be reduced to a 2D plot?
13. Write a procedure to plot 3D spectrogram of a signal.
14. What is a mother wavelet? Write the wavelet and scaling functions for Haar wavelet.

15. Explain how will you decompose a signal using Haar wavelet?
16. Explain the characteristics of orthogonal wavelets.
17. Describe Daubechies wavelet. Generate a basis matrix of size 8×8 using Daubechies filters. Show that Daubechies basis matrix is orthogonal.
18. Explain the meaning of multiresolution analysis. How do we get multiresolution analysis by dilating the mother wavelet or by decimating the signal?
19. Considering our intuitive idea of resolution, what will be the meaning of multi-resolution analysis?
20. Explain the block schematic for two-level wavelet decomposition using wavelet and scaling filters.
21. What are wavelet packets? Why are wavelet packets required?
22. Explain full-wavelet packet decomposition using a block schematic.
23. What are different standard wavelets? Why are we using number of standard wavelets?
24. Explain the time frequency tiling of WT.
25. Calculate the order of operations and time bandwidth product for orthogonal wavelet decomposition.
26. Explain the meaning of gain of decomposition. When will you get high gain of decomposition?
27. What are characteristics of the biorthogonal wavelets?
28. Explain different applications for wavelet transform.
29. Explain the implementation of sampling rate conversion using WT.
30. Explain the use of WT for filtering of speech.
31. How will you achieve compression of the signal by properly coding wavelet coefficients?

Problems

1. Calculate DFT for $x(n) = [1\ 3\ 5\ 7]$.
2. Calculate IDCT of the result of Problem 1 and show that you get the same signal back.
3. Record speech utterance in your voice and find its DCT. Check if energy compaction is done.
4. Consider a sampling frequency of 8000 Hz for a recorded speech signal and take 256 samples of data and take 256-point DCT. Design a low-pass filter with cut-off frequency of 1000 Hz.
5. Consider a sampling frequency of 8000 Hz for a recorded speech signal and take 256 samples of data and take 256-point DCT. Design a band-pass filter to pass a band between 1000 and 2000 Hz.
6. Calculate the WT of $x(n) = [1\ 3\ 5\ 7]$ using Haar wavelet.

7. Calculate two-level wavelet decomposition using Haar wavelet for

$$h(n) = [1\ 2\ 4\ 7\ 8\ 9\ 1\ 0]$$

8. Decompose the signal using Daubechies wavelet.

$$X(n) = [2\ 4\ 1\ 4]$$

9. Using the result of Problem 7, show that Haar wavelet has poor energy compaction.

Suggested Projects

1. Record a speech file in your voice. Write MATLAB program to track the voiced segment, find pitch values, and track a pitch synchronous segment. Plot a pitch synchronous segment.
2. Record a speech file in your own voice and code properly in DCT domain to achieve 50% compression.
3. Record a speech file in your own voice and low-pass filter it in DCT domain.
4. Record a speech file in your own voice and high-pass filter it in DCT domain.
5. Record a speech file in your own voice and implement sampling rate conversion by a factor of 3 using DCT.
6. Record a speech signal in your voice and write a MATLAB program to plot a 3D spectrogram.
7. Record a speech file in your own voice and code properly in WT domain to achieve 50% compression.
8. Record a speech file in your own voice and low-pass filter it in WT domain.
9. Record a speech file in your own voice and high-pass filter it in WT domain.
10. Record a speech file in your own voice and implement sampling rate conversion by a factor of 4 using WT.

Answers

Multiple-Choice Questions

1. (a)
2. (b)
3. (a)
4. (b)

5. (c)
6. (b)
7. (b)
8. (b)
9. (d)
10. (b)
11. (a)
12. (c)
13. (c)
14. (a)
15. (a)
16. (b)
17. (c)
18. (d)

Problems

1. DCT coefficients are 8.0000, −4.4609, 0, −0.3170.
2. —
3. —
4. DFT filter has pass-band up to 64th DFT coefficient.
5. DFT filter has passband between 64th and 128th DFT coefficients.
6. 2.8284, 8.4853, −1.4142, −1.4142.
7. First level:

2.1213	7.7782	12.0208	0.7071
−0.7071	−2.1213	−0.7071	0.7071

Second level:

7.0000	9.0000
−4.0000	8.0000

Third level:

11.3137 −1.4142

8. 3.2185, 2.9597, −2.4507, −1.6848
9. The result of Problem 7 indicates that although the detail function coefficients are less, they have significant value. Hence, the energy compaction is poor.

9

Image Processing Techniques

LEARNING OBJECTIVES

- Basics of image processing
- Edge enhancement techniques
- Spatial filtering and smoothing
- Sobel and Prewitt edge detectors
- Laplacian of Gaussian (LOG)
- Image processing using transformations
- Histogram equalization
- Statistical image processing
- Image parameters and extraction
- Image compression
- Noise cancellation techniques
- Image resizing

This chapter deals with image formation, image storage, and image representation. The filtering techniques used for images, namely, low-pass filters, high-pass filters, edge enhancement filter, high-frequency emphasis filter, and contrast enhancement filters in spatial domain, are discussed in detail. We will describe different spatial masks for image smoothing and edge detectors such as Sobel and Prewitt. The edge detection using derivative of gradient, i.e., Laplacian mask, will be explained. Laplacian of Gaussian (LOG) is discussed. We then go through different image transformations, such as logarithmic, piecewise linear transformations, which are described for image enhancements. We describe different statistical parameters of an image. The different processing techniques like image compression, noise cancellation, and image resizing using transform domain techniques like Discrete cosine transform (DCT) and wavelet transform (WT) will be explained.

9.1 Image Representation and Spatial Filtering

Let us start with the image formation. The image is formed by illuminating the object and analyzing the reflected energy. Thus, the image is basically a multiplicative signal generated by multiplication of illuminance and reflected component. This clearly indicates that the linear digital filters we have studied are of no help for image filtering. For image filtering or processing, we have to make use of homomorphic processing system for multiplication, as discussed in Chapter 5, Section 5.7.1. The homomorphic system for multiplication consists of just the LOG block that will convert the multiplicative signals into additive components, and we can use linear time invariant (LTI) filter in the LOG domain. The inverse system will be exp block.

We will concentrate on digital images. The image is sampled on a two-dimensional (2D) grid. Each sampled value is further quantized to get the matrix of integer values. Let us see how these images are represented. The 512×512 image is a matrix of size 512×512. This means a matrix will have 512 rows and 512 columns, and each element of the matrix will represent the intensity level of the pixel at the position given by its row number and column number. Generally, the intensity of every pixel in the gray scale image is represented as an 8-bit number, with 255 representing white pixel and 0 representing the black pixel. We will deal with gray scale images first. Figure 9.1 represents an image of size 8×8.

The spatial filtering technique uses a 3×3 or 5×5 or 7×7 mask, and the mask convolves with the image under consideration. Let us understand the process of convolution. Figure 9.2 shows image filtering using 3×3 masks in spatial domain. The block representing pixel values as Z_1, Z_2, etc. is the image block. Mask pixels are represented as W_1, W_2, etc., and when the mask convolved with the image block, the center pixel of the image block, namely, Z_5, is replaced by the result R, shown in Figure 9.2 as

$$R = W_1 Z_1 + W_2 Z_2 + \ldots W_9 Z_9 \tag{9.1}$$

After evaluating the replacement value for the centre pixel, the mask slides over the image. Now, the pixel Z_6 will be at the center, and so on. After sliding over the entire row, it shifts down one row and again slides over the row. This will happen until all rows are exhausted. The reader can easily see that in the first and last rows, first and last column pixels are not replaced. If you want to replace these pixels as well, then the extra rows and

62	79	23	119	120	105	4	0
10	10	9	62	12	78	34	0
10	58	197	46	46	0	0	48
176	135	5	188	191	68	0	49
2	1	1	29	26	37	0	77
0	89	144	147	187	102	62	208
255	252	0	166	123	62	0	31
166	63	127	17	1	0	99	30

FIGURE 9.1
Image of size 8×8.

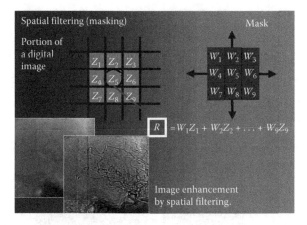

FIGURE 9.2
Details of spatial filtering.

columns with zero pixels are to be added to the image before the convolution starts. We can list the basic image filtering steps as follows:

1. Use a kernel (mostly 3×3)
2. Convolve the kernel with the image
3. Shift the kernel over the image row-wise and then column-wise, and then modify the center pixel every time by the sum of the product of the kernel and image pixel-wise

Let us now consider masks for low-pass filter or smoothing of any image. The typically used masks are as follows.

$$\text{Mask } 1 = \frac{1}{9} \begin{vmatrix} 1 & 1 & 1 \\ 1 & 1 & 1 \\ 1 & 1 & 1 \end{vmatrix} \text{ and Mask } 2 = \frac{1}{16} \begin{vmatrix} 1 & 2 & 1 \\ 2 & 4 & 2 \\ 1 & 2 & 1 \end{vmatrix}$$

Considering the procedure for convolution, the reader can easily verify that using the masks stated earlier, every center pixel will get replaced by the average of the neighboring pixels or the weighted average of the neighboring pixels. The weights are based on the degree of closeness of the concerned pixel to the center one. As every pixel value is the average value of the eight neighboring pixels, the intensity variations will be smoothened. This will implement the smoothing of intensity in the image.

We will take one image now and apply smoothing masks on that. Let us write a program to low-pass filter or smooth the image. Figure 9.3 shows original cameraman image and Figures 9.4 and 9.5 show the filtered image using masks 1 and mask 2, respectively.

Let us write a MATLAB program as follows.

```
clear all;
A=imread('cameraman.jpg');
figure;
imshow(A);title('cameraman image');xlabel('x-coordinate');ylabel('y
coordinate');
```

```
m=1/9*[1 1 1
     1 1 1
     1 1 1];
c=conv2(single(A),single(m),'same');
c=int8(c);
figure;
imshow(c);title('Smoothed cameraman image');xlabel('x-
coordinate');ylabel('y coordinate');
```

Let us now consider different neighboring patterns used for filtering. The typical neighboring patterns are shown in Figure 9.6.

FIGURE 9.3
Original cameraman image.

FIGURE 9.4
Cameraman image filtered using mask 1.

Smoothed cameraman image

y-Coordinate

x-Coordinate

FIGURE 9.5
Cameraman image filtered using mask 2.

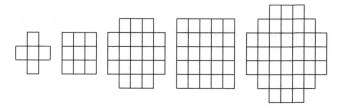

FIGURE 9.6
Different neighboring patterns.

When smoothing of the image is done by simple averaging, typically, 4, 8, 20, 24, 36 neighboring pixels, as shown in Figure 9.5, are used for evaluating the average. Consider that we are using 20 neighbors, for example. The mask will be written as

$$
\text{Mask 3} = \frac{1}{21}
\begin{vmatrix}
0 & 1 & 1 & 1 & 0 \\
1 & 1 & 1 & 1 & 1 \\
1 & 1 & 1 & 1 & 1 \\
1 & 1 & 1 & 1 & 1 \\
0 & 1 & 1 & 1 & 0
\end{vmatrix}
$$

Let us apply mask 3 to cameraman image. The output after putting this mask in the MATLAB program given earlier is shown in Figure 9.7. The image is more blurred.

9.1.1 Edge Detection Using Spatial Filtering

Let us now apply the edge operators to the image. The image operators are basically the differentiators. The mask is specified for detecting the edges in horizontal, vertical, and two diagonal directions, as shown here.

Smoothed cameraman image

y-Coordinate

x-Coordinate

FIGURE 9.7
Smoothed image using mask 3.

$$\text{Sobel edge operator for horizontal edge detection} = \begin{vmatrix} -1 & -2 & -1 \\ 0 & 0 & 0 \\ 1 & 2 & 1 \end{vmatrix}.$$

Let us look more closely to the Sobel edge operator for detecting the horizontal edge. If we look at the first row and the last row, we can see that it takes the difference of the pixel in the first row and the third row. The row for the concerned pixel is not used. This difference is doubled for the center position as these pixels are closer to the center pixel that will be modified. This is seen as the difference operator. If the difference is higher, it is perceived as the edge. Depending on the difference between the pixel in the first row and the third row, we will have to retain or discard the edge. If the difference is high, it is the prominent edge. The threshold is to be properly selected for retaining or discarding edge. If the difference is smaller than the threshold, the edge is not so prominent and it may not be discarded by suitable selection of threshold for the given image under consideration. The reader can remember that when the difference in vertical direction is calculated, we detect the edge in the horizontal direction. Figure 9.8 shows the original cameraman image, Figure 9.9 shows the Sobel edge operated image, and Figure 9.10 shows the image after applying threshold. After proper thresholding, only the prominent edges are seen. The edge operators for detecting vertical edges and edges in diagonal directions are given below.

$$\text{Sobel edge operator for vertical edge detection} = \begin{vmatrix} -1 & 0 & 1 \\ -2 & 0 & 2 \\ -1 & 0 & 1 \end{vmatrix}.$$

$$\text{Sobel operators for diagonal edge detection} = \begin{vmatrix} 0 & 1 & 2 \\ -1 & 0 & 1 \\ -2 & -1 & 0 \end{vmatrix} \text{ and } \begin{vmatrix} -2 & -1 & 0 \\ -1 & 0 & 1 \\ 0 & 1 & 2 \end{vmatrix}$$

Cameraman image

y-Coordinate

x-Coordinate

FIGURE 9.8
Original cameraman image.

Edge detected image using Sobel
operator for horizontal direction

y-Coordinate

x-Coordinate

FIGURE 9.9
Image after Sobel edge detection for horizontal edges.

Figure 9.11 illustrates the curves for the intensity variation in the successive pixels and its derivatives. The edge is detected at the point of maximum of the derivative curve. If the maximum value of the derivative is high, it is a prominent edge. If the slope of the intensity graph is small, i.e., difference in the pixel intensity values is less, it is a nonprominent edge and may be discarded. So, we select a threshold to select prominent edges.

Figure 9.12 shows edge detected and thresholded image for vertical edge detection. Figure 9.13 shows the edge detected and thresholded image for diagonal edges.

Edge detected and thresholded image
using Sobel operator for horizontal direction

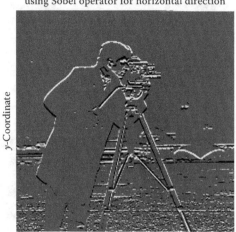

x-Coordinate

FIGURE 9.10
Image after Sobel edge detection and thresholding for horizontal edges.

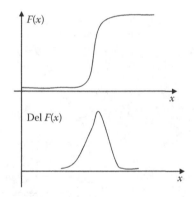

FIGURE 9.11
Use of first derivative for edge detection.

Let us write a MATLAB program as follows.

```
clear all;
A=imread('cameraman.jpg');
figure;
imshow(A);title('cameraman image');xlabel('x-coordinate'); ylabel('y
coordinate');
m=[-1 -2 -1
 0 0 0
 1 2 1];
c=conv2(single(A),single(m),'same');
c=int8(c);
figure;
imshow(c);title('edge detected image using Sobel operator for horizontal
direction');xlabel('x-coordinate');ylabel('y coordinate');
```

Edge detected and thresholded image using
Sobel operator for vertical direction

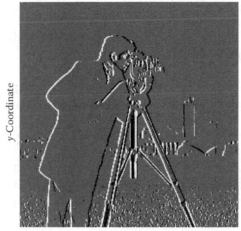

FIGURE 9.12
Sobel edge operated image for detecting vertical edges.

Edge detected and thresholded image using
Sobel operator for diagonal direction

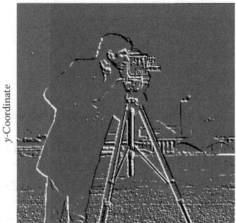

FIGURE 9.13
Edge detected image in diagonal direction using Sobel operator.

```
%now use a threshold.
for i=1:512,
    for j=1:512,
        if abs(c(i,j))<80,
            c(i,j)=0;
        end
    end
end
```

```
figure;
imshow(c);title('edge detected and thresholded image using Sobel operator
for horizontal direction');xlabel('x-coordinate');ylabel('y coordinate');
```

There is one more edge operator, namely, Prewitt operator, with the kernel mask for horizontal, vertical, and diagonal edge detection, given by

$$\text{Prewitt operator for horizontal edge detection} = \begin{vmatrix} -1 & -1 & -1 \\ 0 & 0 & 0 \\ 1 & 1 & 1 \end{vmatrix}.$$

$$\text{Prewitt operator for vertical edge detection} = \begin{vmatrix} -1 & 0 & 1 \\ -1 & 0 & 1 \\ -1 & 0 & 1 \end{vmatrix}.$$

$$\text{Prewitt operators for diagonal edge detection} = \begin{vmatrix} 0 & 1 & 1 \\ -1 & 0 & 1 \\ -1 & -1 & 0 \end{vmatrix} \text{and} \begin{vmatrix} -1 & -1 & 0 \\ -1 & 0 & 1 \\ 0 & 1 & 1 \end{vmatrix}$$

Sobel and Prewitt operators make use of the first derivative or the gradient.
    Consider the image pixels denoted as

$$\begin{vmatrix} Z1 & Z2 & Z3 \\ Z4 & Z5 & Z6 \\ Z7 & Z8 & Z9 \end{vmatrix}.$$

The gradient in horizontal and vertical direction is calculated as

$$\blacktriangledown f = |Gx| + |Gy|, \text{ where } Gx = (Z8 - Z5); \ Gy = (Z6 - Z5).$$

We may also use the cross differences $Gx = (Z9 - Z5)$; $Gy = (Z8 - Z6)$ to detect the edges in the diagonal directions. If we represent the center pixel as $f(x, y)$, then the gradient in x direction is given by $f(x, y + dy) - f(x, y)$ or $f(x, y) - f(x, y - dy)$. The gradient is thus $f(x, y + dy) - f(x, y - dy)$, or if we represent it by the weight of the corresponding pixel, it is [–1 0 1] for all columns. This justifies the mask used for horizontal gradient or for horizontal edge detection. Similarly, we can justify the mask used for vertical edge detection.

9.1.2 Laplacian Mask

Let us consider a Laplacian operator. It is basically an edge detection operator. It is a second order or double differential operator. It can be said to be the divergence (∇) of the gradient operator, (∇f). So, if we have f as a double differentiable real-valued function, then the Laplacian of f is defined by

$$\Delta f = \nabla \cdot \nabla f = \nabla^2 f.$$

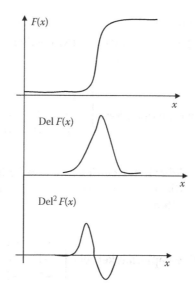

FIGURE 9.14
Plot of the intensity function, its derivative, and double derivative.

We will consider the meaning of the Laplacian of an image. It finds or highlights regions of very rapid changes in the intensity values in the image. Hence, it is used for edge detection. It is usually applied to the smoothed image using a Gaussian filter to reduce its sensitivity to noise. If there is random noise in the image, these noise pixels will also be detected as edge pixels. This is avoided using Gaussian operator. Consider the rapid variation in the intensity represented by the graph in Figure 9.14, which shows the graph of function F with respect to x, its derivative, and a double derivative of function F with respect to x. It is seen that the maximum slope location in the function F graph can be tracked by tracking the zero crossing in the double derivative graph. If we track the maximum and minimum positions in the double derivative graph, we will get double edge for each edge. Hence, zero crossing is detected to find the edge. Detecting zero crossing is more precise compared to detecting the maximum.

Consider the image pixels denoted as

$$
\begin{vmatrix}
Z1 & Z2 & Z3 \\
Z4 & Z5 & Z6 \\
Z7 & Z8 & Z9
\end{vmatrix}
$$

If we denote the center pixel, i.e., Z5, as $f(x, y)$, then the pixels Z4 and Z6 will be represented as $f(x-dx, y)$ and $f(x+dx, y)$, respectively. The pixels Z2 and Z8 will be represented as $f(x, y-dy)$ and $f(x, y+dy)$, respectively. The gradient in x direction can be written as $Z8 - Z5 = f(x, y+dy) - f(x, y), Z5 - Z2 = f(x, y) - f(x, y-dy)$. This is because the denominator for the gradient is one pixel. The gradient of the gradients can be written as

$$G^2x = (Z5 - Z2) - (Z8 - Z5) = 2f(x, y) - f(x, y+dy) - f(x, y-dy)$$
$$G^2y = (Z5 - Z4) - (Z6 - Z5) = 2f(x, y) - f(x-dx, y) - f(x+dx, y)$$

The gradient of gradients in horizontal and vertical direction is calculated as

$$\mathbf{\nabla}^2 f = \left|G^2 x\right| + \left|G^2 y\right| = 4f(x, y) - f(x, y + dy) - f(x, y - dy) - f(x - xd, y) - f(x + dx, y)$$

The Laplacian operator can now be written as

$$\text{Laplacian operator} = \begin{bmatrix} 0 & -1 & 0 \\ -1 & 4 & -1 \\ 0 & -1 & 0 \end{bmatrix}$$

Note that we have used the weights in the earlier equation for the gradient of the gradients. If we use the gradients in the diagonal direction as well, the reader can easily verify that the Laplacian kernel can be written as

$$\text{Alternative Laplacian kernel} = \begin{bmatrix} -1 & -1 & -1 \\ -1 & 8 & -1 \\ -1 & -1 & -1 \end{bmatrix}$$

A 5×5 Laplacian kernel can be written as

$$5 \times 5 \text{ Laplacian kernel} = \begin{bmatrix} 0 & 0 & -1 & 0 & 0 \\ 0 & -1 & -2 & -1 & 0 \\ -1 & -2 & 16 & -2 & -1 \\ 0 & -1 & -2 & -1 & 0 \\ 0 & 0 & -1 & 0 & 0 \end{bmatrix} \text{ or } \begin{bmatrix} 0 & 0 & 1 & 0 & 0 \\ 0 & 1 & 2 & 1 & 0 \\ 1 & 2 & -16 & 2 & 1 \\ 0 & 1 & 2 & 1 & 0 \\ 0 & 0 & 1 & 0 & 0 \end{bmatrix}$$

We will get the kernel with center pixel value of –16 or 16, depending on whether we take the forward difference or the backward difference.

9.1.3 Laplacian of Gaussian

Let us go through Gaussian smoothing operation. The smoothing operator called Gaussian is a mask used for executing a 2D convolution. It is used to blur the image so that it removes the noise pixels. It is basically a low-pass filter, but it uses a kernel that represents a bell-shaped Gaussian function. This kernel is found to have some special properties, which are detailed later in the chapter. The Gaussian distribution in two dimensions can be written as

$$G(x, y) = \frac{1}{2\pi\sigma^2} e^{-(x^2+y^2)/2\sigma^2}$$

The one-dimensional Gaussian distribution with a mean of zero and variance equal to 23 is shown in Figure 9.15.

To implement Gaussian smoothing, we have to use a 2D distribution function specified earlier as a "point-spread" function, and convolve it with the image. To realize the convolution, the Gaussian function must be represented as a collection of digital pixels to obtain a discrete approximation to the Gaussian function. Infinitely long kernel cannot be used. Considering that the Gaussian function tends to become low after three standard

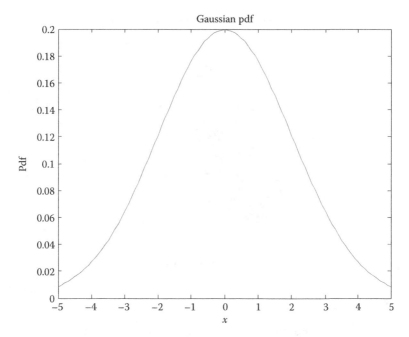

FIGURE 9.15
One-dimensional Gaussian distribution.

deviations from the mean, we can truncate the kernel at this point. The following equation shows a suitable integer-valued convolution kernel that approximates a Gaussian with a standard deviation of 1.0.

$$\text{Integer-valued Gaussian kernel} = \frac{1}{273} \begin{bmatrix} 1 & 4 & 7 & 4 & 1 \\ 4 & 16 & 26 & 16 & 4 \\ 7 & 26 & 41 & 26 & 7 \\ 4 & 16 & 26 & 16 & 4 \\ 1 & 4 & 7 & 4 & 1 \end{bmatrix}$$

Figures 9.16 through 9.18 show the Gaussian filtered, LOG operated and thresholded, and LOG operated images, respectively.

Let us write a MATLAB program to first use Gaussian smoothing and then the Laplacian filter to detect the edges.

```
clear all;
clc;
A=imread('cameraman.jpg');
figure;
imshow(A);title('cameraman image');xlabel('x-coordinate');ylabel('y
coordinate');
m=(1/273)*[1 4 7 4 1
     4 16 26 16 4
7 26 41 26 7
7 16 26 16 4
     1 4 7 4 1];
```

Smoothed image using Gaussian filter

y-Coordinate

x-Coordinate

FIGURE 9.16
Smoothed image using Gaussian filter.

Edge detected image using Laplacian LOG operator

y-Coordinate

x-Coordinate

FIGURE 9.17
Edge detected image using LOG.

```
c=conv2(single(A),single(m),'same');
c=int8(c);
figure;
imshow(c);title('smoothed image using Gaussian filter');xlabel('x-
coordinate');ylabel('y coordinate');
%now use a threshold.
m1=[0 0 1 0 0
    0 1 2 1 0
    1 2 -16 2 1
    0 1 2 1 0
    0 0 1 0 0];
```

Edge detected and thresholded LOG operator

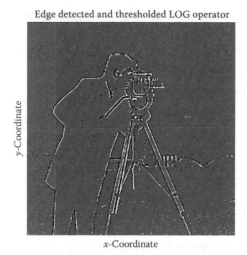

FIGURE 9.18
Edge detected and thresholded image using LOG.

```
d=conv2(single(c),single(m1),'same');
d=int8(d);
figure;
imshow(d);title('edge detected image using Laplacian LOG
operator');xlabel('x-coordinate');ylabel('y coordinate');
for i=1:512,
    for j=1:512,
        if abs(d(i,j))<100,
            d(i,j)=0;
        end
    end
end
figure;
imshow(d);title('edge detected and thresholded LOG operator');xlabel('x-
coordinate');ylabel('y coordinate');
```

Concept Check

- Why is image signal a multiplicative signal?
- What is the meaning of homomorphic processing?
- How will you convert multiplicative signals in the additive form?
- How is the image represented?
- State different spatial masks used for smoothing operation.
- State different mask sizes commonly used.
- In spatial filtering, explain the convolution operation.
- Define Sobel operator for detecting edges in horizontal, vertical, and diagonal directions.

- Define a gradient in horizontal direction and justify the mask used for detecting horizontal edge.
- What is Laplacian operator?
- Why is a Gaussian filter used before Laplacian?
- Why is zero crossing used for detecting the edge?
- Write the kernel used for Laplacian.
- Write the kernel used for Gaussian filter.

9.2 Transformations on Image

We will now consider different transformations on the image for increasing the contrast in the image to highlight the wanted part of the image with a particular range of intensity values. We will deal with different transformations such as thresholding, piece-wise linear transformation, logarithmic transformation, negation of the image, gray-level slicing, and bit-plane slicing. Let us go through these transformations one by one.

9.2.1 Linear Transformations

We will consider different linear transformations for contrast stretching. Let us first consider the thresholding and image negation. Referring to Figure 9.19 for image thresholding, we can see that if the intensity value is greater than the threshold, it is changed to 255, and if it is less than the threshold, it is changed to 0. Thus, thresholding operation converts a gray scale image into a black and white image. Figure 9.20 shows the output image used after thresholding as a result of MATLAB program. For negation, we can see from Figure 9.21 that pixels with 100% intensity values are now changed to 0% intensity values, and vice versa. Unlike the thresholding operation, the negation will keep the image as a gray scale image but will invert the pixel intensity values to about 255. This type of operation is useful for X-rays. Let us apply thresholding and negation to the same cameraman image and see the output. Figure 9.22 shows the negative image.

A MATLAB program to do image thresholding and negation is as follows.

```
clear all;
clc;
A=imread('cameraman.jpg');
figure;
```

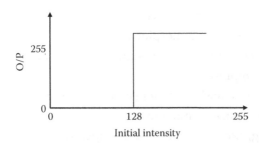

FIGURE 9.19
Transformation used for thresholding.

FIGURE 9.20
Cameraman image after thresholding with threshold of 128.

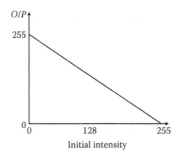

FIGURE 9.21
Transformation graph for image negation.

FIGURE 9.22
Cameraman image after applying negation transformation.

```
imshow(A);title('cameraman image');xlabel('x-coordinate');ylabel('y
coordinate');
for i=1:512,
    for j=1:512,
        if A(i,j)>128,
            A(i,j)=255;
        else A(i,j)=0;
        end
    end
end
figure;
imshow(A);title('thresholded image');xlabel('x-coordinate');ylabel('y
coordinate');
```

A MATLAB program to execute image negation is as follows.

```
clear all;
clc;
A=imread('cameraman.jpg');
figure;
imshow(A);title('cameraman image');xlabel('x-coordinate');ylabel('y
coordinate');
for i=1:512,
    for j=1:512,
    A(i,j)=255-A(i,j);
    end
end
figure;
imshow(A);title('Negative image');xlabel('x-coordinate');ylabel('y
coordinate');
```

Let us go through the MATLAB program for contrast stretching now. The selective contrast stretching transformation is shown in Figure 9.23. The slope of the graph is 0.8 for intensity values below 100 and above 200. The slope is 1.2 from 100 to 200. We will concentrate on the intensity values of 100–200; we will allow the amplification of these values by a factor of 1.2 and reduce the other intensity values by a factor of 0.8. To write a program to implement this transformation, we have to write equations for three straight graphs shown in Figure 9.23. The equation for the line passing through the origin is $y = 0.8x$, the equation for the middle line between $x = 100$ to 200 can be written as $y = 1.2(x - 100) + 80$, an equation for the third line can be written as $y = 0.8(x - 200) + 200$ as the last point for middle line, i.e., for $x = 200$ we get $y = 1.2(200 - 100) + 80$, i.e., $= 200$. The maximum value for y with $x = 255$ can be found using the equation for the third line $y = 0.8(255 - 200) + 200 = 244$ as shown in Figure 9.23. Using this basic knowledge, the reader is encouraged to write a program for any selective enhancement graph.

Refer to the MATLAB program given here.

```
clear all;
clc;
A=imread('cameraman.jpg');
imshow(A);title('cameraman image');xlabel('x-coordinate');ylabel('y
coordinate');
for i=1:512,
    for j=1:512,
        if A(i,j)<100
```

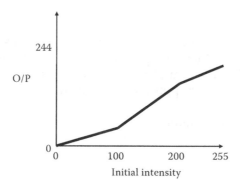

FIGURE 9.23
Transformation graph for selective contrast enhancement.

```
            A(i,j)=(0.8*A(i,j));
        end
        if (A(i,j)>100)&&(A(i,j)<200)
            A(i,j)=(((A(i,j)-100)*1.2)+80);
        end
        if A(i,j)>200
            A(i,j)=(((A(i,j)-200)*0.8)+200);
        end
    end
end
figure;
imshow(A);title('cameraman image after selective enhancement
transformation');xlabel('x-coordinate');ylabel('y coordinate');
```

The result of selective contrast stretching is shown in Figure 9.24. The reader can verify that the picture in the background is clearer now after selective enhancement.

Let us now consider the logarithmic transformation.

FIGURE 9.24
Result of selective contrast enhancement.

9.2.2 Gray-Level Slicing

We will discuss the gray-level slicing and its plane slicing in this section. Consider a typical gray-level slicing method that highlights the pixels intensity levels in the range 40–50%; other levels are reduced to a constant level as shown in Figure 9.25.

When this gray-level slicing is applied on the cameraman image, the result is shown in Figure 9.26.

A MATLAB program is as follows.

```
clear all;
clc;
A=imread('cameraman.jpg');
imshow(A);title('cameraman image');xlabel('x-coordinate');ylabel('y
coordinate');
for i=1:512,
    for j=1:512,
        if (A(i,j)>100)&&(A(i,j)<200)
            A(i,j)=200;
                else A(i,j)=10;
        end
```

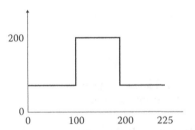

FIGURE 9.25
Gray-level slicing.

Cameraman image after gray-level slicing

x-Coordinate

FIGURE 9.26
Processed image after applying gray-level slicing.

```
      end
end
figure;
imshow(A);title('cameraman image after gray level slicing');xlabel('x-
coordinate');ylabel('y coordinate');
```

Consider a typical gray-level slicing method that highlights the pixel intensity levels from 100 to 200 and other levels that are preserved. The graph of the transformation is shown in Figure 9.27.

Cameraman image after applying gray scale transformation is shown in Figure 9.28.

```
clear all;
clc;
A=imread('cameraman.jpg');
imshow(A);title('cameraman image');xlabel('x-coordinate');ylabel('y
coordinate');
for i=1:512,
    for j=1:512,
        if (A(i,j)>100)&&(A(i,j)<200)
```

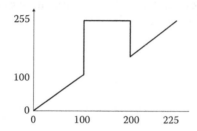

FIGURE 9.27
Gray-level transformation.

Cameraman image after-gray level slicing

x-Coordinate

FIGURE 9.28
Cameraman image after gray scale slicing shown in Figure 9.27.

```
            A(i,j)=200;
        end
    end
end
figure;
imshow(A);title('cameraman image after gray level slicing');xlabel('x-
coordinate');ylabel('y coordinate');
```

9.2.3 Bit-Plane Slicing

Let the intensity of the pixel be represented by 8-bit number. The image will now consist of eight intensity planes corresponding to each bit in the representation. Bit-0 represents the least significant bit, and bit-7 represents the most significant bit. Higher order planes will contain visually significant data, whereas lower order bit planes will contain detail data. Let us decompose the image in 8-bit planes and examine the data in each plane. Different bit-plane images, such as 6th, 5th, 3rd, 2nd bit-plane images, are shown in Figures 9.29 through 9.32, respectively. The 6th plane is closer to most significant bit (MSB) bit. The smooth image is seen, i.e., we see visually significant data, and for bits closer to least significant bit (LSB), we see detail data, which means tit gives edge information.

A MATLAB program to do so is given here.

```
clear all;
clc;
A=imread('cameraman.jpg');
figure;
imshow(A);title('cameraman image');xlabel('x-coordinate');ylabel('y
coordinate');
for i=1:512,
    for j=1:512,
        if A(i,j)>64,
            A(i,j)=255;
        else
```

6th bit plane of cameraman image

x-Coordinate

FIGURE 9.29
6th plane sliced cameraman image.

```
          A(i,j)=0;
      end
   end
end
figure;
imshow(A);title('6th bit plane of cameraman image');xlabel('x-
coordinate');ylabel('y coordinate');
```

9.2.4 Nonlinear Transformations

The nonlinear transformations include logarithmic transformations, exponential transformations, etc. In logarithmic transformation, the low intensities are amplified and

FIGURE 9.30
5th plane sliced cameraman image.

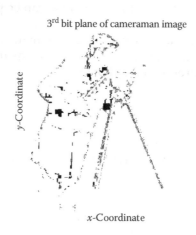

FIGURE 9.31
3rd plane sliced cameraman image.

2nd bit plane of cameraman image

x-Coordinate

FIGURE 9.32
2nd plane sliced cameraman image.

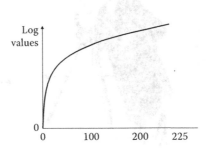

FIGURE 9.33
Log transformation graph.

high-intensity values are compressed as shown in Figure 9.33. The gamma transformation can be written as $y = c \times x^{\gamma}$. Power law transformation can be plotted for different values of γ as shown in Figure 9.33.

Let us apply this gamma transformation to the cameraman image. Figure 9.34 shows the cameraman image after applying gamma transformation. As the low-intensity values are amplified, we see white background instead of gray.

```
clear all;
clc;
A=imread('cameraman.jpg');
figure;
imshow(A);title('cameraman image');xlabel('x-coordinate');ylabel('y
coordinate');
for i=1:512,
    for j=1:512,
        A(i,j)=(gammaln(double(A(i,j))));
    end
end
figure;
```

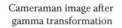

Cameraman image after
gamma transformation

x-Coordinate

FIGURE 9.34
Cameraman image after gamma transformation.

```
imshow(int8(A));title('cameraman image after gamma
transformation');xlabel('x-coordinate');ylabel('y coordinate');
```

Let us now consider the logarithmic transformation given by $y = c \times \log(1+x)$, where y and x represent the intensities of input and output pixels, respectively.

9.3 Histogram Equalization

Histogram equalization is one more method for image enhancement. This section will discuss the method to draw the histogram for the image, mapping probability density function (pdf) values to cumulative distributive function (CDF) and then equalizing the histogram.

9.3.1 Histogram Evaluation

Let us define the histogram. It is defined as the plot of frequency of occurrence on y axis and the value of the bin on x axis. What is a bin? We have to divide the total range of values of the intensity of the image in ranges like 0–4, 5–9, 10–14, and so on, until the value of 255. This divides the range of 0–255 in total of 51 bins, each of five values. We have to count how many times the value of the pixel lies in each range. This is the frequency of occurrence. The procedure for histogram calculation can be written as

1. Find the total range of pixel values.
2. Divide it in small bins of suitable range, say, 5.
3. Count the number of pixels lying in each range. This gives the frequency of occurrence.
4. Plot the frequency of occurrence versus bin values. This plot is a bar graph and is called a histogram. It is in a way the pdf graph of the function.

Figure 9.35 shows the histogram of the image with 51 bins, and Figure 9.36 shows the histogram of the image with 255 bins. We have used the cameraman image throughout the book in order to illustrate various effects.

Let us write a MATLAB program to draw the histogram for the cameraman image.

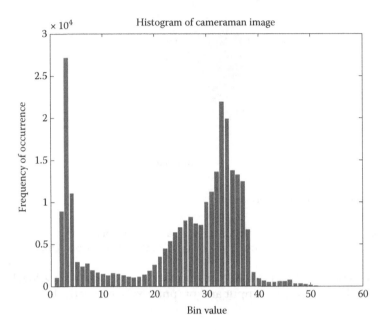

FIGURE 9.35
Histogram of image with 51 bins.

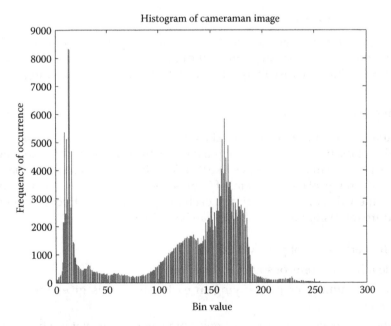

FIGURE 9.36
Histogram of the image with 255 bins.

```
clear all;
clc;
A=imread('cameraman.jpg');
figure;
imshow(A);title('cameraman image');xlabel('x-coordinate');ylabel('y
coordinate');
for i=1:51,
 c(i)=0;
end
for k=1:51,
    for i=1:512,
        for j=1:512,
            if (A(i,j)<=k*5-1)&&(A(i,j)>(k-1)*5-1),
                c(k)=c(k)+1;
            else
            end
        end
    end
end
figure;
bar(c);title('Histogram of cameraman image');xlabel('bin
value');ylabel('frequency of occurrence');
```

9.3.2 Mapping the pdf Value with CDF

Let us now map the frequency of occurrence in the histogram, i.e., the pdf to the CDF. We add the probabilities upto a certain pixel value to find the CDF. Figure 9.37 shows the CDF plot for the histogram.

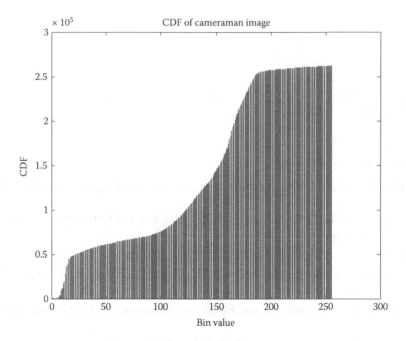

FIGURE 9.37
CDF for the histogram in Figure 9.36.

Let us write a MATLAB program to find the CDF. We have repeated the program here
to draw the histogram.

```
clear all;
clc;
A=imread('cameraman.jpg');
figure;
imshow(A);title('cameraman image');xlabel('x-coordinate');ylabel('y
coordinate');
for i=1:255,
    c(i)=0;
end
for k=1:255,
    for i=1:512,
        for j=1:512,
            if (A(i,j)<=k)&&(A(i,j)>(k-1)),
                c(k)=c(k)+1;
            else
            end
        end
    end
 end
figure;
bar(c);title('Histogram of cameraman image');xlabel('bin
value');ylabel('frequency of occurrence');
for i=1:255,
    d(i)=0;
end
d(1)=c(1);
for k=2:255,
    d(k)=d(k-1)+c(k);

end
figure;

bar(d);title('CDF of cameraman image');xlabel('bin
  value');ylabel('CDF');
```

9.3.3 Histogram Equalization

The nature of the histogram can be examined and the following conclusions can be drawn.

1. If the points in the histogram are concentrated toward the origin, it is a low intensity or dark image.
2. If the points concentrate toward 255, it is a bright image.
3. If the image is a low-contrast image, histogram values will cluster toward the center of intensity range.
4. If image is a high-contrast image, histogram values will spread over all intensities.

Let us consider the steps to improve the contrast in the image.

1. Low-contrast image has a histogram clustered over a small range of intensity values.
2. To improve the contrast, stretch the histogram so that it will spread over the entire range.
3. The image will now have a large dynamic range.

Let us understand how to modify the histogram.

1. It is possible to develop a transformation to stretch the histogram.
2. Transformation design is called histogram equalization.

Histogram equalization means redistribution of the histogram so that it uniformly occupies the entire range of values. We can use the transformation as CDF computation.
    The steps for histogram equalization can be stated as follows.

1. First, normalize the intensity values in the range of 0–255.
2. Map the intensity values using a monotonically increasing function S given by

$S = T(r) = CDF(r)$, where r is the intensity values of the pixels.

To implement histogram equalization, the pixel values in the original image are now replaced by the CDF values obtained earlier. The image is displayed using these new pixel values to show the modified image. Figure 9.38 shows the histogram equalized image. If we try to plot the histogram of the modified image, we will see that it is now equally distributed compared to the previous histogram of the original image. Figure 9.39 shows the histogram of the histogram equalized image.
    Let us write the program to do so.

Histogram equalized cameraman image

y-value

x-value

FIGURE 9.38
Histogram equalized image.

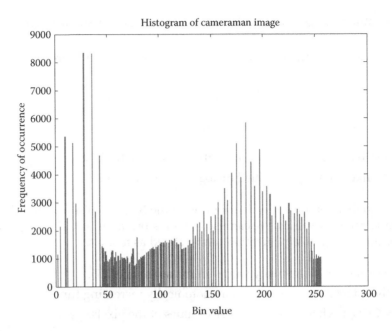

FIGURE 9.39
Histogram of the equalized image.

```
d1=max(d);
d1=d*255/max(d);
A1=0;

for k=1:255,
    for i=1:512,
        for j=1:512,
            if (A(i,j)<=k)&&(A(i,j)>(k-1)),
                A1(i,j)=(d1(k));
            end
        end
    end
end

figure;
imshow(uint8(A1));title('Histogram equalized cameraman image');xlabel('x
value');ylabel('y value ');
for i=1:255,
    c1(i)=0;
end
for k=2:255,
    for i=1:512,
        for j=1:512,
            if (A1(i,j)<=k)&&(A1(i,j)>(k-1)),
                c1(k)=c1(k)+1;
            else
            end
        end
    end
```

FIGURE 9.40
Low- and high-contrast image with histograms.

```
 end
figure;
bar(c1);title('Histogram of cameraman image');xlabel('bin
value');ylabel('frequency of occurrence');
```

The example of histogram of low-contrast and equalized image is shown in Figure 9.40 along with the histograms. For the equalized image, the histogram is uniformly spread over the entire range, whereas for low-contrast image, it is clustered near low-intensity values.

9.3.4 Statistical Image Processing

The statistical image processing methods include the evaluation of mean, standard deviation, variance, skew, and kurtosis, etc. All these parameters are explained and discussed in detail in previous chapters. The reader is encouraged to write a MTLAB program to find the statistical parameters for image.

Concept Check

- What is thresholding?
- What is negation?
- State different types of transformations used for contrast stretching.
- Draw different gray scale transformations.

- What is bit lane slicing?
- Write equation for gamma transformation.
- What is histogram?
- State the procedure for drawing the histogram of the image.
- Write any one transformation for histogram equalization.
- Can you draw conclusions related to contrast in the image from the histogram?

9.4 Transform Domain Image Processing

This section discusses image processing using DCT and WT for image compression, resizing, edge detection, and filtering. The detail theory of the transforms DCT and WT is already discussed in Chapter 8. It is not repeated here. We will go through different applications for images using DCT first.

9.4.1 Image Processing Using DCT

Let us now consider the image coding using DCT. Consider a 4×4 image. The DCT coefficients are usually scanned in a zigzag manner as shown in Figure 9.41. In general, a large section of the tail end of the scan will consist of zeros. The reason for the zigzag scanning is that the higher order DCT coefficients have a small amplitude.

Let us read a 4×4 block of any image and take DCT of the block using a MATLAB program.

```
%program to find dct of 4*4 image block
clear all;
a=imread('cameraman.jpg');
imshow(a);
for i=205:208,
    for j=261:264,
        b(i-204,j-260)=a(i,j);
    end
end
imshow(b);
disp('4*4 image pixel values are');
disp(b);
c=dct2(b);
figure;
imshow(c);
disp('4*4 image dct values are');
disp(c);
```

The results of the execution of this program, that is, the pixel values of 4×4 image block and 4×4 DCT output, are as follows.

4×4 image pixel values are

120	115	105	110
116	112	106	116
43	49	46	55
42	52	47	51

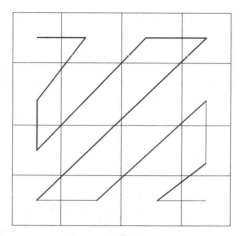

FIGURE 9.41
Zigzag scanning for the image of size 4 × 4.

4 × 4 image DCT values are

321.2500	−0.3459	5.2500	−9.3277
119.0452	11.3336	6.7145	1.5732
−0.2500	5.0581	−3.2500	−0.2010
−49.0395	−1.9268	−1.4283	−0.3336

We can see that the DCT value for the leftmost corner output is the highest. It is the DC value of the image. When we move along the zigzag path as shown in Figure 9.41, the DCT values are found to decrease. By selecting the proper step size for higher order coefficients, the coefficients at the tail of the scan can be coded using zero values. This gives the compression of the image. The last coded nonzero coefficient value will be declared, with the code at the end indicating the receiver that the further coefficients in the tail are zeros. Let us take a 512 × 512 image and its DCT. We will retain only the first 100 rows and 100 columns. It can be seen that the image obtained after processing is almost the same as the original one, indicating that further higher order coefficients contain very little information and we can compress the image by using only 100 × 100 pixel samples. A MATLAB program to do so is as follows below.

```
clear all;
a=imread('cameraman.jpg');
imshow(a);title('original image');xlabel('x co-ordinate');ylabel('y
coordinate');
b=dct2(a);
for i=101:512,
    for j=101:512,
        b(i,j)=0.0;
    end
end
c=((idct2(b)));
figure;
imshow(uint8(c));title('low pass filtered image using 150 rows and 150
columns of DCT');xlabel('x co-ordinate');ylabel('y coordinate');
```

Figures 9.42 and 9.43 depict the original image and processed image, respectively. Both the images are almost the same as far as the visible details are considered.

Let us consider image filtering using DCT. We will read the image and take its 2D DCT. To low-pass filter the image, we have to retain only the lower order coefficients. Refer to zigzag scanning of the image. To make it simple, we will retain the coefficients in the left-most square of size 60×60 and make all other higher order coefficients as zero. We will then take inverse DCT to get the processed image. A MATLAB program to do it is given.

```
clear all;
a=imread('cameraman.jpg');
```

FIGURE 9.42
Original image.

FIGURE 9.43
Processed image.

```
imshow(a);title('original image');xlabel('x co-ordinate');ylabel('y
coordinate');b=dct2(a);
for i=61:512,
    for j=61:512,
        b(i,j)=0.0;
    end
end
c=((idct2(b)));
figure;
imshow(uint8(c));title('low pass filtered/ Blurred image');xlabel('x
co-ordinate');ylabel('y coordinate');
```

Figures 9.44 and 9.45 depict the original image and the low-pass filtered image, respectively.

DCT can also be used for an image edge enhancement. We have used gain of 0.2 for lower order coefficients for 60 rows and 60 columns, gain of 1.4 to enhance middle order coefficients, which contain the edge information, and very high order coefficients are reduced to zero.

A MATLAB program for edge enhancement is given here.

```
%image edge detection using DCT
clear all;
a=imread('cameraman.jpg');
imshow(a);title('original image');xlabel('x co-ordinate');ylabel('y
coordinate');
b=dct2(a);
for i=1:60,
    for j=1:60,
        b(i,j)=0.2*b(i,j);
    end
end
for i=61:200,
    for j=61:200,
```

Original image

x-Coordinate

FIGURE 9.44
Original image.

Low pass filtered/blurred image

y-Coordinate

x-Coordinate

FIGURE 9.45
Low-pass filtered image.

```
        b(i,j)=1.4*b(i,j);
    end
end
for i=201:512,
    for j=201:512,
        b(i,j)=0.0;
    end
end
c=((idct2(b)));
figure;
imshow(uint8(c));title('edge detected image');xlabel('x
co-ordinate');ylabel('y coordinate');
```

The edge-enhanced image is shown in Figure 9.46.

Consider the image resizing using DCT. We will use cameraman image and take its 2D DCT. We will retain only the lower order coefficients, that is, the first 512 rows and first 512 columns. The inverse DCT gives the filtered image and then it will be down-sampled by 2 to get image resizing by a factor of 1/2. A MATLAB program is given here.

```
clear all;
a=imread('cameraman.jpg');
b=dct2(a);
for i=257:512,
    for j=257:512,
        b(i,j)=0.0;
    end
end
c=((idct2(b)));
imshow(uint8(c));title('low pass filtered image');xlabel('x
co-ordinate');ylabel('y coordinate');
figure;
for i=2:2:512,
    for j=2:2:512,
```

Edge detected image

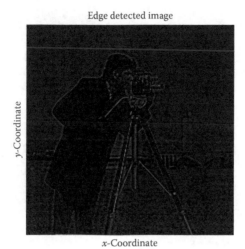

FIGURE 9.46
Edge detected image.

```
        d(i/2,j/2)=uint8(c(i,j));
    end
end
imshow(d);title('resized image');xlabel('x coordinate');ylabel('y
coordinate');
```

Figures 9.47 and 9.48 show the low-pass filtered and resized images, respectively.

9.4.2 Image Processing Using WT

We can obtain compression using WT for an image. We will take its 2D discrete wavelet transform (DWT) and try to get signal compression. A MATLAB program to execute the signal compression is as follows.

Low-pass filtered image

FIGURE 9.47
Low-pass filtered image.

Resized image

y-Coordinate

x-Coordinate

FIGURE 9.48
Resized image.

```
% program to get compression using WT
clear all;
a=imread('cameraman.jpg');
imshow(a);title('original image');xlabel('x co-ordinate');ylabel('y coord
inate');[ca,cv,cs,cd]=dwt2(a,'db1');
for i=1:256,
    for j=1:256,
        cv(i,j)=0.0;
    end
end
for i=1:256,
    for j=1:256,
        cd(i,j)=0.0;
    end
end
for i=1:256,
    for j=1:256,
        cs(i,j)=0.0;
    end
end
d=idwt2(ca,cv,cs,cd,'db1');f
igure;
imshow(uint8(d));title('image recovered by making all detail functions
zero');xlabel('x co-ordinate');ylabel('y coordinate');
```

Figure 9.49 shows a plot of first-level wavelet decomposition of image signal using Haar wavelet with only low-resolution function shown. Detail function and high-low (HL) and low-high (LH) part is displayed separately in Figure 9.50 as it has very small amplitude and is not visible in front of low-resolution part. The detail components, namely, HH, LH, and HL, are made zero as they have a value less than the step size. We will take inverse WT and recover the image. Figure 9.51 shows a plot of the recovered image signal. We can see that the image recovered is almost the same as the original one. This indicates that for the first-level wavelet decomposition, we can select a step size as large so that all detail function coefficients are coded using zero value. This gives 75% signal compression, as we are coding only 25% of the coefficients.

Wavelet decomposed image

y-Coordinate

x-Coordinate

FIGURE 9.49
Low-resolution decomposed image.

Wavelet decomposed image

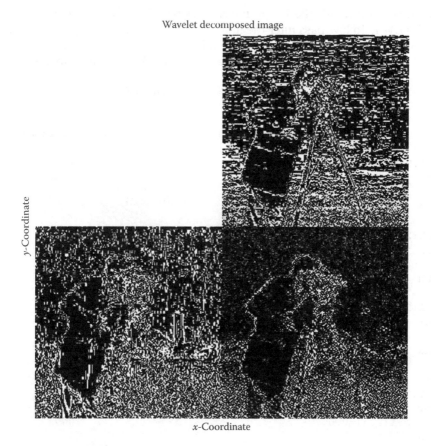

y-Coordinate

x-Coordinate

FIGURE 9.50
HL, LH, and HH part of decomposed image.

Image recovered by making all detail functions zero

x-Coordinate

FIGURE 9.51
Recovered image achieving 75% compression.

Let us now go through the image filtering using DWT. We will read the image and take its 2D DWT. To low-pass filter the image, we have to retain only LL coefficients after second-level decomposition. We will make all detail coefficients zero. We will then take inverse 2D DWT to get the processed image. A MATLAB program to do it is as follows.

```
% A program to low pass filter the image using WT
clear all;
a=imread('cameraman.jpg');
imshow(a);title('original image');xlabel('x co-ordinate');ylabel('y coord
inate');[ca,cv,cs,cd]=dwt2(a,'db1');
for i=1:256,
    for j=1:256,
        cv(i,j)=0.0;
    end
end
for i=1:256,
    for j=1:256,
        cd(i,j)=0.0;
    end
end
for i=1:256,
    for j=1:256,
        cs(i,j)=0.0;
    end
end
[ca1,cv1,cs1,cd1]=dwt2(ca,'db1');
for i=1:128,
    for j=1:128,
        cv1(i,j)=0.0;
    end
end
for i=1:128,
```

```
    for j=1:128,
        cd1(i,j)=0.0;
    end
end
for i=1:128,
    for j=1:128,
        cs1(i,j)=0.0;
    end
end
ca=idwt2(ca1,cv1,cs1,cd1,'db1');
d=idwt2(ca,cv,cs,cd,'db1');
figure;
imshow(uint8(d));title('image low pass filtered');xlabel('x
co-ordinate');ylabel('y coordinate');
```

Figure 9.52 shows the low-pass filtered image.

To detect the edges in the image, we will make LL and HH part of the decomposed image zero and keep only LH and HL components. After taking inverse wavelet transform (IWT), we will get the edge-enhanced image. A MATLAB program to do so is shown.

```
% A program to detect the edge in the image using WT
clear all;
a=imread('cameraman.jpg');
imshow(a);title('original image');xlabel('x co-ordinate');ylabel('y
coordinate');
[ca,cv,cs,cd]=dwt2(a,'db1');
for i=1:256,
    for j=1:256,
        cd(i,j)=0.0;
    end
end

[ca1,cv1,cs1,cd1]=dwt2(ca,'db1');
```

Image low-pass filtered

x-Coordinate

FIGURE 9.52
Low-pass filtered image.

```
for i=1:128,
    for j=1:128,
        cd1(i,j)=0.0;
    end
end
for i=1:128,
    for j=1:128,
        ca1(i,j)=0.0;
    end
end
ca=idwt2(ca1,cv1,cs1,cd1,'db1');
d=idwt2(ca,cv,cs,cd,'db1');
figure;
imshow(uint8(d));title('image -edge detected using WT');xlabel('x
co-ordinate');ylabel('y coordinate');
```

Figure 9.53 shows the edge detected image.

The process of image resizing using WT is illustrated. Let us choose decimation by a factor of 2. We will use an low-pass filter using WT. To implement the LPF, we will make LH, HL, and HH coefficients zero after the first-level decomposition. After taking IWT, we will get the recovered image, which is low-pass filtered. We will down-sample the image by a factor of 2 to get the decimated image. A MATLAB program to do so is given here.

```
% A program to decimate the image by a factor of 2
clear all;
a=imread('cameraman.jpg');
imshow(a);title('original image');xlabel('x co-ordinate');ylabel('y
coordinate');
[ca,cv,cs,cd]=dwt2(a,'db1');
for i=1:256,
    for j=1:256,
        cd(i,j)=0.0;
    end
end
```

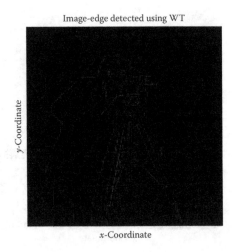

FIGURE 9.53
Edge detected image.

Image-decimated image by a factor of 2

FIGURE 9.54
Decimated image.

```
for i=1:256,
    for j=1:256,
        cv(i,j)=0.0;
    end
end
for i=1:256,
    for j=1:256,
        cs(i,j)=0.0;
    end
end
d=idwt2(ca,cv,cs,cd,'db1');
figure;
for i=2:2:512,
    for j=2:2:512,
        d1(i/2,j/2)= d(i,j);
    end
end
imshow(uint8(d1));title('image -decimated image by a factor of
2');xlabel('x co-ordinate');ylabel('y coordinate');
```

Figure 9.54 shows the decimated image.

Concept Check

- Why is zigzag scanning used for scanning DCT coefficients of an image?
- Can you execute compression using DCT?
- State the different applications for image processing using WT.

Summary

This chapter deals with different processing techniques used for image processing such as spatial filtering, intensity transformations, histogram equalization, transform domain image processing including image compression, enhancement, filtering, resizing, etc.

1. We started with image representation. The spatial filters are introduced and the convolution operation of kernel and the image is explained. The kernels for smoothing operations are stated and its implementation is illustrated. Different kernel sizes and their effect on the smoothing operation are illustrated. The edge detection using Sobel and Prewitt operators for detecting an edge in horizontal, vertical, and diagonal directions is explained. A MATLAB program to detect the edges is given. The double derivative, Laplacian, is introduced, and the need for Gaussian filter before Laplacian is explained. The LOG operator is discussed in detail along with a MATLAB program to implement it. The curves for the derivative and double derivative of the simple rapid intensity variation case are drawn and the edge is shown to lie at zero crossing point. The masks used for Sobel and Laplacian are justified by considering the actual derivatives and double derivatives for concerned pixel points. The use of zero crossing for edge detection is emphasized. The linear operators such as negation and thresholding are dealt with. Selective contrast enhancement using a suitable transformation is described. The LOG transformation and gamma transformation are described and implemented.

2. The histogram evaluation for any image is explained and illustrated using a MATLAB program. The mapping of intensity values using a CDF transformation is discussed and implemented. The histogram equalization is described and implemented. The nature of histogram and the conclusions related to contrast are discussed. The statistical measurements are just stated, and it is indicated that they can be evaluated on similar lines as that of speech signal.

3. The transform domain image processing is described in the next section. The theory of transforms such as DCT and WT is already covered in Chapter 8 and is not repeated. The use of these transforms for image coding, image filtering, image enhancement, and image resizing is explained and implemented using MATLAB programs.

Key Terms

Image representation	Prewitt operator
Kernel	Horizontal edge
Two-dimensional convolutions	Vertical edge
Smoothing mask	Diagonal edge
Low-pass filter	Laplacian
Edge detection	Double derivative
Gradient operator	Gaussian filter
Sobel operator	Laplacian of Gaussian (LOG)

Zero crossing detection	Histogram equalization
Linear transformations	Cumulative distribution function (CDF)
Thresholding	DCT
Negation	WT
Selective contrast enhancement	Image coding
Gamma transformation	Image filtering
Log transformation	Image enhancement
Histogram	Image resizing

Multiple-Choice Questions

1. Smoothing mask is
   a. A spatial filter
   b. A spatial low-pass filter
   c. A spatial high-pass filter
   d. A spatial bandpass filter

2. Sobel operator is
   a. An Integrator
   b. A differentiator
   c. A double derivative
   d. A double integrator

3. A gradient along x direction detects
   a. Vertical edge
   b. Horizontal edge
   c. Diagonal edge
   d. Edge in any direction

4. A Gaussian filter is used before Laplacian
   a. To low-pass filter and to suppress edges due to noise
   b. To pass the edges
   c. To high-pass filter the noise
   d. To detect all possible edges

5. A Laplacian is
   a. Gradient operator
   b. Double derivative
   c. Double integrator
   d. Gradient and differentiator

6. After using LOG, we detect the edge
   a. As a peak in the double derivative graph
   b. As a zero crossing point in double derivative graph

    c. As a peak of the derivative graph

    d. As the zero crossing in derivative graph

7. The nonlinear intensity transformation

    a. Is used to magnify the intensity values

    b. Is used to enhance the selected intensity values and improve the contrast

    c. Is used to reduce the contrast

    d. Is used to detect the edges

8. The gamma transformation

    a. Enhances low intensities and compresses high-intensity values

    b. Compresses all intensity values

    c. Enhances all intensity values

    d. Enhances high intensities and compresses low-pass values

9. Histogram is a graph of

    a. Frequency of occurrence versus probability

    b. Frequency of occurrence versus bin value of the intensity

    c. Probability versus CDF

    d. Intensity versus frequency of occurrence

10. If the histogram is concentrated in a low-intensity area

    a. The image is dark image with low contrast

    b. The image is bright image

    c. The image is a high-contrast image

    d. The image is bright image with high contrast

11. After equalization,

    a. The histogram gets concentrated in low-intensity area

    b. The histogram gets concentrated in high-intensity area

    c. The histogram gets equally spread out

    d. The histogram is peaked

12. A homomorphic system for multiplication has a canonical system consisting of

    a. Log

    b. FFT followed by LOG

    c. LOG followed by DFT

    d. FFT

13. DCT can be used for compression of the image by selecting the proper step size

    a. Because DCT has a property of energy spreading

    b. Because DCT has a property of energy compaction

    c. Because DCT has a property of compression

    d. Because DCT has a property of adjusting the step size

14. We can use DCT for image resizing, i.e., image reduction
    a. Using DCT as a antialiasing filter by selectively retaining lower order DCT coefficients
    b. Using DCT as a high-pass filter by selectively retaining DCT coefficients
    c. Using DCT as a filter by selectively discarding DCT coefficients
    d. Using DCT as an anti-imaging filter by selectively retaining DCT coefficients

15. WT can be used for image compression by selecting proper step size
    a. For LH, HL, and HH components
    b. For LL components
    c. For LL and HH components
    d. For LH and HL components

16. We can use WT for image resizing, i.e., image reduction
    a. Using WT as a antialiasing filter by selectively retaining lower order LL coefficients
    b. Using WT as a high-pass filter by selectively retaining WT coefficients
    c. Using WT as a filter by selectively discarding WT coefficients
    d. Using WT as an anti-imaging filter by selectively retaining WT coefficients

Review Questions

1. Explain the image formation and its representation.
2. Illustrate the use of a 3×3 mask for filtering. Explain the convolution operation.
3. Write different masks or kernels used for smoothing operation. Explain how smoothing takes place.
4. Write Sobel operator kernels for detecting the edges in vertical, horizontal, and diagonal directions. Explain how the edges are detected. Justify the kernel used for edge detection.
5. Draw the curves for derivative operation for the intensity variation at the edge and explain how the edge is detected as the maximum value of the derivative.
6. Define Laplacian mask and justify it using the definition of double derivative in horizontal, vertical, and diagonal directions.
7. Why is a Gaussian filter used before Laplacian? Draw the curves for the intensity variation in an image, its derivative and double derivative, and explain how the edge is detected at the point of zero crossing of the double derivative?
8. Write a digital spatial mask for the Gaussian operation.
9. Draw different possible linear and nonlinear transformations for image enhancement. Explain how enhancement takes place.

10. State the transformations for LOG and gamma transformations.
11. Explain the procedure for drawing the histogram of any image.
12. Explain possible use of CDF as the transformation for histogram equalization.
13. How is the histogram equalized? What is the meaning of equalization?
14. Explain the use of DCT for image compression and filtering.
15. Explain the use of DCT for image resizing and edge enhancement.
16. Explain the use of WT for image compression and filtering.
17. Explain the use of WT for image resizing and edge enhancement.

Problems

1. Write a program to smoothen the image using a smoothing kernel and implementing a convolution operation.
2. Write a program to enhance the edge in vertical direction using Sobel operator. Select proper threshold to suppress the unwanted edges.
3. Write a program to filter an image using a Laplacian operator for detecting the edges.
4. Write a program to use LOG for any blurred image corrupted by paper and salt noise. Select the proper threshold to suppress unwanted edges.
5. Write a program to enhance the selected portion of the image having intensity values in the range of 120–150 and suppress the other intensity levels.
6. Write a program to plot the histogram of any standard image and implement histogram equalization using CDF.
7. Write a program to implement contrast enhancement on the blurred image using DCT.
8. Write a program to increase the size of the image by a factor of 2 using DCT.
9. Write a program to implement contrast enhancement on the blurred image using WT.
10. Write a program to increase the size of the image by a factor of 2 using WT.

Answers

Multiple-Choice Questions

1. (b)
2. (b)
3. (a)
4. (a)

5. (b)
6. (b)
7. (b)
8. (a)
9. (b)
10. (a)
11. (c)
12. (a)
13. (b)
14. (a)
15. (a)
16. (a)

Problems

1. Refer to the programs in the text for smoothing operation.
2. See the Sobel kernel for vertical direction edges and select the proper threshold.
3. Use any Laplacian mask stated in the text.
4. Add salt and paper noise to the image and use Gaussian digital mask stated in the text to remove noise. Then use Laplacian with thresholding to detect edges.
5. Refer to program in the text for selective enhancement using a transformation.
6. Refer to histogram equalization program in the text.
7. Refer to program in the text.
8. This is a question for increasing the size of the image. So, first implement interpolation and then use an anti-imaging filter using DCT.
9. Refer to program in text.
10. This is to be implemented in the same way as in problem 8. In place of DCT, use WT.

10

Applications of Random Signal Processing

LEARNING OBJECTIVES

- Case study—Handwritten Character Recognition—Dr. Sushama Shelke
- Case study—Writer Verification—Dr. Sharada Kore

This chapter deals with two different case studies. One related to Marathi handwritten character recognition and the other related to writer verification using handwritten document. These case studies are written by author's Ph.D. students, namely, Dr. Sushama Shelke and Dr. Sharada Kore, respectively.

10.1 Case Study 1: Handwritten Character Recognition

Optical character recognition (OCR) is the recognition of printed or handwritten text by a computer. This involves photo scanning of the text character by character, analysis of the scanned image, and then translation of the character image into character codes, such as ASCII. Research in OCR is popular for its various application areas such as office automation, bank check verification, post offices for mail sorting, and a large variety of business and data entry applications. Other applications involve reading aid for the blind, library automation, language processing, and multimedia design.

10.1.1 Components of an OCR System

Similar to any pattern recognition system, the character recognition system contains the basic components as shown in Figure 10.1.

The character image is captured using an acquisition device such as flatbed scanner. In preprocessing, the system carries out data cleaning tasks, which may include noise reduction, skew detection and correction, slant detection and correction, normalization, binarization, etc. This is usually followed by a segmentation process to separate the lines, words, and characters in the image. Feature extraction is of vital importance to character recognition systems, particularly handwriting recognition systems. It serves two main purposes: extracting the most representative features that are used by classifiers and reducing redundancies in data. In offline systems, feature extraction is affected by several factors such as different background of documents, nonuniform illumination of

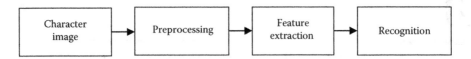

FIGURE 10.1
Components of an OCR.

the scanner, noise introduced by electronics and writing tools, and different qualities of paper and types of ink. Finally, the character recognition task can be carried out by pattern recognition approaches, namely, statistical, structural, artificial neural network, and soft-computing.

Since the mid-1980s, *neural networks* have become popular in handwriting recognition as they learn from examples. They are robust, insensitive to noise, and have generalization ability. Among various architectures, the feed-forward neural networks remain dominant because of the well-known training methods, such as back propagation of errors and approximation function. Multilayer perceptron (MLP) classifiers are able to form complex hyperplane decision regions that can classify a large number of classes.

10.1.2 Challenges in Devanagari Handwriting Recognition

The Devanagari script has a large number of characters and compound characters. There are various issues that make the recognition of handwritten characters a challenging task and affect the recognition rate to a considerable extent. Some examples of such issues are as follows:

- Variations in the writing style of the writers
- Variations in the font size, pen width, pen ink
- Shape changes due to preprocessing parameters
- Variations in the character spacing, skew, and slant
- In the case of compound characters, the strategies for joining two or more consonants are different. Characters may be split during preprocessing
- Some character pairs are very much similar to each other
- The header line may or may not be drawn by the writer
- Segmentation of modifiers
- Segmentation of touching characters.

This demands an efficient system, which takes care of these issues at all the stages in the OCR system, from preprocessing to recognition. The challenges can be met by implementing a multistage recognition scheme. Each stage of classification task is carried out based on features from course level to fine level or global level to local level, finally leading to recognizing the character.

The detailed design of a system to recognize handwritten Devanagari characters is shown in Figure 10.2. This system adopts a multistage recognition scheme.

The system is designed to recognize 39 handwritten Devanagari characters, as shown in Figure 10.3. The characters are scanned at 300 dpi in bmp file format using a flatbed scanner.

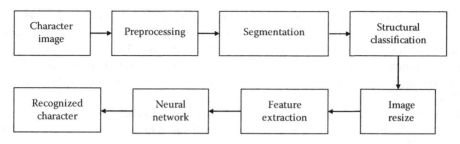

FIGURE 10.2
Multistage recognition system blocks schematic.

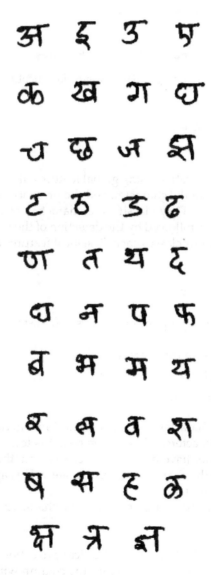

FIGURE 10.3
Characters used in the system.

At first, the handwritten Devanagari character data set is collected from more than 10 writers. A sample set of about 100 samples per character is collected, resulting in about more than 4000 character samples in the database. The characters are preprocessed to remove noise and converted to binary format after thresholding. The characters are segmented from lines and words using horizontal and vertical projection profiles.

The large number of characters in the Devanagari character set with a wide range of variations in the writing style demand a preclassification of the characters before the final recognition. The preclassification is done using a tree classifier based on the course structural features like the presence of vertical line in the character, position of the vertical line, and presence of holes in the character. The tree classifier classifies the characters into 24 different classes based on these parameters.

10.1.3 Tree Classification Based on Structural Features

The preclassification is done using a two-stage classification based on the structural features. These stages are

1. Detection of global features.
2. Detection of local features.

The first stage employs classification using global features like the presence of vertical line in the character, its position in the character, and the presence of holes. These features can be termed as global features. They classify the characters coarsely into six classes. The detection of global features is followed by the detection of the local features, which further classify the six classes into four classes each. The local features are character-specific than the global level features.

- Detection of global features

Global features used for classifying the characters at first stage include

1. Presence of vertical bar in the character
2. Position of the vertical bar
3. Presence of the enclosed region.

About 60% of Devanagari characters exhibit a vertical line in them. This vertical line is at the center in two of the characters, while in the rest, it is toward the end. The remaining 40% of the characters do not have a vertical line. Also, another feature that is enclosed region is present in 56% of the characters approximately. This approximation is due to the writing style of individual writer.

To detect whether these features are present in the character, the following algorithm is implemented.

1. *Detection of vertical bar in the character*: Vertical projection profile of the character image $f(m, n)$ is calculated in order to find the column with maximum number of pixels n_{max}. An average height of the vertical bar is considered to be 85% of the

total height of the image. This value is set as a threshold T_V to find the presence of a vertical bar in a character. Thus, if

$$n_{max} > T_V, \text{vertical bar present} \tag{10.1}$$

The first-stage structural classification is shown in Table 10.1.

2. *Detection of position of vertical bar in the character*: If the presence of a vertical bar is detected, then its location is found so as to further classify the character as per its location within the character. Again an average threshold T_M is set to 30%, for the position of the vertical bar in the character. If

$$T > T_M, \text{vertical bar toward center} \tag{10.2}$$

Else, the vertical bar is at the end. Here,

$$T = \frac{n - n_{max}}{n} \times 100 \tag{10.3}$$

3. *Detection of presence of enclosed region in the character*: Here, eight-neighbor adjacency is used to find the presence of connected components or the enclosed regions. Two foreground pixels p and q are said to be connected if there exists an eight-connected path between them, consisting entirely of foreground pixels. Table 10.1 shows the classification of characters based on these global features.

- Detection of local features

  The local features used here are

a. Presence of endpoints in the lower part of the character.

  To find these features, the binary image $f(m, n)$ is first thinned to yield a single-pixel-wide character. This character is then passed on to hit-or-miss transformation to find the endpoints of the character. Eight structuring elements are used to detect the location of endpoints in all eight directions. The image is then partitioned into

TABLE 10.1

First-Stage Structural Classification

Class	Global Features		
	Mid Bar	End Bar	Enclosed Region
No bar enclosed (NBE)	0	0	1
No bar not enclosed (NBNE)	0	0	0
Mid bar enclosed (MBE)	1	0	1
Mid bar not enclosed (MBNE)	1	0	0
End bar enclosed (EBE)	0	1	1
End bar not enclosed (EBNE)	0	1	0

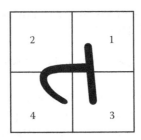

FIGURE 10.4

Character partitioning for endpoint detection.

four quadrants, as shown in Figure 10.4. A vector $V=[V1\ V2\ V3\ V4]$ is defined, where $V1$, $V2$, $V3$, and $V4$ indicate the presence of endpoints in quadrant 1, 2, 3, and 4, respectively, by setting or resetting them accordingly. Here, quadrants 3 and 4 only are of interest. The presence of endpoints in quadrants 3 and 4 sets the values for $V3$ and $V4$. The combination of values in $V3$ and $V4$ classifies the character into four classes, namely, 00, 01, 10, and 11, where class 00 indicates that there is no endpoint in quadrants 3 and 4, whereas 01 indicates that there is an endpoint in quadrant 3, no endpoint in quadrant 4, and so on. Table 10.2 shows this classification.

The entire two-stage structural classification is shown in Figure 10.5. After this classification, 24 classes are obtained for the entire database.

- Feature extraction

  The next stage extracts the normalized pixel density features from the character images $s(x, y)$, as shown in Figure 10.6. Here, 35 features are obtained by partitioning the 70×50 sized characters into 35 nonoverlapping blocks and counting the number of zeros in them.

The number of zero pixels in $s(x, y)$ is found, where $x = 1, 2, \ldots, 7$ and $y = 1, 2, \ldots, 5$. The normalized pixel density features $npd(x, y)$ are calculated using,

$$npd(x, y) = \frac{100 - s(x, y)}{100} \tag{10.4}$$

TABLE 10.2

Second-Stage Structural Classification

Class	Local Features	
	Endpoint in Quadrant 4	**Endpoint in Quadrant 3**
00	Absent	Absent
01	Absent	Present
10	Present	Absent
11	Present	Present

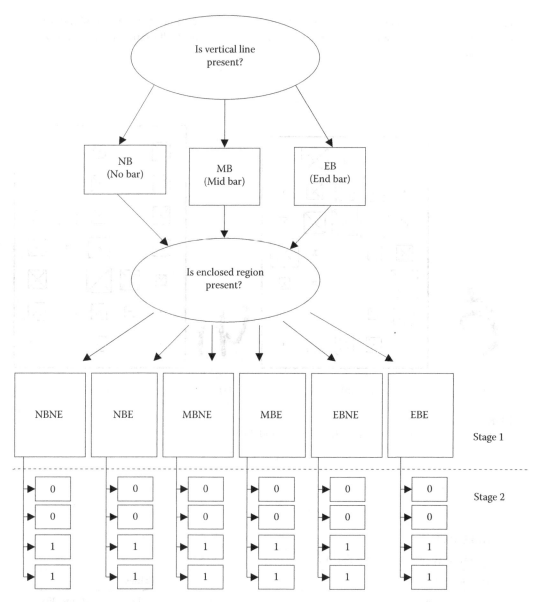

FIGURE 10.5
Two-stage structural classification.

10.1.4 Recognition Using Neural Network

All these features are applied to a feed-forward, back-propagation, neural network built separately for each of the 24 classes. During the training phase, the features are applied to train the network, and during the testing phase, the features are used to recognize the characters. The parameters setting of the neural network are given in Table 10.3.

The system gives a recognition accuracy of 91.54%. The time required to test a character is approximately 0.05 s, when tested on an Intel Core 2 Duo CPU running on 2 GHz with 2 GB RAM.

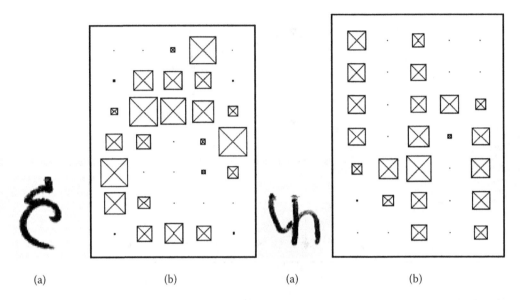

(a)   (b)   (a)   (b)

FIGURE 10.6
Normalized pixel density features: (a) original image and (b) its features.

TABLE 10.3
Neural Network Parameter Settings

Parameters	Values
Number of inputs	With normalized pixel density features: 35
Number of hidden layers	1
Number of neurons in hidden layer	Equal to number of inputs
Hidden layer activation function	Log-sigmoid transfer function
Number of neurons in output layer	Number of characters in the structural class
Output layer activation function	Linear
Learning rate	0.5
Goal	0.001
Error function	Mse
Maximum number of epoch	1000
Training algorithm	Gradient descent back propagation

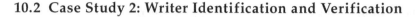

10.2 Case Study 2: Writer Identification and Verification

10.2.1 Introduction

Automatic person identification and verification is very important in the era of automation. Several different ways are used to identify and verify a person. The traditional modes of person identification and verification are based on possession mode or the knowledge mode. In possession mode, a person possesses some identity card, passport, or smart card. The problem with this mode is that the identity card can be stolen or lost. In knowledge mode, person identification and verification is done based on a person's knowledge on a piece of information, e.g., password. The problem with this mode is the password can be copied by an unauthorized person. So, the traditional methods are more prone to forgery and are less secure.

Most popular methods at present are based on biometrics. The biometrics methods are based on who are you. These methods are more secure, easy to use, and there is no repudiation compared to traditional methods of person identification and verification.

The biometrics modalities are broadly classified as physiological biometrics and behavioral biometrics. Physiological biometrics methods use physical property of the human body, e.g., face, iris, ear, fingerprints, and DNA. Behavioral biometrics methods use individual traits of person's behavior, e.g., how you speak, your style of utterance, your accent of utterance, how you write, i.e., your style of writing. Behavioral biometrics methods are less accurate, but they are less invasive.

Handwriting pertains to a class of behavioral biometrics. It is a unique characteristic of a person and can be used for person identification and verification task. Automatic person identification and verification based on handwriting is known as writer identification and verification. Writer identification and verification began to thrive in the 1990s. Owing to the rapid growth in computer technologies and advanced research in pattern recognition area, person identification and verification based on handwriting is a hot research topic in recent years.

10.2.2 Importance of Writer Identification and Verification

Writer verification is gaining in importance due to its applicability, mainly in the field of forensic area, to determine the identity of a writer in ransom notes, threat letters, suicide notes, document forgeries, etc. Advantages of handwriting biometrics over other forms are

1. Identification in conjunction with the intentional aspects of a crime and
2. Evidence material and the details of an offense can be quite remote,

The traditional method of person identification and verification used by forensic document examiners is more tedious, time-consuming, and human dependent. To overcome several of these limitations, automatic writer identification and verification approach has emerged as a promising alternative. The main advantages of computer-based automatic writer recognition are less human intervention, noninvasive, user-friendly, and well-accepted legally.

The other applications of writer verification are biometric security systems based on handwritten password. The advantages of handwritten password over typed one are that they more secure, flexible, and less invasive.

10.2.3 Main Factors Discriminating Handwritings

The writers are found to use different writing styles. The main factors discriminating handwriting are the size of writing (width and height), shape of characters used in the writing, the writing slant, the vertical strokes present in the writing, roundness or curvature, and consistency in the writing. These factors discriminating handwriting are depicted in Figure 10.7.

10.2.4 Factors Affecting Handwriting

The above-mentioned parameters are affected by many factors such as physical conditions, mental conditions of writer, training provided to writer, writing environment, time

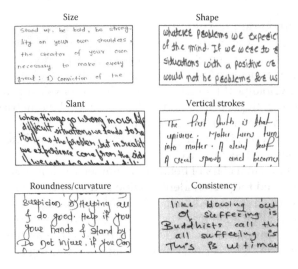

FIGURE 10.7
Factors discriminating different styles of handwriting.

provided for writing, and age of writer. Handwriting is a dynamic characteristic, hence, is a signal with some random component. It is found to vary for person for every page being written and is also found to vary from person to person. Because of its randomness, it is very difficult to achieve high accuracy for writer verification. It is still a useful biometrics modality due to its flexibility and noninvasiveness.

10.2.4.1 System for Writer Identification and Verification

The implementation of writer identification and verification system involved handwriting image acquisition, preprocessing, feature extraction, and writer identification and verification task. Let us go through these steps one by one.

10.2.4.2 Handwriting Acquisition

Handwriting sample can be taken offline or online. In the offline system, scanner or camera is used for handwriting digitization. The information provided is more or less static. It is used in forensic area. Online scan system uses special digitizers such as pen tablet device. It provides dynamic information such as pen tip movement, pen pressure, and sequencing order. It can be used for low-cost access control systems. The problem with this system is noise introduced by the capturing device.

10.2.4.3 Preprocessing

The acquired image can be processed for separation of foreground image (ink information) from background image (ruled lines, graphics, images, etc.), noise removal, and normalization with respect to scaling, rotation, and translation.

10.2.4.4 Feature Extraction

The most commonly used features for the existing methods of writer recognition are explained in this section.

1. *Entropy*: It gives the information about the average ink distribution in the handwritten document image [6]. It is a statistical measure of randomness that can be used to characterize the texture of the input image. Entropy (E) is defined as

$$E = \sum_i p_i \times \log p_i \tag{10.5}$$

   where p is the probability or the frequency of occurrence of the pixel intensity in the image. The pen control and applied pressure in the handwritten document are different for different person. The average ink distribution in the document image gives the information about the pen pressure of a writer and the type of writing instrument used. Therefore, we used entropy as a feature for writer verification.

2. *Run-length*: It gives the information about the intraword and interword spacing in the handwritten document image. The histograms of number of black and white pixels in the horizontal and vertical directions were used as features for writer

verification. The gray-scale image is converted into binary by Otu's thresholding. Figure 10.8 shows run-length features in horizontal and vertical directions for a given original image. The size of feature vector is 100.

3. *Edge-Based Directional Feature*: The original image is converted into gray scale. Feature extraction starts with conventional edge detection using Sobel, followed by thresholding that generates a binary image in which only the edge pixels are detected. Consider each edge pixel in the middle of a square neighborhood, as shown in Figure 10.9. Then, check the presence of edge fragments (4 pixels wide). The histogram of all instances of edge fragments emerging from center pixel is counted. The dominating angle p(θ) shown in Figure 10.9 is the writing slant, which is one of the discriminating features between two handwritings.

4. *Chain Code Method*: The histogram of chain code feature extracts the information about the pixel distribution in different directions, which captures the shape of characters produced in the writing, slant, vertical strokes present, and roundness. The implementation steps are given in Figure 10.10a. The original image is converted into binary image. Then, the connected components and their boundaries are detected. The boundaries are labeled with freeman chain code using directions and values given in table as shown in Figure 10.10b. The example of boundary labeling is shown in Figure 10.10c.

Check each pixel on boundary with respect to next pixel. If the next pixel is in horizontal direction with respect to current pixel, then the value assigned to current pixel is 0, as $dy=0$ and $dx=+1$. If the next pixel is in the right diagonal direction, $dy=-1$, and $dx=+1$, the value assigned to that pixel is 7. Thus, the tracing of the

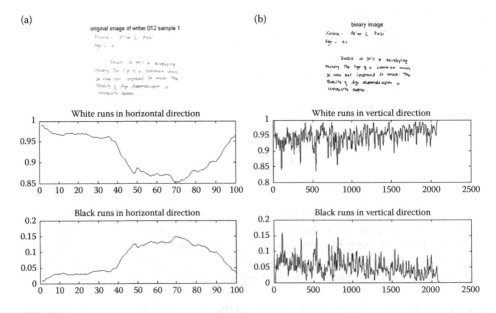

FIGURE 10.8
Run-length features of given image in (a) horizontal directions, left-hand graph; (b) vertical directions, right-hand graph.

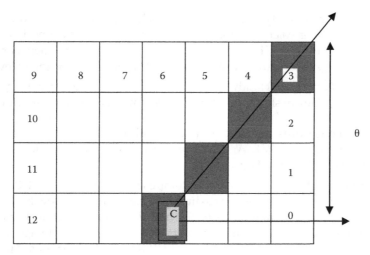

FIGURE 10.9
Edge-based directional feature.

entire boundary is done. We will obtain one code for each connected pixel having values in the range 0–7 representing either of eight directions. This code is called as chain code, and it represents the pixel distribution in different direction, which captures ink distribution information in eight directions.

If the chain code is traced starting from different pixel, its sequence will change for each starting pixel. The chain code obtained must be independent of the starting pixel. It is achieved by evaluating the frequency of occurrences of chain code in eight

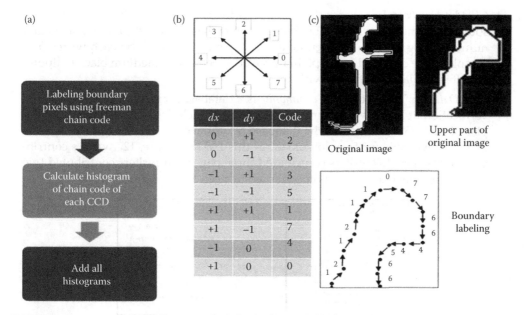

FIGURE 10.10
Feature extraction using chain code: (a) implementation steps, (b) boundary labeling directions, and (c) example of boundary labeling.

directions, i.e., probability of occurrence of each direction. It is called histogram of the chain code. It is calculated for all the boundaries of each connected component in the document image. The histograms of all connected components are added to calculate average histogram of chain code as a feature at the document level.

5. *Wavelet Transform (WT)*: Handwriting can be viewed as a texture image as shown in Figure 10.11. Figure shows handwritten document for two different writers.

Texture-based features are used for writer verification. The texture-based features can be extracted using WT. Wavelet transform of an image results into four sub-band images as shown in Figure 10.12.

Different bands in the wavelet decomposed image represent the following information. Low band gives approximation image; image in low high band gives pixel distribution information in horizontal direction, image in high low band gives pixel distribution information in vertical direction, and image in high high band gives pixel distribution in diagonal left and right directions. Thus, wavelet decomposition gives information in three different directions.

To extract feature using WT, the single-level wavelet decomposition, is performed on a given input image. It results into four sub-bands of wavelet coefficients. The histogram of each sub-band is calculated to get a feature vector at the document level. The normalized value of histogram is used as a feature vector.

10.2.5 Off-Line English Handwriting Databases

The information related to different databases includes Center of Excellence in Document Analysis and Recognition (CEDAR) database, IAM database, IBM_UB_1 database, and the SLK database generated by the authors.

1. *CEDAR Database*: It includes three handwriting samples of each 900 writers. The source document was the CEDAR letter. which has 150 words, all alphabets and numerals. A given source document was copied three times by each writer. The samples were collected using plain unreeled sheets and a medium black ballpoint pen. It is available for free download.

2. *IAM Database*: This is the *most commonly used* database for writer identification and verification in English handwriting document. It includes two handwriting samples of each 657 writers. It is the database of offline cursive handwritten Western text. It is available for free download. In this IAM database, 127 writers contributed four pages, 159 writers contributed three pages, 301 writers contributed two pages, and only one page was contributed by each 356 writers.

FIGURE 10.11
Two different handwritings represent different texture images.

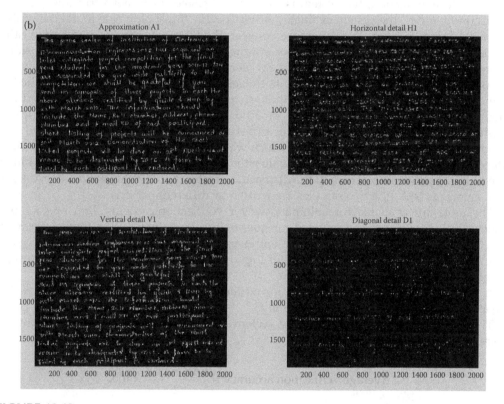

FIGURE 10.12
Result of single-level wavelet decomposition: (a) gray image after thresholding, and (b) wavelet decomposed sub-band images.

3. *IBM_UB_1 Database*: It is a dual-mode (online and offline) unconstrained English handwriting database. Handwriting samples were collected and processed in both modes, online and offline, simultaneously. It contains 2000 pairs of 43 writers. Each pair has one page of continuous writing on a particular topic, while the other page contains a few sampled, core words that encapsulate the written document.

4. *SLK Dataset*: The SLK database is the first offline English handwriting database of Indian people. It is made available publically for researchers working in writer identification and verification area. SLK database is made available for free

download from Web site: www.sharadakore.com. The SLK database has three databases. Database 1 includes source documents of printed text and handwritten text. Database 2 includes handwritten text only. Database 3 includes handwritten text limited to 50% lines of set 2.

The writers were asked to write a given English text of five to six lines using ball pen and sketch pen on A4 size plain paper. The details of handwriting data sample collection is provided in Table 10.4.

10.2.5.1 Writer Verification System

After feature extraction, the system is trained and tested for writer verification task. In the training mode, the system evaluates writer-specific threshold value for all the writers in the training data set. The difference between the feature vectors of two handwriting samples is used as a threshold value for a writer. The threshold is calculated and stored along with the template feature for all the writers in the data set.

After training the system, it enters into the testing mode. In the testing mode, the two handwriting samples are presented to the system and checked whether the two handwriting samples are written by the same person or not. Comparison is done on one-to-one basis, and decision is taken based on the threshold value. The distance is compared with the threshold value of claimed writer, which was calculated during the training mode. If the distance is less than the threshold value, the system answer is YES—the two samples are written by the same person. If the distance is greater than the threshold value, the system answer is NO—the two samples are written by different persons.

The tools are available for automatic writer verification systems. But, no tool provides 100% accuracy. It can be used for quantitative measures by the forensic examiners. The tools used by forensic document examiners are Forensic Information System Handwriting, WANDA, Trimodel Writer Identification Project, CEDAR, and Groningen Intelligent Writer Identification System.

A new tool, SLKSDA, was designed for writer verification in 2016 by the authors. The snapshot of the tool is shown in Figure 10.13.

The performance of writer verification system can be evaluated in terms of the following parameters.

$$\text{Average verification accuracy} = \frac{\text{FAR} + \text{FRR}}{2} \times 100, \tag{10.6}$$

TABLE 10.4

Details of Handwriting Sample Collection

Category	Age Group	No. of Writers
Children	8–16	103
Women	17–25	840
	26–35	12
	36–45	05
Men	17–25	31
	26–35	04
	36–45	05
Total	8–45 men and women	1000

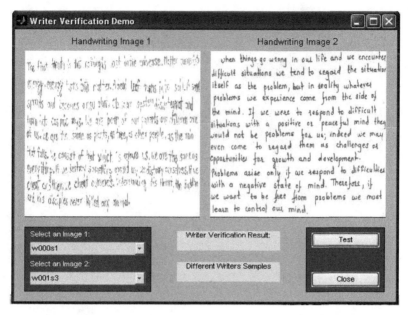

FIGURE 10.13
SLKSDA, writer verification demo tool.

where

$$\text{False acceptance rate (FAR)} = \frac{\text{Falsely accepted samples}}{\text{Total samples tested}} \times 100 \qquad (10.7)$$

$$\text{False rejection rate (FRR)} = \frac{\text{Falsely rejected samples}}{\text{Total samples tested}} \times 100 \qquad (10.8)$$

This section provides a summary of existing methods and tools available for the writer identification and verification. The system developed by the authors gives an FAR of 1.2%.

Future research may concentrate on the following aspects. Finding features for discrimination of handwritten documents is a challenging problem. The methods are based on natural handwriting samples and do not detect forgeries or disguised handwriting. The effect on system performance can be studied when samples are from different sources such as scanner, camera, and pen tablet. Writer identification and verification in multiscript environment can be done.

Index